D0208467

STEEL FRAMED STRUCTURES

Stability and Strength

Related titles

BEAMS AND BEAM COLUMNS: STABILITY AND STRENGTH

edited by R. Narayanan

1. Elastic Lateral Buckling of Beams D. A. NETHERCOT
2. Inelastic Lateral Buckling of Beams N. S. TRAHAIR
3. Design of Laterally Unsupported Beams D. A. NETHERCOT and N. S. TRAHAIR
4. Design of I-Beams with Web Perforations R. G. REDWOOD
5. Instability, Geometric Non-Linearity and Collapse of Thin-Walled Beams T. M. ROBERTS
6. Diaphragm-Braced Thin-Walled Channel and Z-Section Beams T. PEKÖZ
7. Design of Beams and Beam Columns G. C. LEE and N. T. TSENG
8. Trends in Safety Factor Optimisation N. C. LIND and M. K. RAVINDRA

PLATED STRUCTURES: STABILITY AND STRENGTH

edited by R. Narayanan

1. Longitudinally and Transversely Reinforced Plate Girders H. R. EVANS
2. Ultimate Capacity of Plate Girders with Openings in Their Webs R. NARAYANAN
3. Patch Loading on Plate Girders T. M. ROBERTS
4. Optimum Rigidity of Stiffeners of Webs and Flanges M. ŠKALOUD
5. Ultimate Capacity of Stiffened Plates in Compression N. W. MURRAY
6. Shear Lag in Box Girders V. KŘISTEK
7. Compressive Strength of Biaxially Loaded Plates R. NARAYANAN and N. E. SHANMUGAM
8. The Interaction of Direct and Shear Stresses on Plate Panels J. E. HARDING

SHELL STRUCTURES: STABILITY AND STRENGTH

edited by R. Narayanan

1. Stability and Collapse Analysis of Axially Compressed Cylindrical Shells A. CHAJES
2. Stiffened Cylindrical Shells under Axial and Pressure Loading J. G. A. CROLL
3. Ring Stiffened Cylinders under External Pressure S. KENDRICK
4. Composite, Double-Skin Sandwich Pressure Vessels P. MONTAGUE
5. Fabricated Tubular Columns used in Offshore Structures W. F. CHEN and H. SUGIMOTO
6. Collision Damage and Residual Strength of Tubular Members in Steel Offshore Strucures T. H. SØREIDE and D. KAVLIE
7. Collapse Behaviour of Submarine Pipelines P. E. DE WINTER, J. W. B. STARK and J. WITTEVEEN
8. Cold-Formed Steel Shells G. ABDEL-SAYED
9. Torispherical Shells G. D. GALLETLY
10. Tensegric Shells O. VILNAY

STEEL FRAMED STRUCTURES

Stability and Strength

Edited by

R. NARAYANAN

M.Sc. (Eng.), Ph.D., D.I.C., C.Eng., F.I.C.E., F.I.Struct.E., F.I.E.
Reader in Civil and Structural Engineering,
University College, Cardiff, United Kingdom

ELSEVIER APPLIED SCIENCE PUBLISHERS
LONDON and NEW YORK

ELSEVIER APPLIED SCIENCE PUBLISHERS LTD
Crown House, Linton Road, Barking, Essex IG11 8JU, England

Sole Distributor in the USA and Canada
ELSEVIER SCIENCE PUBLISHING CO., INC.
52 Vanderbilt Avenue, New York, NY 10017, USA

British Library Cataloguing in Publication Data

Steel framed structures: stability and strength.
1. Structural frames 2. Steel, Structural
I. Narayanan, R.
624.1′773 TA660.F7

ISBN 0-85334-329-2

WITH 22 TABLES AND 173 ILLUSTRATIONS

The selection and presentation of material and the opinions expressed in this publication are the sole responsibility of the authors concerned.

Filmset and printed in Northern Ireland by The Universities Press (Belfast) Ltd.

PREFACE

In producing this fourth volume in the series on stability and strength of structures, we have continued the policy of inviting several expert contributors to write a chapter each, so that the reader is presented with authoritative versions of recent ideas on the subject. Sufficient introductory material has been included in each chapter to enable anyone with a fundamental knowledge of structural mechanics to become familiar with various aspects of the subject.

Each of the ten chapters in the book highlights the techniques developed to solve a selected facet of frame instability. Thus the earlier chapters deal principally with the instability of entire frames, as influencing the design of multi-storey structures; the chapters which follow cover a wider range of instability problems, such as those connected with joints, braces, etc., as well as special problems associated with thin-walled structures, arches and portals. The topics chosen for the various chapters reflect the diversity of problems within the general area of frame instability and the range of analytical tools developed in recent years.

As editor, I wish to express my gratitude to all the contributors for the willing cooperation they extended in producing this volume. I sincerely hope that the many ideas and experimental data included in the book will meet the needs of the engineer and researcher alike, and provide a stimulus to further progress in our understanding of structural behaviour.

R. NARAYANAN

CONTENTS

LIST OF CONTRIBUTORS

D. ANDERSON

Senior Lecturer, Department of Engineering, University of Warwick, Coventry CV4 7AL, UK.

W. F. CHEN

Professor and Head of Structural Engineering, School of Civil Engineering, Purdue University, West Lafayette, Indiana 47907, USA.

K. H. GERSTLE

Professor of Civil Engineering, University of Colorado, Campus Box 428, Boulder, Colorado 80309, USA.

G. J. HANCOCK

Senior Lecturer, School of Civil and Mining Engineering, University of Sydney, NSW 2006, Australia.

M. R. HORNE

Formerly Beyer Professor of Civil Engineering, Simon Engineering Laboratories, University of Manchester, Oxford Road, Manchester M13 9PL, UK.

S. KOMATSU

Professor, Department of Civil Engineering, Osaka University, Yamadaoka 2–1, Osaka 565, Japan.

E. M. LUI

Lecturer, School of Civil Engineering, Purdue University, West Lafayette, Indiana 47907, USA.

I. C. MEDLAND
Senior Lecturer, Department of Theoretical and Applied Mechanics, University of Auckland, Private Bag, Auckland, New Zealand.

L. J. MORRIS
Senior Lecturer, Simon Engineering Laboratories, University of Manchester, Oxford Road, Manchester M13 9PL, UK.

K. NAKANE
Research Fellow, Department of Civil Engineering, University of Canterbury, Christchurch 1, New Zealand.

T. M. ROBERTS
Lecturer, Department of Civil and Structural Engineering, University College, Newport Road, Cardiff CF2, 1TA, UK.

C. M. SEGEDIN
Emeritus Professor, Department of Theoretical and Applied Mechanics, University of Auckland, Private Bag, Auckland, New Zealand.

G. J. SIMITSES
Professor, School of Engineering Science and Mechanics, Georgia Institute of Technology, Atlanta, Georgia 30332, USA.

A. S. VLAHINOS
School of Engineering Science and Mechanics, Georgia Institute of Technology, Atlanta, Georgia 30332, USA.

Chapter 1

FRAME INSTABILITY AND THE PLASTIC DESIGN OF RIGID FRAMES

M. R. Horne

Formerly Beyer Professor of Civil Engineering, University of Manchester, UK

SUMMARY

Idealised approximations to material stress–strain relationships lead to corresponding idealised limit loads—in particular, the rigid–plastic collapse load and the least elastic critical load. The real behaviour, allowing for stability and change of geometry, causes a reduction of carrying capacity below the rigid–plastic collapse value. The extent of the reduction depends on the slendernesses of the members and may be related to the value of the elastic critical load. Justifications for the Merchant–Rankine load, and for the modification suggested by Wood, are discussed, and applications to unbraced multi-storey frames show the usefulness of this procedure. An example of the design of a multi-storey frame involving frame stability effects is given. Special frame stability problems in single-storey pitched roof frames are discussed, and safeguards are described and illustrated by reference to two design examples.

NOTATION

a_i	Coefficients applied to eigenvectors $f_{\Delta i}$
D	Minimum depth of rafter
E	Elastic modulus
F_{Δ}	Deflection function of a frame
$F_{\Delta L}$	Linear deflection function

$f_{\Delta i}$	Eigenvectors for critical deflection modes
$f_{\Delta L}$	Linear deflection function of a frame under unit load
h	Height of columns in a portal frame
h_i	ith storey height in a multi-storey frame
I_C, I_R	In-plane second moments of area of columns and rafters in a portal frame
L	Span of a portal frame
l	Length of a member
M_P	Plastic moment of resistance
m	Bay width in a multi-storey frame
p_y	Specified design strength
p'_y	Effective design strength
R	Axial force in a member
S	Plastic modulus
s	Displacement of frame in direction of an applied load
U	Total potential energy
U_E	Elastic strain energy
U_N	Modified total potential energy (including plastic work)
U_P	Energy absorbed in plastic deformation
U_W	Potential energy of applied loads
u_i	Sway deflection within ith storey
W	Factored vertical load on a portal frame
W_i	Vertical loads applied to a multi-storey frame
W_0	Uniformly distributed vertical load capacity of a portal frame roof in absence of thrust
x	Sway deflection of a frame
α	Ratio of lowest critical load to factored applied load or to failure load
β	Rotation of any member in a rigid–plastic mechanism
ε_y	Elastic strain at yield
ε_s	Strain at beginning of strain-hardening
θ	Rotation of a member in a plastic collapse mechanism: angle of pitch of a rafter
λ	Load factor
λ_C	Lowest elastic critical load factor
λ_{Ci}	ith elastic critical load factor
λ_F	Failure load factor
λ_{MR}	Merchant–Rankine load factor
λ_P	Rigid–plastic collapse load factor

λ_{WMR}	Merchant–Rankine load factor as modified by Wood
μ	Deflection parameter
ρ	$(I_C/I_R)(L/h)$
σ_L	Lower yield stress
σ_U	Upper yield stress
σ_{ult}	Ultimate yield stress
Φ	Hinge rotation
Φ_i	Sway index in the ith storey
Φ_{max}	Maximum value of Φ_i
Ω	'Arching' ratio in a portal frame

1.1 INTRODUCTION: ELASTICITY, PLASTICITY AND STABILITY

The stability and strength of a framed structure may be explored in relation to the various approximations that may be made to the real stress–strain behaviour of the material of which the structure is composed. Mild steel has a stress–strain curve in tension or compression of the form shown in Fig. 1.1(a), in which an elastic phase OA up to an

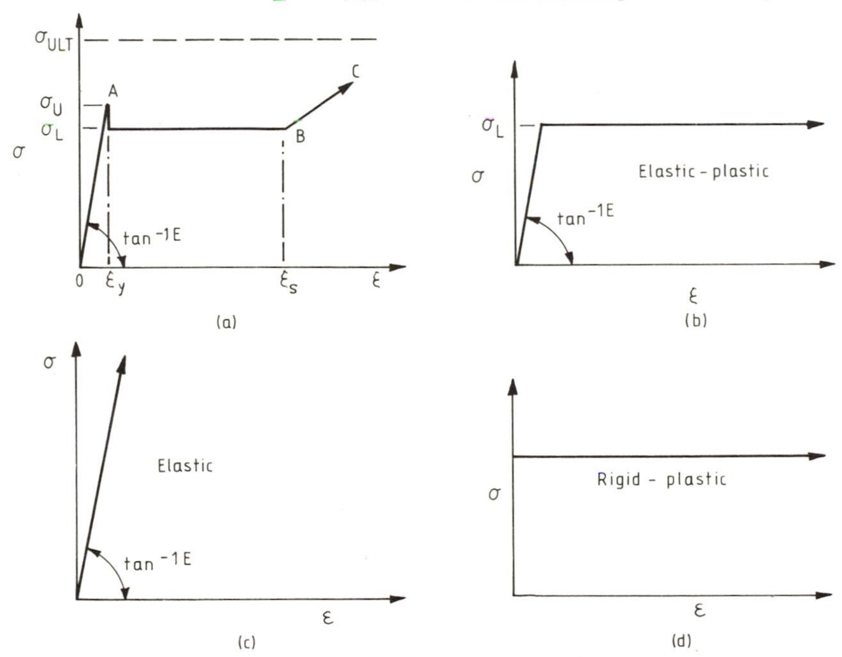

FIG. 1.1. Idealised stress–strain relations.

upper yield stress σ_U and a strain $\varepsilon_y = \sigma_U/E$ (where E is the elastic modulus) is followed by plastic deformation at a lower yield stress σ_L up to a strain ε_s of the order of $10\varepsilon_y$. Beyond a strain of ε_s, strain-hardening occurs, the strain ultimately becoming non-uniform due to necking, leading to fracture at an ultimate stress σ_{ult} some 25–40% above σ_L and at an elongation of some 30%. Within the range of structural interest, the idealised elastic–plastic stress–strain relation in Fig. 1.1(b) is a sufficiently close approximation.

With the aid of computers, it is possible to follow analytically the behaviour of entire structures on the basis of any assumed stress–strain relationship. However, even if the idealised elastic–plastic stress–strain relation of Fig. 1.1(b) is used, limits of computer capacity can soon be reached. Further idealisations of behaviour facilitate design and may be used if their respective limitations are recognised. Provided the effects of the deformations of the structure on the equations of equilibrium are neglected, the *idealised elastic behaviour* (Fig. 1.1(c)) leads to a linear relationship between intensity of loading and the deformations and stresses induced in the structure, and is the basis for many design procedures. However, the analysis is valid only up to the stage at which yield is reached somewhere in the structure—a point which has no consistent relationship to the ultimate strength of the structure. For this purpose, the *rigid–plastic* stress–strain relationship in Fig. 1.1(d), leading to the concept of *plastic collapse mechanisms*, gives for many structures a close estimate of the actual load at which collapse would occur.

While such idealised analyses are useful, they can lead to significant errors when the deformations of the structure are sufficient to change significantly the equations of equilibrium. In elastic analysis, this leads to non-linear behaviour, resulting in theoretically indefinitely large deflections as the elastic critical load is approached. If deflections calculated by a full elastic–plastic analysis significantly affect the equations of equilibrium, then the rigid–plastic collapse load ceases to be a good estimate of the failure load, and needs to be modified for use in design.

In arriving at an understanding of the strength and stability of real framed structures, the results of *idealised perfectly elastic* and of *idealised rigid–plastic* analyses can be combined to give a good estimate of the actual behaviour, even when change of geometry and stability effects are important. These idealised analyses will therefore first be discussed.

1.2 IDEALISED ELASTIC BEHAVIOUR

When instability and change of geometry effects are ignored, the deflections and stresses are linearly related to the intensity of loading. Under proportionate loading, characterised by a load factor, the deflection function $F_{\Delta L}$ is given by

$$F_{\Delta L} = \lambda f_{\Delta L} \tag{1.1}$$

where $f_{\Delta L}$ is the deflection function under unit load factor. Instability effects are introduced by the presence of axial load components in the members of the frame. If these axial loads are assumed to be proportionate to the applied loading (a reasonable approximation for most framed structures unless gross deformations are involved) there are directly calculable reductions in the stiffnesses of members carrying compressive loads. At certain successive critical load factors, $\lambda_{C1} \leqslant \lambda_{C2} \leqslant \lambda_{C3} \cdots$ (the 'eigenvalues'), the stiffness matrix vanishes, leading theoretically to the possibility of infinitely large deflections in the corresponding critical modes (the 'eigenvectors') at those load factors. If the eigenvectors are represented by respective deflection functions, $f_{\Delta 1}, f_{\Delta 2}, f_{\Delta 3}, \ldots$, etc., then these have orthogonal properties, and it is possible to express the linear deflected form $f_{\Delta L}$ in terms of the eigenvectors as

$$f_{\Delta L} = a_1 f_{\Delta 1} + a_2 f_{\Delta 2} + a_3 f_{\Delta 3} + \cdots \tag{1.2}$$

When instability and change of geometry effects (other than gross changes of geometry) are allowed for in a complete elastic analysis, the resulting deflection function F_Δ under load factor λ differs from the linear function $F_{\Delta L}$, and may be expressed as

$$F_\Delta = \frac{\lambda}{1 - \dfrac{\lambda}{\lambda_{C1}}} a_1 f_{\Delta 1} + \frac{\lambda}{1 - \dfrac{\lambda}{\lambda_{C2}}} a_2 f_{\Delta 2} + \frac{\lambda}{1 - \dfrac{\lambda}{\lambda_{C3}}} a_3 f_{\Delta 3} + \cdots \tag{1.3}$$

Hence the deflections theoretically approach very large values as λ approaches the lowest critical value λ_{C1}. In practice, the structure will cease to behave elastically at some stage, usually well before the lowest elastic critical load is reached. Apart from this consideration, however, eqn (1.3) would become invalid at large deflections because of two main considerations:

(1) When deflections become gross, it is no longer possible to express any deflected form in terms of the eigenvectors.

(2) When deflections become large, the pattern of axial forces in the members changes, and the critical load factors λ_{Ci} are no longer relevant.

Nevertheless, for structural deflections within the limits of practical interest, eqn (1.3) retains fully sufficient accuracy.

The calculation of the lowest critical load factor is important, both because it is the asymptotic value for very large elastic deflections and because it has particular significance in approximate methods for the estimation of elastic–plastic failure loads (see Section 1.5). While computer programs are available for the estimation of critical loads, they are neither so commonly available nor so easily applied and trouble-free as programs for linear analysis. For this reason, various approximate methods of estimating elastic initial loads have been proposed (Wood, 1974; Anderson, 1980). One of the simplest and most reliable is that due to Horne (1975), and uses the results of a linear elastic analysis, as follows.

Considering any rigid-jointed frame (Fig. 1.2(a)), let the loads W_1, W_2, etc., represent the total vertical loads (live plus dead) acting on the structure under factored loading, and suppose it is desired to find the

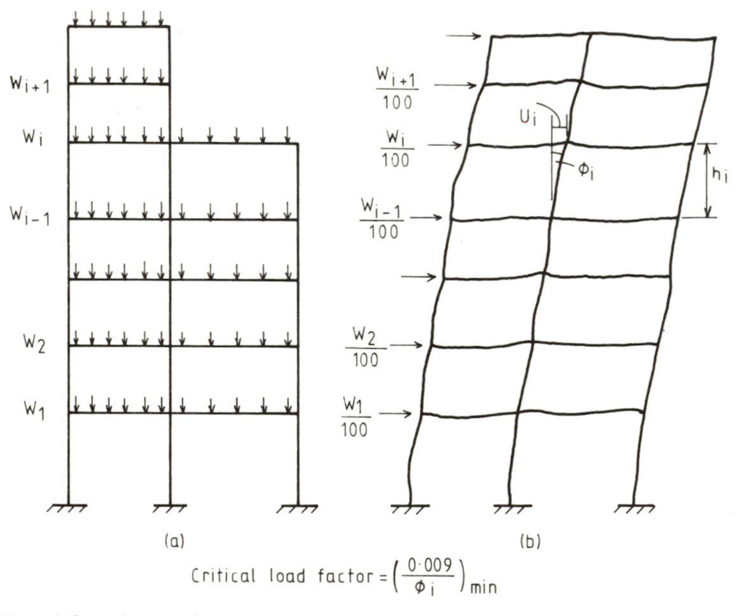

FIG. 1.2. Approximate calculation of the critical load of a rigid frame.

further common load factor α by which these factored loads are to be multiplied to produce elastic critical conditions. The loads W_i at any storey level i include the dead loads for the frame and the walls added at the level, as well as the dead and superimposed loads for the floor itself.

A linear elastic analysis is now performed with horizontal concentrated loads $W_i/100$ applied at each storey level i, as shown in Fig. 1.2(b), and the sway index $\Phi_i = u_i/h_i$ found for each storey (where u_i is the sway at any one floor relative to the floor immediately below and h_i is the storey height). Then, if Φ_{max} is the maximum value of Φ_i, considering all storeys,

$$\alpha = \frac{0{\cdot}009}{\Phi_{max}} \tag{1.4}$$

This method can be applied to a frame with any number of bays or storeys, and is applicable to frames in which the number of bays and/or the number of internal columns varies with the height. A proof that it represents an approximate lower bound to the critical load is given by Horne (1975).

1.3 STABILITY AND PLASTIC COLLAPSE

The plastic theory of structures is now well understood by engineers, and offers a straightforward means of assessing the ultimate load of a continuous structure. For many single-storey and low-rise frames, plastic design can form the basis of the design procedure, but it may be necessary to estimate the effect of instability on the plastic collapse load. The reason why instability affects the collapse load lies in the effect of deformation on the calculated internal forces—deformations either within the length of a member, or of the frame as a whole. The problem is therefore best introduced by considering the effect of deformations on plastic collapse loads.

The fundamental theorems of plasticity refer strictly to rigid–plastic materials, that is, to materials with an infinitely high modulus of elasticity (Fig. 1.1(d)). The structure is assumed to have no deformations at the collapse load. In exploring the effect of the finite deformations induced by elastic behaviour before collapse occurs, it is instructive first to consider the effect of finite deformations in a rigid–plastic structure.

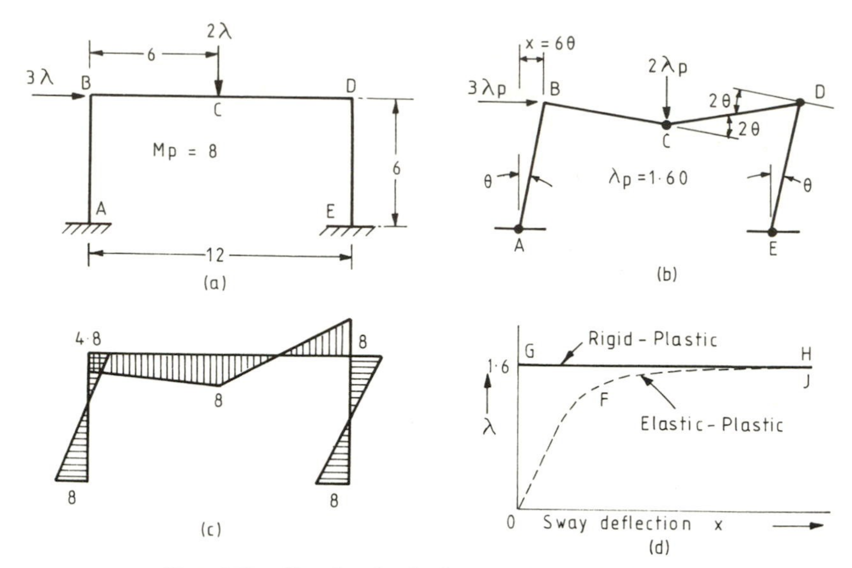

FIG. 1.3. Simple plastic theory of a portal frame.

The work equation at collapse, assuming infinitely small deformations, for the fixed base portal frame in Fig. 1.3(a) is derived from a general equation of the form

$$\lambda_P \sum Ws = \sum M_P \Phi \tag{1.5}$$

where the first summation is for all loads W (multiplied by the plastic collapse load factor λ_P) with their associated small displacements s, while the second summation is for all plastic hinges with full plastic moments M_P and hinge rotations Φ. It follows for the numerical example in Fig. 1.3 that $\lambda_P = 1{\cdot}60$. Confirmation that the correct mechanism is that shown in Fig. 1.3(b) is obtained by constructing the bending moment diagram at collapse, Fig. 1.3(c). This shows that the condition of not exceeding the full plastic moment has been satisfied. The rigid–plastic load–deflection relation is *OGH* in Fig. 1.3(d). If the elastic–plastic idealisation of the stress–strain relation is used to obtain a more accurate estimate of the true load–deflection curve for the structure, the curve *OFJ* is obtained, rising ultimately to meet the rigid–plastic line *GH*.

Considering now the effect of finite deformation, we take a finite rigid–plastic deformation of the portal frame in Fig. 1.3 according to the collapse mechanism, the column *AB* having rotated through an

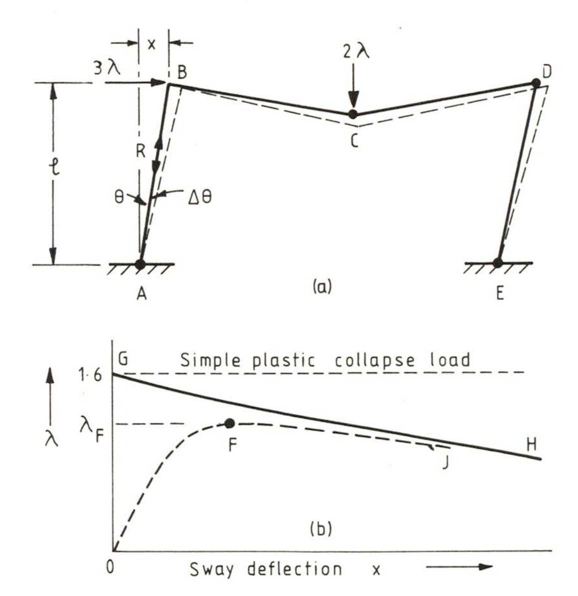

FIG. 1.4. Plastic theory of portal frame allowing for effect of change of geometry.

angle θ. The work equation for an incremental angle change $\Delta\theta$ (Fig. 1.4(a)) is then obtained, so that the derived load factor λ will correspond to the equilibrium state of the structure in the deformed position. It may be shown (Horne, 1963) that, if loads λW produce an axial compressive load R in a member of length l, and the total rotation of that member in the deformed state of the structure is β, then

$$\lambda \sum Ws + \sum R\beta^2 l = \sum M_P \Phi \tag{1.6}$$

The first and last summations in eqn (1.6) correspond to those in eqn (1.5), while the summation $\sum R\beta^2 l$ extends over each rigid length of member between plastic hinges or joints. The quantities s, β and Φ are expressible in terms of the rotations θ of one of the members of the frame in the simple collapse mechanism (Fig. 1.3(b)). The axial loads R may be assumed proportional to the load factor λ, and may be derived from the axial loads appropriate to the simple collapse mechanism at $\lambda = \lambda_P$.

The values of R at $\lambda = 1$ for the portal frame in Fig. 1.3 are found to be 1/3, 5/3, 5/3 and 5/3 for AB, BC, CD and DE, respectively (that is, 1/1·6 of the values appropriate to Fig. 1.3(b)). Hence eqn (1.6) becomes

$$\lambda\{3(6\theta)+2(6\theta)\}+\lambda\{(1/3)(6\theta^2)+(5/3)(6\theta^2)+(5/3)(6\theta^2)+(5/3)(6\theta^2)\}$$
$$=8\{\theta+2\theta+2\theta+\theta\} \quad (1.7)$$

and

$$\lambda=\frac{1\cdot6}{1+1\cdot07\theta}=\frac{1\cdot6}{1+0\cdot178x}$$

where x is the sway deflection. This gives the load–deflection relation *GH* in Fig. 1.4(b). There is thus a reduction of the load factor as the deflections of the structure increase, and this is generally true of all structures in which axial compressive loads preponderate over tensile loads.

The interest of Fig. 1.4(b) lies in the fact that the elastic–plastic behaviour, as represented by *OFJ*, must have a peak value below the rigid–plastic collapse load, represented by point *G*. Hence the rigid–plastic collapse load is to be regarded only as an upper bound on the true failure load of the structure. The extent to which the failure load factor λ_F for elastic–plastic behaviour falls below the rigid–plastic load factor λ_P depends on the slenderness of the structure. The criterion by which one may ascertain the stage at which the peak load has been reached may be obtained by the consideration of 'deteriorated critical loads'.

1.4 THE CONCEPT OF DETERIORATED CRITICAL LOADS

Consider any elastic structure, Fig. 1.5(a), and suppose that at any given load factor the total potential energy U is calculated as a

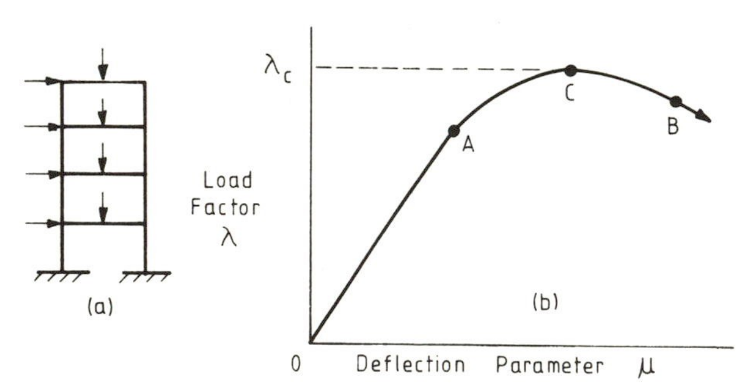

Fig. 1.5. Elastic load–deflection curve for a rigid frame.

function of some deflection parameter μ. The total potential energy U is the sum of the potential energy U_W of the external loads and the elastic strain energy U_E of the structure, so that $U = U_W + U_E$.

We now suppose that, as the loads on the structure are changed, the load–deflection curve takes the form *OACB* in Fig. 1.5(b). It is assumed that the structure remains elastic at all stages. Each point on the load–deflection curve represents a state of the structure in which the external loads are in equilibrium with the internal forces, and hence

$$\frac{\partial U}{\partial \mu} = 0 \tag{1.8}$$

It may be shown that, on the rising part of the load–deflection curve *OAC*, the potential energy U is a minimum with respect to small deviations from the equilibrium state, whence

$$\frac{\partial^2 U}{\partial \mu^2} > 0$$

as shown in Fig. 1.6(a). On the falling part of the curve the structure is unstable, $\partial^2 U/\partial\mu^2 < 0$ (Fig. 1.6(c)), while at the maximum load $\partial^2 U/\partial\mu^2 = 0$ (Fig. 1.6(b)) and the structure is in neutral equilibrium for small displacements. The structure is then at its elastic critical load.

Consider now the load–deflection curve *OAFD* (Fig. 1.7) for an elastic–plastic structure. We include in the total potential energy not only potential energy of their applied loads U_W and the elastic strain energy U_E, but also the energy absorbed in plastic deformation U_P. The energy absorbed in plastic deformation depends on the loading path for the structure, and it is assumed that the path taken is that

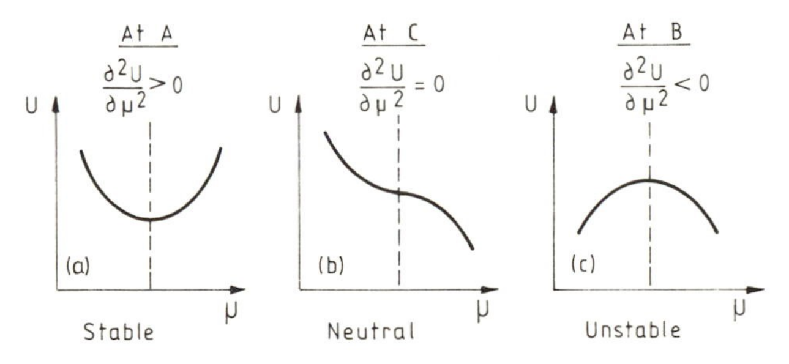

FIG. 1.6. Stability criterion for elastic structures.

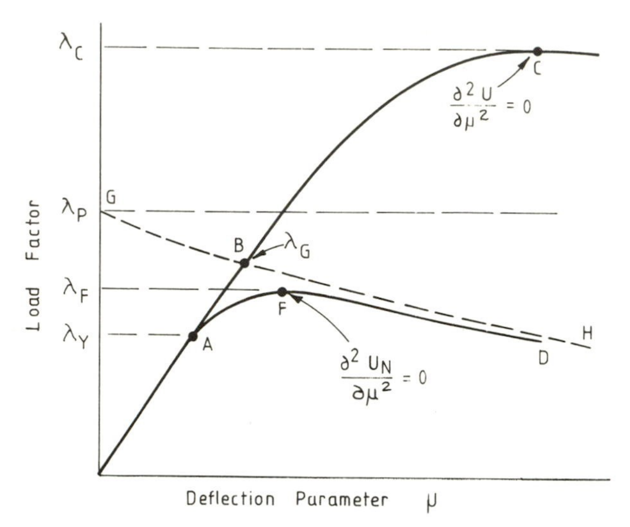

FIG. 1.7. Load–deflection curve for an elastic–plastic structure.

represented by the load–deflection curve. With this proviso, and denoting this modified total energy by U_N, so that

$$U_N = U_W + U_E + U_P \tag{1.9}$$

the condition $\partial U_N/\partial\mu = 0$ is satisfied at all points on the load–deflection curve. Moreover, $\partial^2 U_N/\partial\mu^2 > 0$ before the failure load factor λ_F is reached, $\partial^2 U_N/\partial\mu^2 < 0$ on the falling part of the curve and $\partial^2 U_N/\partial\mu^2 = 0$ at the failure load. However, in the plastic zones of the structure the stress is constant, whence $\partial^2 U_P/\partial\mu^2 = 0$, and hence the condition at failure becomes

$$\frac{\partial^2(U_W + U_E)}{\partial\mu^2} = 0 \tag{1.10}$$

Since the elastic strain energy in the plastic zones of the structure is constant, it follows that the failure criterion of the elastic–plastic structure is identical with the elastic critical load criterion for the same structure, but with parts of the structure—those that are plastically deforming—eliminated. The structure in this depleted condition is termed the deteriorated structure, and the corresponding elastic critical load is called the deteriorated critical load. If, for example, a plastic hinge has formed in an elastic–plastic structure, the hinge provides zero increase of resisting moment for an increase of rotation, and is to

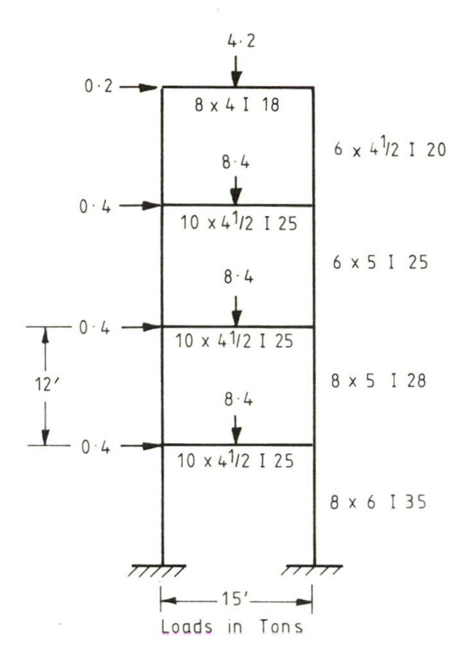

FIG. 1.8. Frame analysed by Wood (1957).

be regarded as a pin joint when computing the deteriorated critical load. As the load on a structure increases and plastic zones spread, the deteriorated critical load factor decreases continuously until it is depressed down to the load corresponding to the actual load applied. It is at this stage that the structure becomes unstable and the failure load factor λ_F is attained.

The concept of deteriorated critical loads has been developed by Wood (1957), who gives the following example. The frame in Fig. 1.8 is composed of I-section members with webs in the plane of the frame, the 'working' values of the loads (in tons, load factor $\lambda = 1$) being as shown. Wood obtained the following values for λ_C, λ_P and λ_F:

Lowest elastic critical load factor	$\lambda_C = 12{\cdot}9$
Rigid–plastic load factor	$\lambda_P = 2{\cdot}15$
Elastic–plastic failure load	$\lambda_F = 1{\cdot}90$

The elastic critical load was calculated for beam loads equally divided between joints, as shown in Fig. 1.9. The rigid–plastic collapse mechanism is shown in Fig. 1.9(a), and the theoretical state of the

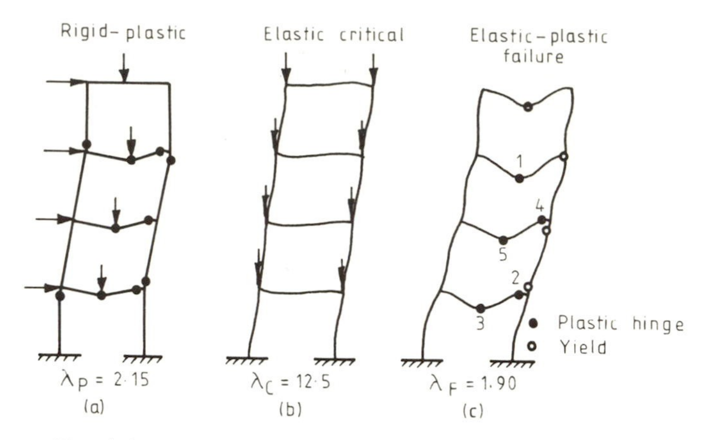

FIG. 1.9. Behaviour of frame analysed by Wood (1957).

elastic–plastic structure at collapse in Fig. 1.9(c). At the theoretical failure load factor of $\lambda_F = 1{\cdot}90$, plastic hinges had formed at positions 1, 2, 3 and 4, and a fifth hinge has practically formed at position 5. A certain amount of plastic deformation had also occurred at the other sections at which yield is indicated.

Wood calculated the elastic critical loads of his frame with pin joints assumed at various sections, in order to obtain general guidance in his full analysis. Some of his results are shown in Figs. 1.10(a)–(e), the circles representing the positions of pin joints. The load factors λ_D quoted represent the elastic critical loads of these deteriorated structures for vertical loads applied at the joints, as in Fig. 1.9(b). None of the deteriorated structures correspond exactly to the deteriorated

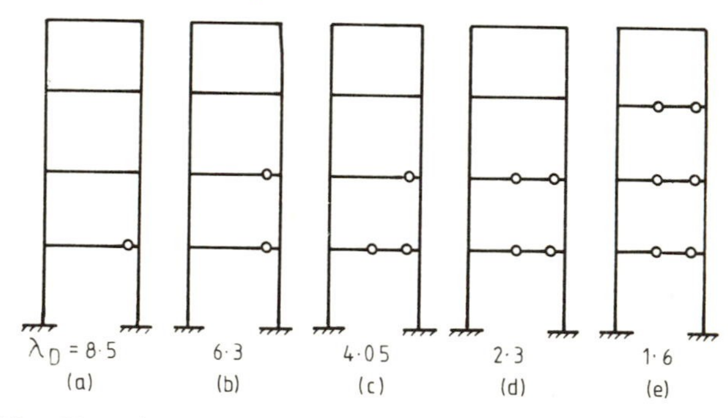

FIG. 1.10. Deteriorated critical loads of frame analysed by Wood (1957).

structure at the failure load in the full analysis ($\lambda_F = 1{\cdot}90$, Fig. 1.8(c)). A deteriorated structure intermediate between those in Figs. 1.10(d) and (e) is, however, seen to be appropriate (cf. Fig. 1.9(c)), and the deteriorated critical load factors 2·30 and 1·60 lie either side of the failure load 1·90.

The concept of the deteriorated critical load clarifies thinking on the subject of elastic–plastic instability of structures. Deteriorated critical loads are also of importance in elastic–plastic computer methods for analysis of structures up to collapse (Majid and Anderson, 1968; Majid, 1972). It is, however, only of very limited assistance in the actual calculation of failure loads, since the deteriorated structure cannot be obtained unless a full elastic–plastic analysis has in any case been derived. Moreover, if deteriorated critical loads are discussed in isolation from a complete analysis, misleading results may be obtained. Thus, it may be seen that, although ten hinges are required for the rigid–plastic collapse of the frame in Fig. 1.8, two hinges only are sufficient to halve the critical load (Fig. 1.10(b)). Despite this alarming reduction, the failure load was only 12% below the rigid–plastic collapse load. Hence a more reliable means of estimating approximately the effect of instability on failure loads is required, and this is discussed in the next section.

1.5 EMPIRICAL APPROACHES TO THE ESTIMATION OF FAILURE LOADS: THE MERCHANT–RANKINE LOAD

It has been seen that, if the material of a structure is assumed to be rigid–plastic, a drooping load–deflection curve *GH* is obtained (Fig. 1.7), descending from the rigid–plastic collapse load factor λ_P. If ideal elastic behaviour is assumed, the load–deflection curve *OBC* rises to the elastic critical value λ_C. The actual behaviour *OAFD* follows the elastic curve up to the load factor λ_y at which yield first occurs, rising to a peak at *F*, before approaching the rigid–plastic mechanism line *GH* at large deflections. Merchant (1954, 1958) suggested that it might be possible to consider the failure load factor λ_F as some function of the load factors λ_y, λ_P and λ_C, and also of the purely abstract load factor λ_G (Fig. 1.7) obtained at the intersection of the elastic curve *OAC* with the rigid–plastic mechanisms line *GH*. The advantage of such an approach is that these load factors are much easier to calculate than the failure load factor λ_F itself.

Merchant (1954) tested, for a large number of theoretical structures, the following formula, which may be regarded as a generalisation of Rankine's formula for struts. The failure load factor λ_F is approximated to by the Merchant–Rankine load factor λ_{MR}, where

$$\lambda_{MR} = \frac{\lambda_P \lambda_C}{\lambda_P + \lambda_C} \tag{1.11}$$

Since for many practical structures λ_C is large compared with λ_P, the requirement $\lambda_F \simeq \lambda_P$ for such structures is satisfied. If λ_F/λ_P is plotted vertically and λ_F/λ_C horizontally, eqn (1.11) with $\lambda_F = \lambda_{MR}$ is simply given by the straight line *AB* in Fig. 1.11. The points plotted on Fig. 1.11 were obtained theoretically by Salem (1958) for one- and two-storey frames loaded as shown. Lines corresponding to various ratios of λ_P/λ_C have been drawn, and it is readily seen that the Merchant–Rankine formula (eqn (1.11)) is most successful when λ_P/λ_C is small and the collapse load is close to the rigid–plastic collapse value. When $\lambda_P/\lambda_C > 0{\cdot}3$ the scattering of the points away from the Merchant–Rankine formula is considerable. Merchant suggested the formula as a safe (that is, lower) limit for the collapse load. Its theoretical significance has been discussed by Horne (1963).

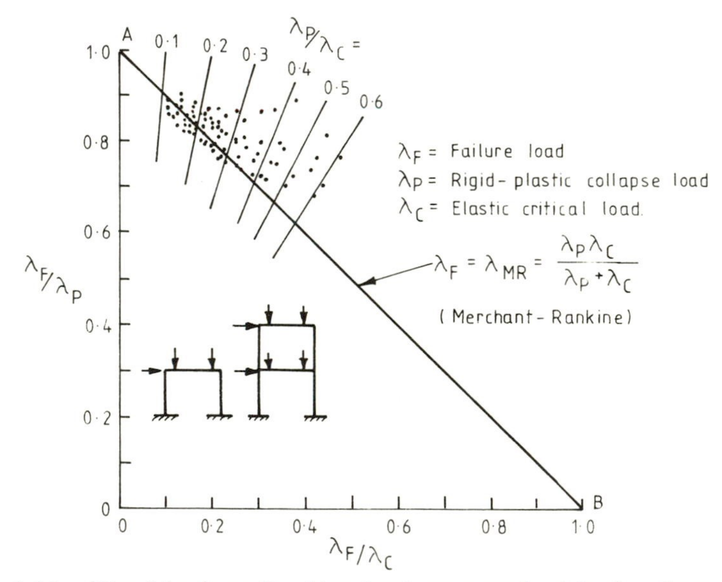

FIG. 1.11. The Merchant–Rankine load compared with the theoretically obtained failure load.

1.6 MODIFICATION OF MERCHANT–RANKINE LOAD

In a series of papers, Wood (1974) suggested a modification of the Merchant–Rankine load to allow for the minimum beneficial effects that must always be present from strain-hardening and restraint provided by cladding. He suggested that, provided

$$\frac{\lambda_C}{\lambda_P} > 10, \quad \text{then} \quad \lambda_F = \lambda_P \tag{1.12}$$

and when

$$10 > \frac{\lambda_C}{\lambda_P} > 4, \quad \text{then} \quad \lambda_F \simeq \lambda_{WMR} = \frac{\lambda_P \lambda_C}{\lambda_P + 0{\cdot}9\lambda_C} \tag{1.13}$$

These proposals may be expressed graphically by the lines *ACD* in Fig. 1.12, and may be compared with the Merchant–Rankine relationship given by the straight line *AB*.

When applying condition (1.12) to derive design requirements, we may note that the required minimum value of λ_F is that corresponding to the required factored loading. We may conveniently put

$$\frac{\lambda_C}{\lambda_F} = \alpha = \frac{\text{Elastic critical load}}{\text{Factored load}}$$

The simple plastic collapse load λ_P is derived by using the design strength p_y of the steel. Suppose we express the required minimum failure load factor of the structure λ_F as a plastic collapse load factor,

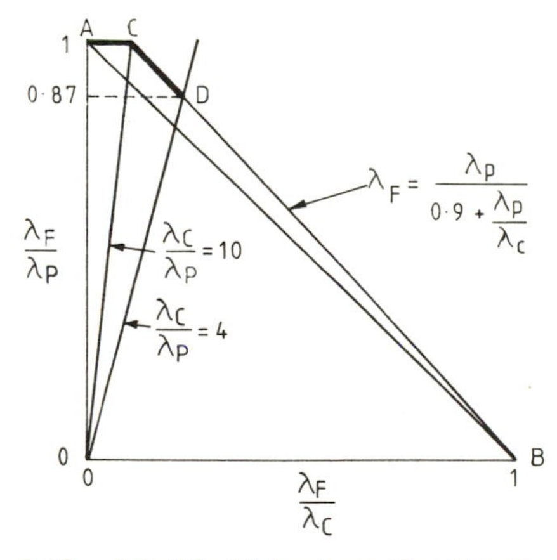

FIG. 1.12. Modified Merchant–Rankine formula.

using, instead of the specified design strength p_y, an *effective* design strength p'_y, so that

$$\frac{\lambda_F}{\lambda_P} = \frac{p'_y}{p_y}$$

whence

$$\frac{\lambda_C}{\lambda_P} = \alpha \frac{p'_y}{p_y}$$

On substituting the above values of $\lambda_F/\lambda_P \simeq \lambda_{WMR}/\lambda_P$ and λ_C/λ_P in eqn (1.12), it is found that, in design, we may assume that when $\lambda_C/\lambda_F > 10$, $p'_y = p_y$ and when

$$10 > \frac{\lambda_C}{\lambda_F} > 4{\cdot}6, \qquad p'_y = \frac{\alpha - 1}{0{\cdot}9\alpha} p_y \tag{1.14}$$

This proposal is the basis for the adoption of plastic design for continuous low-rise frame in the 'Draft British Standard for the Use of Structural Steelwork in Building' BS 5950 (1983). When $\lambda_C/\lambda_F < 4{\cdot}6$, it is recommended that, if elastic–plastic methods of ultimate load design are to be employed, then a special analysis allowing for elastic–plastic behaviour and change of geometry effects should be undertaken. It will, of course, also be necessary to satisfy the deflection limits imposed by the provisions of the appropriate clauses in the Standard.

It should be noted that, if the more conservative unmodified Merchant–Rankine load λ_{MR} (eqn (1.11)) is used in design rather than eqns (1.12) and (1.13), then a modified yield stress P'_y must always be used, where

$$p'_y = \frac{\alpha - 1}{\alpha} p_y \tag{1.15}$$

1.7 ACCURACY OF MERCHANT–RANKINE AND MODIFIED MERCHANT–RANKINE LOADS

Numerical examples based on second-order elastic–plastic analyses led Merchant (1954) to the conclusion that the Merchant–Rankine load (eqn (1.11)) represented an approximate lower bound to the failure load. He did not claim it as a *strict* lower bound (see Fig. 1.11). However, if it is to be used as the basis for design, one needs some assurances that no serious error on the unsafe side will arise. Some evidence has been given emphasising the extent to which the Merchant–Rankine load is not a lower bound (Adam, 1979) and for this reason Anderson and Lok (1983) examined a number of 4-, 5- and 10-storey unbraced frames. These frames were realistic in that they were designed economically for wind and floor loadings specified in codes, with sway deflections at unit factored wind loading limited to 1/300 of each storey height. Some of their results are summarised in Table 1.1. Two loading conditions are considered—one with maximum wind loading and the other with maximum vertical loading. Values are quoted for the ratios of the lowest elastic critical load factor λ_C, the Merchant–Rankine load factor λ_{MR} and the modified Merchant–Rankine load factor λ_{WMR} to the accurately calculated 'second-order' (i.e. allowing for change of geometry) elastic–plastic failure load factor λ_F. It will be seen that, without exception, the Merchant–Rankine load is a safe estimate of the failure load. Anderson and Lok found that only when the storey height exceeded the span of the beams, and when the wind loading was exceptionally low, was the Merchant–Rankine load higher than the elastic–plastic failure load. Such frames are not realistic—and if they occurred, would certainly be braced.

Anderson and Lok recommend that the Merchant–Rankine formula should not be used when the bay width is less than the greatest height

TABLE 1.1

RESULTS OBTAINED BY ANDERSON AND LOK (1983) FOR UNBRACED MULTI-STOREY FRAMES

Number of storeys	*Number of bays*	*Bay width (m)*	*Min. vertical–Max. wind*			*Max. vertical–Min. wind*		
			$\frac{\lambda_C}{\lambda_P}$	$\frac{\lambda_{MR}}{\lambda_F}$	$\frac{\lambda_{WMR}}{\lambda_F}$	$\frac{\lambda_C}{\lambda_P}$	$\frac{\lambda_{MR}}{\lambda_F}$	$\frac{\lambda_{WMR}}{\lambda_F}$
4	2	7·5	9·15	0·96	1·07	5·75	0·91	1·00
	5	7·5	5·34	0·89	0·97	4·78	0·90	0·98
	2	5	11·38	0·95	1·03	6·56	0·89	0·98
	5	5	5·34	0·95	1·03	3·19	0·86	0·93
7	2	7·5	12·82	0·97	1·05	5·66	0·93	1·01
	5	7·5	5·06	0·93	1·01	4·36	0·90	0·97
	2	5	16·59	0·96	1·02	7·98	0·94	1·04
	5	5	5·78	0·90	0·98	3·98	0·89	0·97
10	2	7·5	15·33	0·97	1·04	6·02	0·95	1·04
	4	7·5	6·57	0·93	1·02	3·87	0·90	0·98
	2	5	14·90	0·95	1·01	8·07	0·91	1·07
	4	5	8·23	0·96	1·06	4·56	0·90	0·98

of one storey. For multi-bay frames with unequal bays, the average bay width should be compared with the storey height.

The generally conservative nature of the Merchant–Rankine load for practical frames (Table 1.1) encourages the use of the modified Merchant–Rankine load, although in some cases the predicted load factor at collapse λ_{WMR} may exceed the elastic–plastic failure load λ_F. Anderson and Lok (1983) point out, however, that the cases were not very significant because the excess load capacity was small (not more than 7%), and the computer program used to calculate λ_F ignored the beneficial effects of strain-hardening and strong composite action (see Wood, 1974). For these reasons, and provided the restriction on bay width is observed, there would seem to be no case for not adopting the modified Merchant–Rankine formula for design.

1.8 EXAMPLE OF UNBRACED MULTI-STOREY FRAME DESIGNED WITH ALLOWANCE FOR FRAME STABILITY EFFECTS

The four-storey frame in Fig. 1.13(a) is to be designed for the factored load combination shown in Figs. 1.13(b), (c) and (d), using grade 43 steel ($p_y = 245\ \text{N/mm}^2$). Suitable sections, derived by applying simple plastic theory, are as shown in Table 1.2. The most critical combination of loads is for dead plus superimposed plus wind, and the plastic collapse mechanism is shown in Fig. 1.14. The factor λ is that by which the required factored loads are to be multiplied to cause collapse in the frame as designed, and this has the value 1·018.

Applying Horne's method to the calculation of the elastic critical load, the applied horizontal loads and resulting sway index values are given in Table 1.3. Hence $\Phi_{max} = 0{\cdot}001\,37$ and from eqn (1.4),

$$\alpha = \frac{\lambda_C}{\lambda_F} = \frac{0{\cdot}009}{0{\cdot}001\,37} = 6{\cdot}56$$

and

$$p'_y = \frac{6{\cdot}56 - 1}{0{\cdot}9(6{\cdot}56)}\,240 = 226\ \text{N/mm}^2$$

The effective design stress required to support the factored loads in Fig. 1.13(d) is $245/1{\cdot}018 = 241\ \text{N/mm}^2$, so that some strengthening of the frame is required.

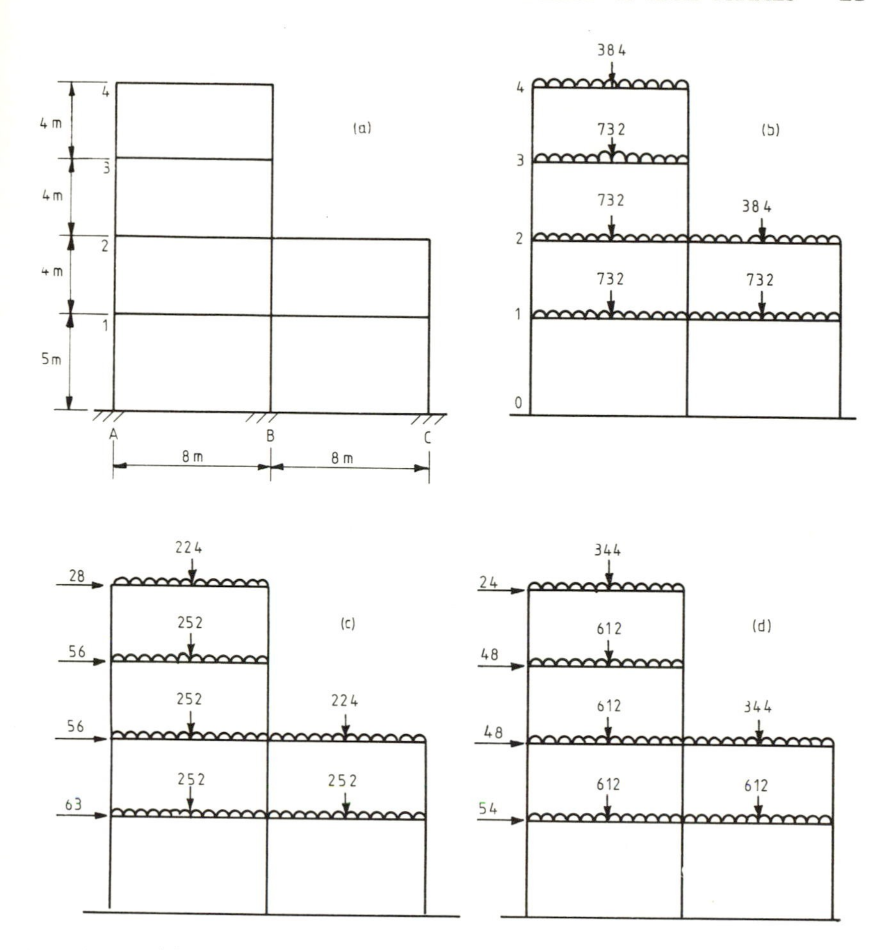

FIG. 1.13. (a) Four-storey frame; (b) equivalent factored dead plus superimposed vertical loads (kN); (c) equivalent factored dead plus wind load (kN); (d) equivalent factored dead plus superimposed plus wind loads (kN).

By inspection of the collapse mechanism in Fig. 1.14, the best section to increase is that of the two-storey column length C02 on the right-hand side. This is therefore changed from 203×203 UC 52 to 254×254 UC 73, giving $\lambda = 1{\cdot}110$ in Fig. 1.14. Hence, the applied factored loads require a minimum effective design stress of $245/1{\cdot}110 = 221$ N/mm^2, which is satisfactory compared with the attained value (from Horne's method) of 226 N/mm^2. (Strictly speaking,

TABLE 1.2
SECTIONS FOR FRAME IN FIG. 1.13, CHOSEN BY PLASTIC THEORY

Member		*Section*
Columns	A02	254×254 UC 89
	A24	203×203 UC 52
	B02	305×305 UC 118
	B24	203×203 UC 52
	C02	203×203 UC 52
Beams	AB1	533×165 UC 73
	AB2	533×165 UB 66
	AB3	610×178 UB 82
	AB4	457×152 UB 52
	BC1	533×165 UB 73
	BC2	356×171 UB 51

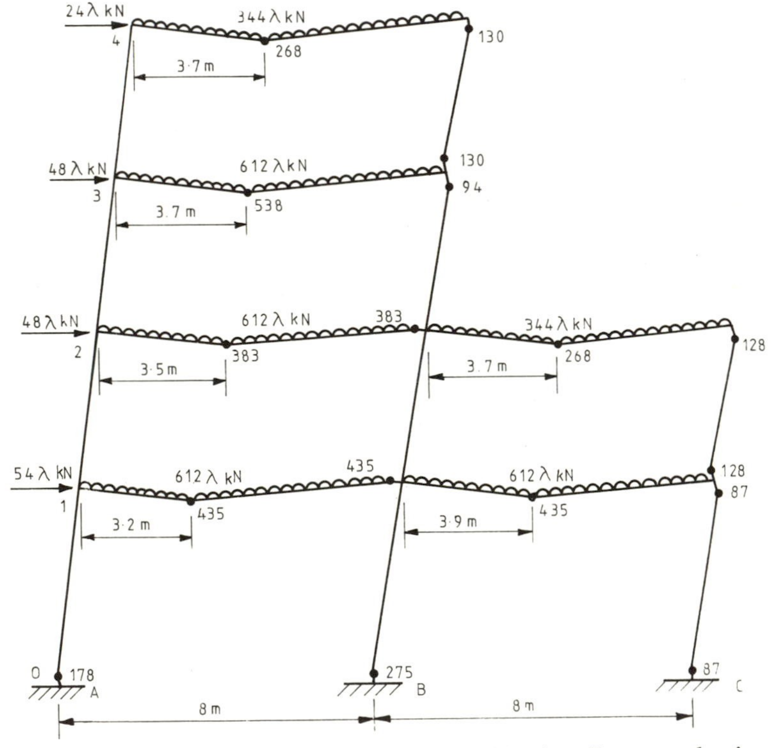

FIG. 1.14. Dead plus superimposed plus wind loads collapse mechanism.

TABLE 1.3
CALCULATION OF CRITICAL RATIO FOR FRAME IN FIG. 1.13 (HORNE'S METHOD)

Storey	*1% Vertical loads (kN)*	*Sway index ($10^{-3}x$)*
1	14·97	1·37
2	11·74	0·87
3	7·80	1·15
4	5·12	0·50

the α values have been derived for the originally designed frame and are not applicable to the revised frame. However, this latter would give slightly higher α values than the original frame, so that the error is on the safe side.)

1.9 OVERALL STABILITY PROBLEMS IN THE PLASTIC DESIGN OF SINGLE-STOREY FRAMES

1.9.1 General Considerations

The assumption has usually been made in the past that overall stability problems do not affect the design of single-storey frames, the argument being that the mean axial stresses in the columns are generally small. While this latter statement is correct, it is also true that such frames may be quite slender in the plane of bending, and this may bring down the ratio of critical load to plastic collapse load to an unacceptably low level.

1.9.2 Single-Bay Frames

In single-bay frames the usual deflection limitations when the frame is subjected to wind loading will usually ensure that overall stability is not a controlling factor. This is certainly the case when the horizontal deflections at the tops of the stanchions are limited to height divided by 300, under unfactored loads. However, if it can be shown that greater deflections would not impair the strength or efficiency of the structure, or lead to damage to cladding, then this deflection limit may be allowed to be exceeded, and there will undoubtedly be a desire to take advantage of this in many single-storey frames. A safeguard against deflections which could affect strength and safety is therefore needed, and is provided by the following requirement (see Horne, 1977).

The horizontal deflection δ, in millimetres, at the top of a column due to horizontal loads applied in the same direction at the top of each column, and equal to 1% of the vertical load in the column due to factored loads, must not exceed $1{\cdot}8h$, where h is the height of the column in metres. In calculating δ, allowance may be made for the restraining effect of cladding.

As a simple check on the requirement, the condition may be shown (Horne, 1977) to be satisfied, provided

$$\frac{h}{D} \ngtr \frac{50}{\Omega}\left(\frac{\rho}{4+\rho \sec \theta}\right)\left(\frac{240}{p_y}\right) \tag{1.16}$$

where

$$\rho = \left\{\frac{2I_C}{I_R}\right\}\left\{\frac{L}{h}\right\}$$

and L = span of frame (m), D = minimum depth of rafter (m), h = column height (m), I_C, I_R = minimum second moments of area of columns and rafters, respectively for bending in the plane of the frame, p_y = design strength of rafter (N/mm^2), θ = angle of pitch of a ridged portal, $\Omega = W/W_0$ = arching ratio, W = factored vertical load on frame, W_0 = maximum uniformly distributed load which could be supported by a horizontal roof beam of span L continuous with the columns with the same distribution of cross-section in plan as the actual rafter.

In the above formula the assumption is made that the columns are rigidly fixed to the rafters, but it is assumed that they have pinned bases. If the columns are fixed-based, then the above treatment is conservative.

1.9.3 Multi-Bay Frames

It might at first be thought that the overall stability problems of multi-bay frames would be less than those of single-bay frames. It is, however, possible to reduce considerably the size of internal rafters by making use of the 'arching effect', whereby the plastic collapse of an internal rafter (Fig. 1.15(a)) is assumed to occur by beam mechanisms, each involving only a rafter to one side of a ridge. Moreover, the balancing of the rafter moments at the tops of the internal columns may allow them to become quite small in section—and even (although this is deprecated) allow them to become pin-ended struts. If such highly 'competitive' design is indulged in, or even approached, two

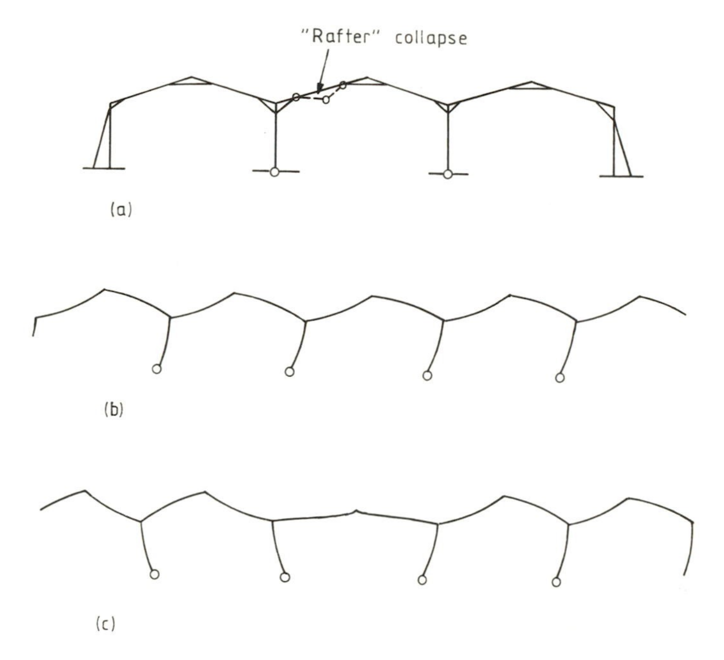

FIG. 1.15. (a) Propped frame with minimum internal members; (b) interior sway buckling; (c) interior snap-through buckling.

forms of instability may arise, as shown in Figs. 1.15(b) and (c). The first is internal sway instability, and is liable to occur in multi-bay frames with slender internal columns. The second is 'snap-through' instability, and is particularly dangerous when considerable use is made of the arching effect with low-pitch rafters.

Safeguards against both forms of instability have been discussed by Horne (1977), and this work has led to the following requirements being recommended for inclusion in the 'Draft British Standard for the Use of Structural Steel in Building' (1983).

Snap-through instability is safeguarded against by requiring that, for each internal bay of a multi-bay frame,

$$\frac{L}{D} \ngtr \frac{25(4+L/h)}{\Omega(\Omega-1)}\left\{1+\frac{I_C}{I_R}\right\}\left\{\frac{240}{p_y}\right\}\tan 2\theta \qquad (1.17)$$

where the symbols have the same meanings as previously. It is assumed in general that the columns are pinned to the foundations and rigidly connected to the rafters. The same treatment may, however, be applied to frames with columns pinned to rafters, by replacing I_C by zero, but it is of course still assumed that the rafters are rigidly connected together above the columns.

Sway instability is controlled by requirements similar to those for single-bay frames. The horizontal deflection at the top of any column due to 1% of vertical factored loads applied simultaneously in the same direction to the tops of all columns should not exceed $1{\cdot}8h$ (millimetres) where h (metres) is the height of the column. The algebraic limitation (eqn (1.16)) on the value h/D ensures the satisfaction of this requirement for columns rigidly connected to the rafters but with $\rho = \{I_C/I_R\}\{L/h\}$.

1.10 EXAMPLE OF DESIGN OF SINGLE-BAY PITCHED PORTAL FRAME WITH ALLOWANCE FOR FRAME STABILITY EFFECTS

Portal frames, at 6 m spacing and of span 12 m, have a height to eaves of 3 m and an angle of pitch of 22·5°. The total factored vertical load is 2550 N/m^2. Establish a suitable design for a frame of uniform section, using grade 50 steel.

Plastic analysis shows a required plastic moment of 99·4 kN m.

$$\text{Required plastic modulus } S = \frac{99{\cdot}4 \times 10^3}{340} = 292 \text{ cm}^3.$$

<u>Try 254×102 UB 25,</u> $S = 305{\cdot}3 \text{ cm}^3$.

The mean axial stress in the rafter is 15 N/mm^2, and the effect on plastic modulus is negligible.

$$W_0 = \text{load capacity for flat roof}$$

$$= \frac{16M_P}{L} = \frac{16 \times 305{\cdot}3 \times 340}{12 \times 10^3} = 138{\cdot}4 \text{ kN}$$

where M_P = plastic moment of resistance

$$W = \text{actual factored load}$$

$$= \frac{2550 \times 6 \times 12}{10^3} = 183{\cdot}6 \text{ kN}$$

Hence

$$\Omega = \text{arching factor} = \frac{183{\cdot}6}{138{\cdot}4} = 1{\cdot}327$$

$$\rho = \frac{2I_C}{I_R}\frac{L}{h} = \frac{2\times 12}{3} = 8{\cdot}00$$

Hence

$$\frac{h}{D} \ngtr \frac{50}{1{\cdot}327}\cdot\frac{8{\cdot}00}{4+8{\cdot}00\sec 22{\cdot}5}\cdot\frac{240}{340} = 16{\cdot}8$$

But

$$D = 257\text{ mm} \quad \text{and} \quad \frac{h}{D} = \frac{3000}{257} = 11{\cdot}7$$

Hence frame is satisfactory for sway stability.

1.11 EXAMPLE OF DESIGN OF MULTI-BAY PITCHED PORTAL FRAME WITH ALLOWANCE FOR FRAME STABILITY EFFECTS

Multi-bay frames consisting of spans of 30 m have rafters of constant pitch 10° and columns of height 4 m, rigidly fixed to the rafters and pinned to foundations. The frames are at 9 m spacing and support a total factored load of 2550 N m^2. Establish whether or not it would be permissible to design the internal rafters to a minimum uniform section, derived by considering the plastic collapse of a single rafter member as a fixed-ended beam.

Total factored load per frame is

$$W = \frac{2550\times 9\times 30}{10^3} = 688{\cdot}5\text{ kN}$$

$$\text{Load/rafter} = 344{\cdot}2\text{ kN}$$

$$\text{Required plastic moment} = \frac{344{\cdot}2\times 15}{16} = 323\text{ kN m}$$

$$\text{Required } S = \frac{323\times 10^3}{340} = 950\text{ cm}^3.$$

Try 457×152 UB 52, $S = 1095\text{ cm}^3$.

Ignoring effect of axial stress on M_P,

$$W_0 = \frac{16M_P}{L} = \frac{16 \times 1095 \times 340}{30 \times 10^3} = 198{\cdot}4 \text{ kN}$$

and

$$\Omega = \frac{W}{W_0} = 3{\cdot}470$$

$$I_R = 21\,345 \text{ cm}^4 \text{ and } D = 450 \text{ mm}$$

For the columns, a 254×254 UC 89 is found to give adequate stability against failure about the minor axis. This gives $I_C = 14\,307 \text{ cm}^4$.

Considering snap-through stability, from condition (1.17),

$$\frac{L}{D} \not> \frac{25(4+7{\cdot}5)}{3{\cdot}47 \times 2{\cdot}47}\left\{1+\frac{14\,307}{21\,345}\right\}\frac{240}{340}\tan 20° = 14{\cdot}4$$

Actual $L/D = 30\,000/450 = 66{\cdot}7$.

Hence the frame cannot be designed to take full advantage of maximum arching action due to danger of snap-through instability.

The large margin by which the permissible slenderness is exceeded indicates that plastic design is not ideally suited to this problem. It also indicates high sensitivity to non-uniform loading. If maximum assistance from arching action is desired, elastic design should be used, in conjunction with an elastic analysis using stability functions. Alternatively, the frame may be overdesigned plastically to the degree required for eqn (1.17) to be satisfied.

Following this latter course, try 610×178 UB 82 for rafters.

$$S = 2194 \text{ cm}^3$$

$$W_0 = \frac{16 \times 2194 \times 340}{30 \times 10^3} = 397{\cdot}8 \text{ kN}$$

$$\Omega = \frac{688{\cdot}5}{397{\cdot}8} = 1{\cdot}73$$

Application of condition (1.17) gives $L/D \not> 73{\cdot}5$, compared with actual $L/D = 50{\cdot}2$, so the section is satisfactory.

1.12 CONCLUSIONS

In order for plastic theory to be safely applied to the design of multi-storey frames and some slender single-storey portal frames, checks need to be made to see whether frame stability effects need to be taken into account. The Merchant–Rankine load, both in its original form and as modified by Wood, are suitable means of making allowance for frame stability effects. These approximate estimates of failure loads require approximate calculations of elastic critical loads. Hence, the rigid–plastic collapse load and the lowest elastic critical load are clearly established as the two most important theoretical quantities in the practical estimation of the ultimate carrying capacity of a rigid framed structure.

REFERENCES

ADAM, V. (1979) Traglastberechnung beliebiger ebener unausgesteifter Rahmentragwerke aus Stahl (The calculation of the bearing capacity of plane unbraced steel steel frames). Dissertation, Technische Hochschule, Darmstadt.

ANDERSON, D. (1980) Simple calculation of elastic critical loads for unbraced, multistorey steel frames. *Structural Engineer*, **58A,** 243.

ANDERSON, D. and LOK, T. S. (1983) Design studies on unbraced, multistorey steel frames. *Structural Engineer*, **61B,** 29.

HORNE, M. R. (1963) Elastic–plastic failure loads of plane frames. *Proc. Roy. Soc. A*, **274,** 343.

HORNE, M. R. (1975) An approximate method for calculating the elastic critical loads of multi-storey plane frames. *Structural Engineer*, **53,** 242.

HORNE, M. R. (1977) Safeguards against frame instability in the plastic design of single storey pitched-roof frames. Presented at Conference on the Behaviour of Slender Structures, The City University, London (14–16 Sept.).

MAJID, K. I. (1972) *Non-Linear Structural Matrix Methods of Analysis and Design*, Butterworths, London.

MAJID, K. I. and ANDERSON, D. (1968) Elastic–plastic design of sway frames by computer. *Proc. Instn. Civ. Engrs.*, **41,** 705.

MERCHANT W. (1954) The failure load of rigid jointed frameworks as influenced by stability. *Structural Engineer*, **32,** 185.

MERCHANT, W., RASHID, C. A., BOLTON, A. and SALEM, A. (1958) The behaviour of unclad frames. *Proc. Fiftieth Anniv. Conf. Instn. Struct. Engrs.*, Institution of Structural Engineers, London.

SALEM, A. (1958) Structural frameworks. PhD Thesis, University of Manchester.

WOOD, R. H. (1957) The stability of tall buildings. *Proc. Instn. Civ. Engrs.*, **11,** 69.

WOOD, R. H. (1974) Effective lengths of columns in multi-storey buildings. Parts 1, 2 and 3. *Structural Engineer*, **52,** 235, 295 and 341.

Chapter 2

MATRIX METHODS OF ANALYSIS OF MULTI-STOREYED SWAY FRAMES

T. M. Roberts

Department of Civil and Structural Engineering, University College, Cardiff, UK

SUMMARY

The ultimate limit state design of plane multi-storeyed sway frames involves consideration of nonlinear effects induced by changes of geometry and the influence of member axial forces. In recent years, nonlinear matrix methods of analysis have been developed and it is now possible to analyse the complete loading history of such frames up to collapse. This chapter deals with the elastic linear, nonlinear and instability analysis of frames using matrix methods. Rigorous formulations of the problem are presented first so that the approximations frequently incorporated in such analyses can be identified. Approximate methods for determining elastic critical loads and the magnitude of the nonlinear effects are then presented.

NOTATION

A	Cross-section area
a	Length of an element or a member of a frame
c	Stability function
E	Young's modulus
f	Shear force
H	Horizontal force
I	Second moment of area about centroid
m	Moment

n	Scalar
P_0	Base load
P_i	External force
q_i	Displacements of external forces
r	Stability function
s	Stability function
t	Axial force
U	Displacement in global X direction
UE	Strain energy
u	Axial displacement in x direction
$\bar{u}$	Axial shortening of element or member
V	Potential energy
V_0	Initial value of V
VP	Potential energy for stationary external forces
W	Displacement in global Z direction
w	Bending displacement in z direction
w_0	Initial value of w
$\bar{w}$	Sway of column or storey of frame (Figs 2.4 and 2.5)
X, Z	Global coordinate axes
x, z	Local element coordinate axes
z	Distance from centroid in z direction
β	Angle between global and local coordinates
β_i	Ratio of second order sway displacement to linear elastic sway displacement
γ	Defined by eqn (2.42)
δ	First variation
δ^2	Second variation
ε	Axial strain
ε_i	Strain
λ	Load factor
λ_{cr}	Critical load factor
ρ	Uniformly distributed load
ρ_E	Defined by eqn (2.42)
σ	Axial stress
σ_i	Stress

Matrices

$[KGA], [\overline{KGA}]$	Geometric stiffness matrix for an element and for a complete structure (tangent stiffness)

$[KL], [\overline{KL}]$	Linear elastic stiffness matrix for an element and for a complete structure
$[KSC], [\overline{KSC}]$	Secant stiffness matrix using stability functions for an element and a complete structure
$[tKGB], [\overline{tKGB}]$	Geometric stiffness matrix for an element and for a complete structure (secant stiffness)
$\{p\}, \{P\}, \{\bar{P}\}$	Nodal force vector for element in local and global coordinates and for complete structure
$\{p^*\}$	Equivalent element nodal force vector
$\{p\rho\}$	Distributed nodal force vector for an element
$\{q\}, \{Q\}, \{\bar{Q}\}$	Nodal displacement vector for element in local and global coordinates and for complete structure
$\{\alpha\}$	Polynomial coefficients
$[\beta]$	Coordinate transformation matrix

2.1 INTRODUCTION

Current design practice for structural steelwork (Horne, 1978) permits elastic analysis of plane, multi-storey sway frames provided that the analysis takes account of nonlinear effects due to changes of geometry and the influence of axial forces. If ordinary linear elastic analysis is used to calculate frame moments, a check should be made on the sway deflection at the factored load and an allowance made, if necessary, for the additional moments thereby induced. The columns are then designed in the usual way in terms of their effective lengths (Wood, 1974; Horne, 1978).

Over the past 20–30 years, considerable research effort has been directed towards the analysis of the instability and nonlinear behaviour of frames (Merchant, 1955; Smith and Merchant, 1956; Bowles and Merchant, 1958; McMinn, 1962; Majid, 1972). This has been aided by the development of high speed electronic digital computers and matrix methods of analysis, which now enable problems of extreme size and complexity to be solved. Theories have been developed and computer programs written which are capable of analysing the geometrically nonlinear, elasto-plastic behaviour of frames up to collapse (Majid, 1968, 1972).

This chapter describes the elastic linear, nonlinear and instability analysis of plane multi-storey sway frames using matrix methods. Element stiffness matrices are derived throughout in accordance with general energy principles (Washizu, 1968) and finite element procedures (Zienkiewicz, 1971) so that the approximations inherent in many other, probably more familiar, formulations can be identified.

2.2 ENERGY PRINCIPLES

The total potential energy V of a structural system (Washizu, 1968) can be defined by the equation

$$V = V_0 - \int P_i \, dq_i + \int \left(\int \sigma_i \, d\varepsilon_i \right) d \, \text{vol} \tag{2.1}$$

where P_i and q_i represent the external forces and corresponding displacements, and σ_i and ε_i represent the internal stresses and corresponding strains. V_0 is the potential energy of the system prior to application of external forces. The integrals in eqn (2.1) represent the work done by the external forces and the strain energy of the structure, which are equal. Therefore, along any equilibrium path, V is constant and hence the first and second variations of V along the equilibrium path, denoted by δV and $\delta^2 V$, are zero.

Assuming that V can be expressed in terms of a number of prescribed displacements, q_i, δV and $\delta^2 V$ are defined as

$$\delta V = \frac{\partial V}{\partial q_i} \cdot \delta q_i, \qquad \delta^2 V = \frac{1}{2} \frac{\partial^2 V}{\partial q_i \partial q_j} \cdot \delta q_i \, \delta q_j \tag{2.2}$$

Hence from eqn (2.1)

$$\delta V = -P_i \, \delta q_i + \int \sigma_i \, \delta \varepsilon_i \, d \, \text{vol} = 0 \tag{2.3}$$

$$\delta^2 V = -P_i \, \delta^2 q_i - \tfrac{1}{2} \delta P_i \, \delta q_i + \int \left(\sigma_i \, \delta^2 \varepsilon_i + \tfrac{1}{2} \delta \sigma_i \, \delta \varepsilon_i \right) d \, \text{vol} = 0 \tag{2.4}$$

Since the structure as a whole, and individual parts of the structure, are in equilibrium, eqn (2.3) is valid for all δq_i, not just variations along an equilibrium path. Equation (2.3) provides a basis for linear and nonlinear iterative analysis while equation (2.4) provides a basis for nonlinear incremental analysis.

Variational principles can also be used to investigate the stability of structures. For equilibrium $\delta V = 0$. Stability of the system requires that positive work be done to move the system from the equilibrium state and hence $\delta V=0$ corresponds to a minimum value. Conversely, if, as the system moves slightly from the equilibrium state, energy is given out, which can be manifest only as kinetic energy, the system is unstable. Hence for stable equilibrium, $\delta V=0$ and the second variation of V for stationary values of the external forces, denoted by $\delta^2 VP$, is positive definite. From equation (2.4), $\delta^2 VP$ is given by

$$\delta^2 VP = -P_i\,\delta^2 q_i + \int \left(\sigma_i\,\delta^2 \varepsilon_i + \tfrac{1}{2}\delta\sigma_i\,\delta\varepsilon_i\right) \mathrm{d\,vol} \tag{2.5}$$

Critical conditions occur when $\delta^2 VP$ changes from positive definite to zero, indicating a possible transition from stable equilibrium to instability.

2.3 LINEAR ELASTIC ANALYSIS

A typical beam column element is shown in Fig. 2.1. The external forces at the nodes A and B are denoted by f, m and t and the displacements in the local x and z directions are denoted by u and w. The total axial strain ε (tensile positive) due to displacements u and w is given by

$$\varepsilon = -zw_{\mathrm{xx}} + u_{\mathrm{x}} \tag{2.6}$$

In eqn (2.6), z is the distance from the centroid in the z direction and subscripts x and xx denote differentiation. The first variation of ε with respect to displacements u and w is simply

$$\delta\varepsilon = -z\,\delta w_{\mathrm{xx}} + \delta u_{\mathrm{x}} \tag{2.7}$$

For linear elastic material having Young's modulus E, the axial stress σ is related to the axial strain by

$$\sigma = E\varepsilon \tag{2.8}$$

Substituting eqns (2.6)–(2.8) into eqn (2.3) and integrating over the

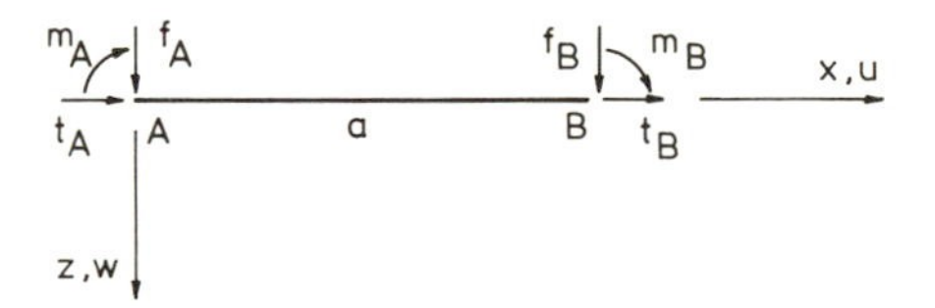

FIG. 2.1. Element of a frame.

area of the cross-section gives

$$\{\delta q\}^{\mathrm{T}}\{p\} = \int [\delta w_{xx}\ \delta u_x]\begin{bmatrix} EI & 0 \\ 0 & EA \end{bmatrix}\begin{bmatrix} w_{xx} \\ u_x \end{bmatrix} \mathrm{d}x \tag{2.9}$$

A is the cross-section area, I the second moment of area about the centroid and

$$\{p\}^{\mathrm{T}} = [f_{\mathrm{A}} m_{\mathrm{A}} t_{\mathrm{A}} f_{\mathrm{B}} m_{\mathrm{B}} t_{\mathrm{B}}] \tag{2.10}$$

$$\{\delta q\}^{\mathrm{T}} = \delta[w_{\mathrm{A}} w_{x\mathrm{A}} u_{\mathrm{A}} w_{\mathrm{B}} w_{x\mathrm{B}} u_{\mathrm{B}}] \tag{2.11}$$

Following standard finite element procedures (Zienkiewicz, 1971), the stiffness matrix for an element can be derived from eqn (2.9) by assuming suitable displacement functions for w and u. Neglecting interaction between bending and axial displacements, the exact expressions for a member loaded by nodal forces only are

$$w = [1\ \ x\ \ x^2\ \ x^3\ \ 0\ \ 0]\{\alpha\} \tag{2.12}$$

$$u = [0\ \ 0\ \ 0\ \ 0\ \ 1\ \ x]\{\alpha\} \tag{2.13}$$

The polynomial coefficients $\{\alpha\}$ are completely specified by the nodal values of w, w_x and u. It is worth noting that these functions are in accordance with established convergence criteria in that they provide nodal continuity of derivatives of order one less than appears in the variational equation (eqn (2.9)). This validates assembly of elements which is equivalent to integration of the energy over the entire structure.

Substituting the nodal coordinates of the element, length a, into eqns (2.12) and (2.13) gives

$$\{q\} = \begin{bmatrix} w_{\mathrm{A}} \\ w_{x\mathrm{A}} \\ u_{\mathrm{A}} \\ w_{\mathrm{B}} \\ w_{x\mathrm{B}} \\ u_{\mathrm{B}} \end{bmatrix} = \begin{bmatrix} 1 & 0 & 0 & 0 & 0 & 0 \\ 0 & 1 & 0 & 0 & 0 & 0 \\ 0 & 0 & 0 & 0 & 1 & 0 \\ 1 & a & a^2 & a^3 & 0 & 0 \\ 0 & 1 & 2a & 3a^2 & 0 & 0 \\ 0 & 0 & 0 & 0 & 1 & a \end{bmatrix}\{\alpha\} = [A]\{\alpha\} \tag{2.14}$$

From eqn (2.14)

$$\{\alpha\} = [A]^{-1}\{q\} = \frac{1}{a^3}\begin{bmatrix} a^3 & 0 & 0 & 0 & 0 & 0 \\ 0 & a^3 & 0 & 0 & 0 & 0 \\ -3a & -2a^2 & 0 & 3a & -a^2 & 0 \\ 2 & a & 0 & -2 & a & 0 \\ 0 & 0 & a^3 & 0 & 0 & 0 \\ 0 & 0 & -a^2 & 0 & 0 & a^2 \end{bmatrix}\{q\} \tag{2.15}$$

The derivatives of w and u can be expressed as

$$\begin{bmatrix} w_{xx} \\ u_x \end{bmatrix} = \begin{bmatrix} 0 & 0 & 2 & 6x & 0 & 1 \\ 0 & 0 & 0 & 0 & 0 & 1 \end{bmatrix} \{\alpha\} = [S][A]^{-1}\{q\} \tag{2.16}$$

Substituting in eqn (2.9) gives

$$\{\delta q\}^{\mathrm{T}}\{p\} = \{\delta q\}^{\mathrm{T}} \int [A]^{-1\mathrm{T}}[S]^{\mathrm{T}} \begin{bmatrix} EI & 0 \\ 0 & EA \end{bmatrix} [S][A]^{-1}\,\mathrm{d}x\{q\} \tag{2.17}$$

Integrating over the length of the element, eqn (2.17) can be expressed simply as

$$\{\delta q\}^{\mathrm{T}}\{p\} = \{\delta q\}^{\mathrm{T}}[KL]\{q\} \tag{2.18}$$

$[KL]$ is the linear elastic stiffness matrix for bending and axial displacements, which is given by

$$[KL] = \begin{bmatrix} 12 & 6a & 0 & -12 & 6a & 0 \\ 6a & 4a^2 & 0 & -6a & 2a^2 & 0 \\ 0 & 0 & Aa^2/I & 0 & 0 & -Aa^2/I \\ -12 & -6a & 0 & 12 & -6a & 0 \\ 6a & 2a^2 & 0 & -6a & 4a^2 & 0 \\ 0 & 0 & -Aa^2/I & 0 & 0 & Aa^2/I \end{bmatrix} \frac{EI}{a^3} \tag{2.19}$$

Prior to assembly of the individual element stiffness matrices it is necessary to transform the element nodal forces and displacements to a global set of axes. From Fig. 2.2, $\{q\}$ is related to global displacement $\{Q\}$ by the equation

$$\{q\} = \begin{bmatrix} \cos\beta & 0 & -\sin\beta & 0 & 0 & 0 \\ 0 & 1 & 0 & 0 & 0 & 0 \\ \sin\beta & 0 & \cos\beta & 0 & 0 & 0 \\ 0 & 0 & 0 & \cos\beta & 0 & -\sin\beta \\ 0 & 0 & 0 & 0 & 1 & 0 \\ 0 & 0 & 0 & \sin\beta & 0 & \cos\beta \end{bmatrix} \{Q\} = [\beta]\{Q\} \tag{2.20}$$

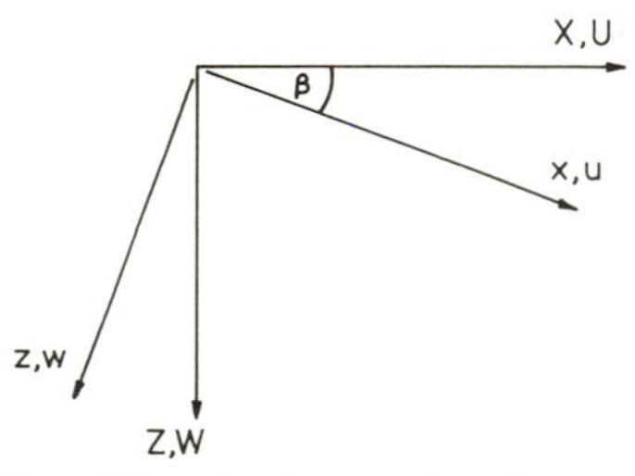

FIG. 2.2 Global and local coordinates.

The same transformation matrix $[\beta]$ relates $\{p\}$ to the global forces $\{P\}$. Noting also that $[\beta]^{-1} \equiv [\beta]^{\mathrm{T}}$, eqn (2.16) can be expressed in terms of global forces and displacements as

$$\{\delta q\}^{\mathrm{T}}\{P\} = \{\delta q\}^{\mathrm{T}}[\beta]^{\mathrm{T}}[KL][\beta]\{Q\} \tag{2.21}$$

Assembly of elements to form the overall structural stiffness matrix is simply a matter of superimposing the energy contributions of the individual elements. The $\{\delta q\}^{\mathrm{T}}$ can then be cancelled since the resulting scalar energy equation is valid for all $\{\delta q\}$. Hence, the stiffness equations for the complete structure reduce to the form

$$\{\bar{P}\} = [\overline{KL}]\{\bar{Q}\} \tag{2.22}$$

After application of the prescribed boundary conditions, eqn (2.22) can be solved for the unknown displacements to give

$$\{\bar{Q}\} = [\overline{KL}]^{-1}\{\bar{P}\} \tag{2.23}$$

The member bending moments and axial forces m and t are then given by

$$m = EIw_{xx} = EI[0\ \ 0\ \ 2\ \ 6x\ \ 0\ \ 0][A]^{-1}[\beta]\{Q\} \tag{2.24}$$

$$t = EAu_{x} = EA[0\ \ 0\ \ 0\ \ 0\ \ 0\ \ 1][A]^{-1}[\beta]\{Q\} \tag{2.25}$$

So far, only concentrated nodal forces have been considered. Beam members in frames are, however, often subjected to distributed loading, of intensity ρ per unit length. This can be incorporated in the analysis by equating the potential energy of the distributed loads to that of an equivalent nodal force vector $\{p\rho\}$ containing moments m and transverse forces f only. Hence, for an element in local coordinates

$$\{\delta q\}^{\mathrm{T}}\{p\rho\} = \int \rho\,.\,\delta w\,\mathrm{d}x = \{\delta q\}^{\mathrm{T}} \int \rho[A]^{-1\mathrm{T}}[1\ \ x\ \ x^2\ \ x^3\ \ 0\ \ 0]^{\mathrm{T}}\,\mathrm{d}x \tag{2.26}$$

2.4 NONLINEAR ELASTIC ANALYSIS

Nonlinear elastic analysis of frames can be performed either incrementally or iteratively.

Incremental analysis involves the determination of the incremental or tangent stiffness matrix relating small increments in external forces and corresponding displacements; this depends on the current geometry and state of stress. Complete solutions for the entire loading history can then be obtained by incrementing either forces or displacements (Roberts, 1970; Roberts and Ashwell, 1971). Incrementing displacements has the advantage that solutions do not break down at horizontal tangents on load–deflection curves, which is one form of critical load condition.

Iterative solutions are based on the determination of a secant stiffness matrix, which is derived assuming that the current geometry and state of stress is known (Majid, 1972). The first cycle then gives new values for the current geometry and state of stress which are then included in the second cycle, and the sequence is repeated until the assumed values are consistent with the calculated values.

In general, incremental analysis is theoretically more sound than iterative analysis. Incremental analysis follows the complete loading history and is able to detect bifurcations or branching points along equilibrium paths. This is not true of iterative solutions which may not converge to the lowest equilibrium path which is of interest in practice. However, for multi-storey frames, such complexities seldom exist and either form of solution appears satisfactory. Incremental analysis is considered first since it illustrates the full interaction between bending and axial displacements for individual elements and will help to indicate the approximations often made in iterative solutions.

2.4.1 Incremental Analysis

The incremental stiffness matrix for a finite element of a frame can be derived as follows (Roberts, 1970). It is assumed that all displacements are small so that nonlinear strains can be related to the initial geometry; this is a satisfactory assumption for the majority of practical frames. For an element having an initial transverse imperfection w_0, the nonlinear expression for the axial strain ε is (Timoshenko and Gere, 1961)

$$\varepsilon = -z(w_{xx} - w_{0xx}) + u_x + 0{\cdot}5(w_x^2 - w_{0x}^2) \tag{2.27}$$

The first and second variations of ε with respect to displacements u and w are

$$\delta\varepsilon = -z\,\delta w_{xx} + \delta u_x + w_x\,\delta w_x \tag{2.28}$$

$$\delta^2\varepsilon = 0{\cdot}5\delta w_x\,.\,\delta w_x \tag{2.29}$$

Substituting eqns (2.27)–(2.29) into eqn (2.4) and noting that $\delta^2 q_i$ vanishes, since the nodal displacements q_i are assumed to be linear functions, gives

$$0{\cdot}5\delta P_i\,\delta q_i = \int E\{-z(w_{xx} - w_{0xx}) + u_x + 0{\cdot}5(w_x^2 - w_{0x}^2)\}0{\cdot}5\delta w_x\,\delta w_x\ \mathrm{d\,vol} + \int 0{\cdot}5E\{-z\,\delta w_{xx} + \delta u_x + w_x\,\delta w_x\}^2\ \mathrm{d\,vol} \qquad (2.30)$$

Integrating over the area of the cross-section, eqn (2.30) can be arranged in matrix form as

$$\{\delta q\}^{\mathrm{T}}\{\delta p\} = \int [\delta w_{xx}, \delta u_x]\begin{bmatrix} EI & 0 \\ 0 & EA \end{bmatrix}\begin{bmatrix} \delta w_{xx} \\ \delta u_x \end{bmatrix}\mathrm{d}x + \int EA[\delta w_x, \delta u_x]\left[\begin{array}{c|c} u_x + 0.5(w_x^2 - w_{0x}^2) + w_x^2 & w_x \\ \hline w_x & \end{array}\right]\begin{bmatrix} \delta w_x \\ \delta u_x \end{bmatrix}\mathrm{d}x \qquad (2.31)$$

Assuming suitable displacement functions for w, w_0 and u (see eqns (2.12) and (2.13)) and proceeding as in Section 2.3, eqn (2.31) can be reduced to the form

$$\{\delta q\}^{\mathrm{T}}\{\delta p\} = \{\delta q\}^{\mathrm{T}}[[KL] + [KGA]]\{\delta q\} \qquad (2.32)$$

$[KL]$ is as defined in Section 2.3 and $[KGA]$ is referred to as the geometric stiffness matrix for incremental analysis. The derivation of $[KGA]$ from the second integral in eqn (2.31), assuming w and w_0 to be cubic polynomials, is complex. The derivation of $[KGA]$ can be simplified however by assuming linear polynomials for w and w_0, defined by the nodal values of w and w_0 only (Roberts and Azizian, 1983). All terms in the second integral of eqn (2.31) then become constants and integration is simply a matter of multiplying by the length of the element. This procedure, as well as simplifying the derivation considerably, is advantageous for convergence of finite element solutions.

After transformation to global coordinates and assembly of elements, the incremental equations for the whole structure take the form (see eqn (2.22))

$$\{\delta\bar{P}\}=[[\overline{KL}]+[\overline{KGA}]]\{\delta\bar{Q}\} \qquad (2.33)$$

Solutions can be obtained by incrementing either loads or displacements. The geometric stiffness matrix has to be reformed prior to each increment to account for the current geometry or state of stress and the total forces and displacements accumulated from the increments $\{\delta\bar{P}\}$ and $\{\delta\bar{Q}\}$. Stresses at any stage can be determined from eqn (2.27) and the accumulated displacements.

The accuracy of incremental solutions will depend upon the size of each increment. Also, if the simple linearisation techniques described are not employed, it may prove necessary to use a large number of elements to model each member, beam or column, of the frame.

2.4.2 Iterative Analysis

Derivation of the secant stiffness matrix for use in iterative methods of analysis can be based on eqn (2.3). Assuming the nonlinear expression for the axial strain, as defined by eqn (2.27), and substituting in eqn (2.3) gives (Roberts, 1970)

$$P_i\,\delta q_i=\int E\{-z(w_{xx}-w_{0xx})+u_x+0{\cdot}5(w_x^2-w_{0x}^2)\}$$
$$\times\{-z\,\delta w_{xx}+\delta u_x+w_x\,\delta w_x\}\,\mathrm{d\,vol} \qquad (2.34)$$

Integrating over the area of the cross-section, eqn (2.34) can be arranged in matrix form as

$$\{\delta q\}^{\mathrm{T}}\{p\}=\int[\delta w_{xx},\,\delta u_x]\begin{bmatrix}EI & 0\\ & EA\end{bmatrix}\begin{bmatrix}w_{xx}\\ u_x\end{bmatrix}\mathrm{d}x$$
$$+\int[\delta w_x,\,\delta u_x]EA\begin{bmatrix}0{\cdot}5(u_x+w_x^2-w_{0x}^2) & 0{\cdot}5w_x\\ 0{\cdot}5w_x & 0\end{bmatrix}\begin{bmatrix}w_x\\ u_x\end{bmatrix}\mathrm{d}x$$
$$-\int EI\,w_{0xx}\,\delta w_x\,\mathrm{d}x-\int 0{\cdot}5EAw_{0x}^2\,\delta u_x\,\mathrm{d}x \qquad (2.35)$$

Following the previously described finite element procedures, eqn (2.35) can be reduced to a form similar to eqn (2.32) with the last two terms contributing a constant column vector on the right hand side. As in the previous section, eqn (2.35) indicates full interaction between axial and bending displacements.

However, the major nonlinear influence in the behaviour of frames is the influence of axial forces on the flexure of members. Assuming, therefore, that the axial force t is simply EAu_x, as given by a linear elastic analysis, and assuming $w_0 = 0$ for simplicity, eqn (2.35) reduces to the form (Chajes, 1974)

$$\{\delta q\}^T\{p\} = \int [\delta w_{xx}, \delta u_x] \begin{bmatrix} EI & 0 \\ 0 & EA \end{bmatrix} \begin{bmatrix} w_{xx} \\ u_x \end{bmatrix} dx + \int \delta w_x \,.\, t \,.\, w_x \,.\, dx \tag{2.36}$$

Equation (2.36) can be simplified further by linearising w when deriving the geometric stiffness matrix, as discussed in Section 2.4.1.

Assuming suitable displacement functions for w and u, eqn (2.36) can be reduced to the form

$$\{\delta q\}^T\{p\} = \{\delta q\}^T[[KL] + [tKGB]]\{q\} \tag{2.37}$$

$[KL]$ is as defined by eqn (2.19) and $[tKGB]$ is the geometric stiffness matrix for iterative analysis. Assuming w to be a cubic polynomial

$$[tKGB] = \frac{t}{10a} \begin{bmatrix} 12 & a & 0 & -12 & a & 0 \\ a & 4a^3/3 & 0 & -a & -a^2/3 & 0 \\ 0 & 0 & 0 & 0 & 0 & 0 \\ -12 & -a & 0 & 12 & -a & 0 \\ a & -a^2/3 & 0 & -a & 4a^2/3 & 0 \\ 0 & 0 & 0 & 0 & 0 & 0 \end{bmatrix} \tag{2.38}$$

After transformation to global coordinates and assembly of elements, the secant stiffness equations for the complete structure take the form

$$\{\bar{P}\} = [[\overline{KL}] + [\overline{tKGB}]]\{\bar{Q}\} \tag{2.39}$$

Solutions of eqn (2.39) can be obtained iteratively. For the first cycle $[\overline{tKGB}]$ is assumed zero and a linear elastic solution is obtained. The member axial forces given by the first cycle are then used to determine $[\overline{tKGB}]$ for the second cycle, and so on until the assumed and calculated member axial forces are consistent. Solutions of this type usually converge rapidly since the member axial forces do not vary much from those given by the first linear elastic analysis.

2.4.3 Iterative Analysis Using Stability Functions

An alternative way of expressing the influence of axial forces on the flexural behaviour of members is in terms of so-called 'stability func-

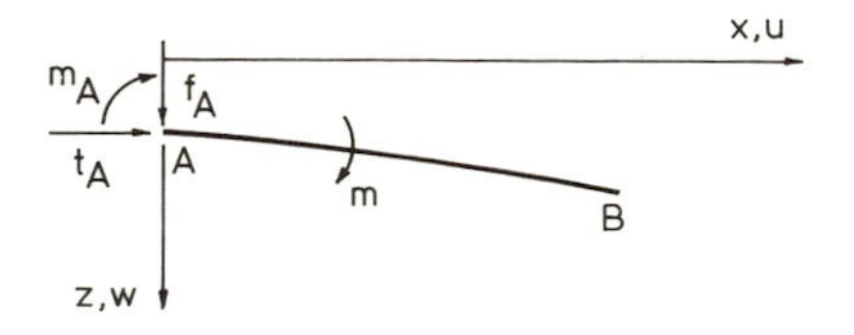

FIG. 2.3. Member of a frame.

tions' (Livesley and Chandler, 1956; Horne and Merchant, 1965). The differential equation governing the flexure of the member shown in Fig. 2.3 is

$$m = EIw_{xx} = f_A \,.\, x - m_A - t_A(w - w_A) \tag{2.40}$$

Solutions of eqn (2.40) subject to various boundary conditions lead to a secant stiffness matrix $[KSC]$ of the form

$$[KSC] = \frac{EI}{a^3}$$
$$\begin{bmatrix} 2s(1+c)/r & s(1+c)a & 0 & -2s(1+c)/r & s(1+c)a & 0 \\ s(1+c)a & sa^2 & 0 & -s(1+c)a & sca^2 & 0 \\ 0 & 0 & Aa^2/I & 0 & 0 & -Aa^2/I \\ -2s(1+c)/r & -s(1+c)a & 0 & 2s(1+c)/r & -s(1+c)a & 0 \\ s(1+c)a & sca^2 & 0 & -s(1+c)a & sa^2 & 0 \\ 0 & 0 & -Aa^2/I & 0 & 0 & Aa^2/I \end{bmatrix} \tag{2.41}$$

In eqn (2.41) s, c and r are defined as follows (eqns (2.42)–(2.44)):

$$\rho_E = \left| \frac{ta^2}{\pi^2 EI} \right| \qquad \gamma = \frac{\pi}{2}\sqrt{\rho_E} \tag{2.42}$$

For compressive axial forces

$$s = \frac{(1-2\gamma \cot 2\gamma)\gamma}{\tan \gamma - \gamma}, \qquad c = \frac{2\gamma - \sin 2\gamma}{\sin 2\gamma - 2\gamma \cos 2\gamma}, \qquad r = \frac{2s(1+c)}{2s(1+c) - \pi^2 \rho_E} \tag{2.43}$$

For tensile axial forces

$$s = \frac{(1-2\gamma \coth 2\gamma)\gamma}{\tanh \gamma - \gamma}, \qquad c = \frac{2\gamma - \sinh 2\gamma}{\sinh 2\gamma - 2\gamma \cosh 2\gamma} \tag{2.44}$$

In eqn (2.41), the terms relating axial forces and displacements are as for linear elastic analysis and terms which represent the interaction between flexural and axial displacements (see Section 2.4.2) are omitted.

After transformation to global coordinates and assembly of elements, the secant stiffness equations for the complete structure take the form

$$\{\bar{P}\} = [\overline{KSC}]\{\bar{Q}\} \tag{2.45}$$

Solutions of eqn (2.45) can be obtained iteratively as described in Section 2.4.2, noting that when member axial forces are assumed zero, $[\overline{KSC}] \equiv [\overline{KL}]$.

2.5 ELASTIC INSTABILITY

As mentioned previously, the most significant nonlinear influence in the elastic behaviour of frames is the influence of axial forces on the flexural stiffness of members. Tensile axial forces can be considered as increasing the flexural stiffness while compressive forces decrease the flexural stiffness. If a set of compressive member axial forces is increased to the extent that the bending stiffness of the frame as a whole reduces to zero, the frame becomes unstable.

There are a number of ways in which the elastic instability of frames can be analysed, many of which reduce to the solution of the same basic set of equations after simplifying assumptions are made.

2.5.1 Vanishing of the Second Variation of Total Potential Energy

The most general approach to the analysis of the elastic instability of structures is based on the vanishing of the second variation of total potential energy, defined by eqn (2.5) (Roberts and Azizian, 1983). Assuming that the nodal displacements q_i are linear functions of displacement variables, $\delta^2 q_i$ vanishes and critical conditions are defined by the equation

$$\delta^2 VP = \int (\sigma_i\, \delta^2 \varepsilon_i + 0{\cdot}5\delta\sigma_i\, \delta\varepsilon_i)\, \mathrm{d\,vol} = 0 \tag{2.46}$$

Substituting the nonlinear expression for the axial strain ε defined by eqn (2.27) gives

$$\delta^2 VP = \{\delta q\}^{\mathrm{T}}[[KL] + [KGA]]\{\delta q\} = 0 \tag{2.47}$$

$\delta^2 VP$ is identical to the right hand side of eqns (2.31) and (2.32) and is a complete quadratic form which changes from positive definite to zero, indicating critical conditions, when the determinant of $[KL]+[KGA]$ vanishes. Hence critical conditions for the complete structure occur when

$$\det|[\overline{KL}]+[\overline{KGA}]|=0 \tag{2.48}$$

An alternative way of interpreting eqn (2.48) is that critical conditions occur when the incremental or tangent stiffness matrix becomes singular.

Although eqn (2.48) is of general applicability, it requires a knowledge of the axial and flexural deformations at the critical points on the loading path. The analysis can be simplified if it is assumed that prior to the frame becoming unstable, only axial deformations occur. The axial forces t are then simply equal to u_x and eqn (2.46) reduces to

$$\delta^2 VP=\{\delta q\}^{\mathrm{T}}[[KL]+[tKGB]]\{\delta q\}=0 \tag{2.49}$$

$\delta^2 VP$ is now identical to the right hand side of eqn (2.37) if $\{q\}$ is replaced by $\{\delta q\}$. Hence, critical conditions for the complete structure occur when

$$\det|[KL]+[\overline{tKGB}]|=0 \tag{2.50}$$

In solving eqn (2.50) it is usually assumed that the critical set of member axial forces can be related to a base set, determined for example from a preliminary linear elastic analysis, by a scalar load factor λ, which is given a negative sign to denote compression. Equation (2.50) then takes the form

$$\det|[\overline{KL}]-\lambda[\overline{tKGB}]|=0 \tag{2.51}$$

Equation (2.51) represents a standard eigenvalue problem. The lowest eigenvalue defines the critical load factor λ_{cr} and the corresponding eigenvector defines the buckled mode.

2.5.2 Singularity of Secant Stiffness Matrix

The secant stiffness equations defined by eqns (2.39) and (2.45) are indeterminate when the secant stiffness matrix becomes singular. Critical conditions occur therefore when

$$\det|[\overline{KL}]+[\overline{tKGB}]|=0 \tag{2.52}$$

$$\det|[\overline{KSC}]|=0 \tag{2.53}$$

Equation (2.52) is identical to eqn (2.50) and can be reduced to the standard eigenvalue problem defined by eqn (2.51). Solution of eqn (2.53) is accomplished by assuming a load factor λ and evaluating the determinant. The process is repeated until the load factor at which the determinant vanishes is found.

2.5.3 Horne's Method

Horne (1975) proposed an approximate method for determining the elastic critical loads of plane multi-storey sway frames, the only analytical requirement being that of performing a standard linear elastic analysis of the frame. The method can be illustrated by considering the instability of the column of length a shown in Fig. 2.4(a).

Assuming that the column buckles into a state of neutral equilibrium, i.e. zero kinetic energy, the potential energy remains constant and the loss of potential energy of the external forces is equal to the increase in strain energy of the column. If, due to buckling, the end A of the column sways by $\delta\bar{w}$ with a corresponding axial shortening $\delta\bar{u}$, the governing energy equation can be written

$$P\,.\,\delta\bar{u} = UE \tag{2.54}$$

in which UE represents the change in strain energy of the column.

To determine UE it is assumed that the buckled shape δw is the same as that produced by a concentrated horizontal force H as shown in Fig. 2.4(b). UE is then equal to $0{\cdot}5H\,.\,\delta\bar{w}$ and eqn (2.54) becomes

$$P\,.\,\delta\bar{u} = 0{\cdot}5H\,\delta\bar{w} \tag{2.55}$$

If the critical load P_{cr} is related to a base load P_0 by a scalar load factor λ_{cr} and H is assumed equal to nP_0 (n being a scalar load factor

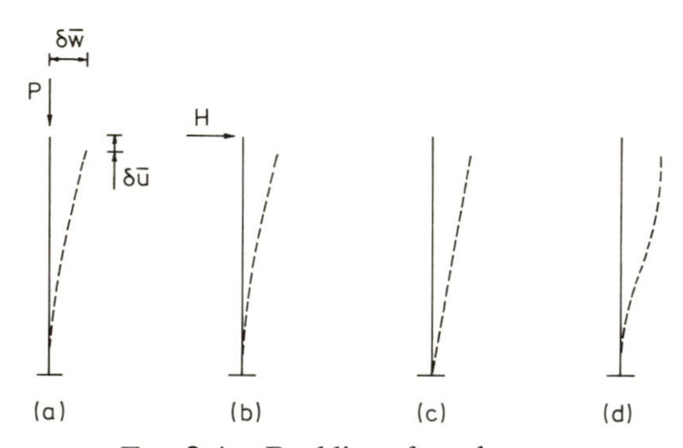

FIG. 2.4. Buckling of a column.

less than unity) eqn (2.55) can be rewritten as

$$\lambda_{cr} P_0 \delta\bar{u} = 0{\cdot}5 n P_0 \, \delta\bar{w} \tag{2.56}$$

or

$$\lambda_{cr} = 0{\cdot}5 n \frac{\delta\bar{w}}{\delta\bar{u}} \tag{2.57}$$

Two extreme cases are now considered for the buckled shape, as shown in Fig. 2.4(c) and (d). The first, which is a simple rigid body rotation, represents the case in which the columns of a frame are stiff compared with the beams. The second, which is a pure sway mode, represents the case in which the beams are stiff compared with the columns. Assuming simple polynomials to represent the buckled shape δw, the axial shortening is given by

$$\delta\bar{u} = \int 0{\cdot}5 \delta w_x \, . \, \delta w_x \, \mathrm{d}x \tag{2.58}$$

For the two extreme cases $\delta\bar{u}$ is given by

$$\delta\bar{u} = 0{\cdot}5 \frac{\delta\bar{w}^2}{a}, \qquad \delta\bar{u} = 0{\cdot}6 \frac{\delta\bar{w}^2}{a} \tag{2.59}$$

Since these two values differ by only approximately 20% it is convenient to take the average, and substituting in eqn (2.57) gives

$$\lambda_{cr} = 0{\cdot}9 n \frac{a}{\delta\bar{w}} \tag{2.60}$$

The procedure can now be summarised as follows. Apply a horizontal force of nP_0 to the end of the column and determine $\delta\bar{w}$. The critical load factor is then given by eqn (2.60).

Now consider the generalisation of this procedure for frames. The frame shown in Fig. 2.5 has $N = 4$ storeys. P_{0i} represents the vertical loads applied at the ith storey and $H_i = nP_{0i}$ are the corresponding horizontal forces assumed applied to the frame. The governing energy equation for the frame (see eqn (2.56)) is then

$$\sum_{i=1,N} \lambda_{cr} P_{0i} \sum_{j=i,N} \delta\bar{u}_j = 0{\cdot}5 \sum_{i=1,N} n P_{0i} \sum_{j=i,N} \delta\bar{w}_j \tag{2.61}$$

Assuming that the rth storey of the frame becomes unstable before all

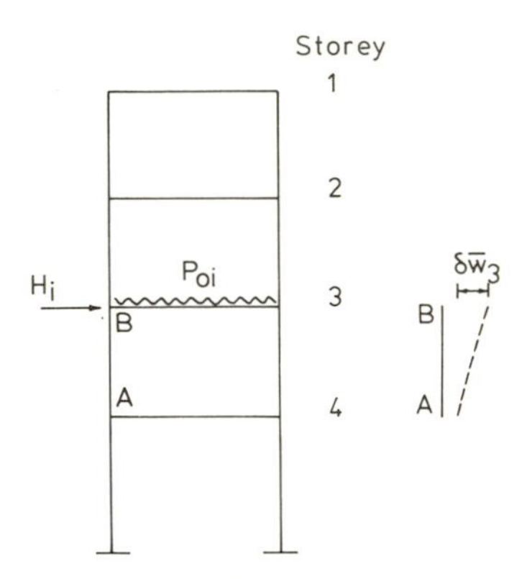

FIG. 2.5. Multi-storey sway frame.

the others, eqn (2.61) can be simplified as

$$\sum_{i=1,r} \lambda_{cr} P_{0i}\, \delta\bar{u}_r = 0{\cdot}5 \sum_{i=1,r} nP_{0i}\, \delta\bar{w}_r \tag{2.62}$$

The summation can now be cancelled to give

$$\lambda_{cr} = 0{\cdot}5n \frac{\delta\bar{w}_r}{\delta\bar{u}_r} \tag{2.63}$$

and proceeding as for the simple column

$$\lambda_{cr} = 0{\cdot}9n \frac{a_r}{\delta\bar{w}_r} \tag{2.64}$$

The procedure for frames can now be summarised. Load the frame with horizontal forces at each storey H_i equal to nP_{0i}. Determine the sway deflections and locate the storey for which $\delta\bar{w}_i/a_i$ is a maximum and assume equal to $\delta\bar{w}_r/a_r$. The critical load factor is then calculated from eqn (2.63).

2.6 SECOND ORDER EFFECTS AND ELASTIC CRITICAL LOADS

Although considerable effort has been devoted to determining the elastic critical loads of plane multi-storey sway frames, elastic critical loads have found little direct application in practice. Sway limitations (Anderson and Islam, 1979; Majid and Okdeh, 1982) to prevent serious damage to non-structural cladding ensure that so-called 'instability effects' or second order nonlinear effects are relatively minor. Apart from this, elastic critical loads do not provide a direct measure of the magnitude of the second order effects and further complicated calculations are required to provide information of use to designers.

Roberts (1981) proposed a procedure for estimating the magnitude of the second-order effects, based on a standard linear elastic analysis. The results can also be used to estimate, very simply, the elastic critical loads of frames if required.

A linear elastic analysis of a plane, multi-storey sway frame gives displacement and member forces at all the joints of the frame. For any member, it is possible to deduce the displacements w, rotation w_x, moments m, shear forces f and axial forces t at the joints or nodes relative to the local coordinate axes of the member.

Each member is in equilibrium, only in the absence of the axial forces which produce the second order effects. It is assumed that the member axial forces produce an additional transverse deflection δw of the member. The corresponding axial shortening $\delta \bar{u}$ is then given by

$$\delta \bar{u} = \int w_x \, . \, \delta w_x \, . \, dx \tag{2.65}$$

In eqn (2.65) w_x is as given by the linear elastic analysis. The work done by the axial forces t during the displacement $\delta \bar{u}$ is $-t\,\delta \bar{u}$ (t is tensile positive) and this is taken as equal to the work done by an equivalent set of nodal forces $\{p^*\} = [f_A^*, m_A^*, 0, f_B^*, m_B^*, 0]^T$ during the displacement δw. Hence

$$\{\delta q\}^T \{p^*\} = -t \int w_x \, . \, \delta w_x \, dx \tag{2.66}$$

Assuming w and δw to be represented by cubic polynomials, and proceeding as in Section 2.3, gives

$$\{\delta q\}^T \{p^*\} = -\{\delta q\}^T [tKGB]\{q\} \tag{2.67}$$

$[tKGB]$ is as defined by eqn (2.38) and $\{q\}$ is the nodal displacements given by the linear elastic analysis.

The equivalent load vectors for each member of the frame can now be determined and a second linear elastic analysis performed with the frame loaded by all the member equivalent load vectors to give a first approximation for the second-order effects. Since this is only a first approximation, the final solution should ideally be obtained iteratively.

A new set of equivalent loads should be calculated from member axial forces and displacements $w+\delta w$ and the process repeated until the calculated values of $w+\delta w$ are consistent with those assumed. An approximation to the iterative procedure can however be made as follows. Consider any of the non-zero displacements, for example the sway of the ith storey of the frame $\bar{w}_i$ (see Fig. 2.5). Assuming that δw is proportional to w, the final value of $\delta \bar{w}_i$, which is denoted by $\Delta \bar{w}_i$, is given approximately by

$$\Delta \bar{w}_i = \bar{w}_i(\beta_i + \beta_i^2 + \beta_i^3 + \cdots + \beta_i^n) \tag{2.68}$$

in which

$$\beta_i = \frac{\delta \bar{w}_i}{\bar{w}_i} \tag{2.69}$$

In practice, the series $(\beta_i + \beta_i^2 + \cdots)$ converges rapidly and only the first two or three terms need be considered. The complete solution corresponding to displacements $\bar{w}_i + \Delta \bar{w}_i$ can now be obtained by superposition.

Equation (2.68) was expressed in terms of storey sway for a particular reason. Provided that the primary loading produces sway in each storey, an estimate of the sway critical load can be made using the principle (Horne and Merchant, 1965) that member axial forces at a load factor λ have the effect of increasing the deformations corresponding to the lowest elastic critical load by a factor $1/(1-\lambda/\lambda_{cr})$. If λ is taken as unity for the primary frame loads

$$1 + \frac{\Delta \bar{w}_i}{\bar{w}_i} = (1 + \beta_i + \beta_i^2 + \cdots \beta_i^n) = 1/(1 - 1/\lambda_{cr}) \tag{2.70}$$

The storey for which $\beta_i = \delta \bar{w}_i/\bar{w}_i$ is a maximum will be the storey in which elastic critical conditions occur first, and the elastic critical load for that storey can be calculated from eqn (2.70).

2.7 ILLUSTRATIVE EXAMPLES

Meaningful comparisons of the various nonlinear methods of analysis discussed are limited, since detailed solutions seldom appear in the literature and so-called 'exact' solutions of nonlinear problems are often questionable. However, all the methods discussed have one aspect in common. After making certain simplifying assumptions, they all lead to a prediction of the elastic critical loads of frames and it is this aspect which will be used, herein, for comparison.

Details of the frames analysed are shown in Fig. 2.6. A single element was used to represent each member (beam or column) of the frame and the axial forces in members were assumed either statically determinate or as given by a preliminary linear elastic analysis. With

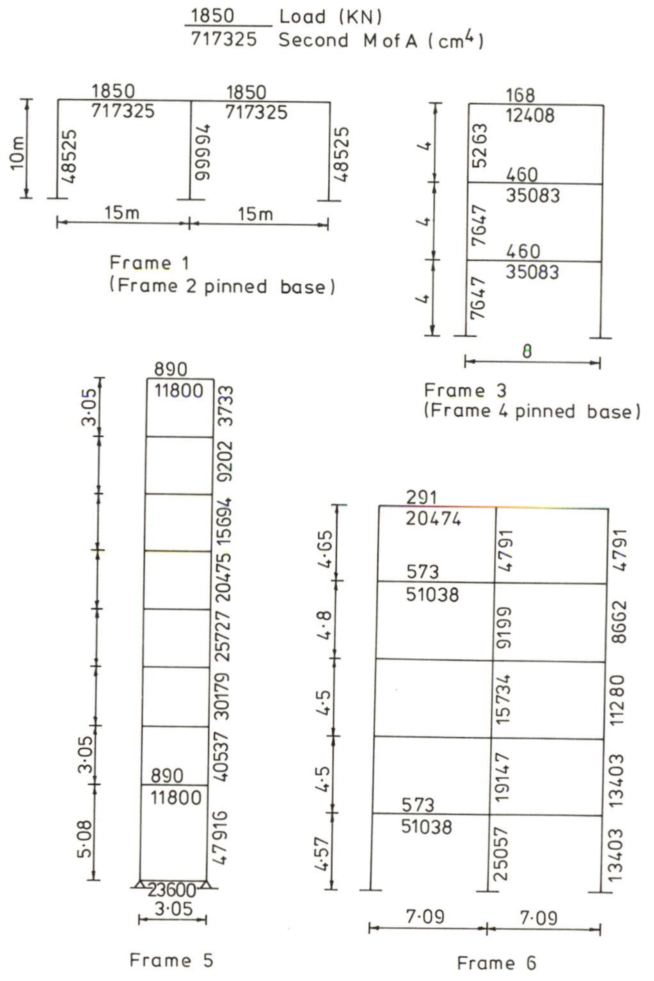

FIG. 2.6. Frame and loading details.

the exception of methods incorporating stability functions, improved accuracy is achieved by using more than one element to represent each member, due to the approximate displacement functions assumed in deriving the element stiffness matrices. In applying the method discussed in Section 2.6, the frames were also loaded with horizontal forces at each storey equal to 10% of the vertical load applied at that storey, to excite sway deformations.

Results of the analysis discussed in Section 2.6 are presented in Table 2.1, in which β_i values for each storey are given, the maximum from which the elastic critical load factor is calculated being underlined. Also given in Table 2.1 are critical load factors determined in accordance with Section 2.5.1 (eigenvalue solution), Section 2.5.2 (stability functions) and Section 2.5.3 (Horne's method).

TABLE 2.1
VALUES OF β_i FOR EACH STOREY AND ELASTIC CRITICAL LOAD FACTORS λ_{cr} OF THE FRAMES SHOWN IN FIG. 2.6

	Frame					
	1	*2*	*3*	*4*	*5*	*6*
β_i values for storey						
1	<u>0·097</u>	<u>0·390</u>	0·034	0·052	0·175	0·069
2			0·056	0·125	0·183	0·103
3			<u>0·070</u>	<u>0·266</u>	0·196	0·116
4					0·213	<u>0·134</u>
5					0·231	0·131
6					0·250	
7					0·267	
8					<u>0·272</u>	
Elastic critical load factors λ_{cr}						
λ_{cr} (Section 6) (Roberts, 1981)	10·3	2·57	14·3	3·76	3·68	7·44
λ_{cr} eigenvalues (Section 5.1) (Virgin, 1982)	10·8	2·68	15·6	3·99	3·49	6·47
λ_{cr} stability functions (Section 5.2) (Horne, 1975)	10·5	2·66	14·7	3·78	3·95	7·70
λ_{cr} Horne's method (Section 5.3) (Horne, 1975)	10·9	2·73	14·5	3·74	3·60	7·36

The results are reasonably consistent, which indicates that all the nonlinear methods of analysis discussed adequately represent the influence of axial forces on the flexural behaviour of frames.

2.8 CONCLUDING REMARKS

Matrix formulations for the elastic linear, nonlinear and instability analysis of frames have been presented. Nonlinear analysis can be performed either incrementally using the tangent stiffness matrix or iteratively using the secant stiffness matrix. Rigorous derivations of both the tangent and secant stiffness matrices indicate full interaction between bending and axial displacements. Critical loading conditions occur when the second variation of the total potential energy changes from positive definite to zero, indicating a transition from stable equilibrium to instability, this condition being defined by the vanishing of the determinant of the incremental or tangent stiffness matrix.

For practical multi-storeyed sway frames, sway limitations to prevent damage to non-structural cladding generally ensure that nonlinear effects are of only minor significance. This enables simplifying approximations to be introduced in matrix analysis, the most significant of which is the assumption that member axial forces are either statically determinate or as given by a preliminary linear elastic analysis. This assumption is equivalent to considering only the influence of axial forces on the flexural behaviour of members and neglects axial shortening due to flexure. Based on this assumption, approximate methods of estimating second order effects and elastic critical loads can be established, and several alternative approaches for determining elastic critical loads reduce to the solution of the same set of equations.

REFERENCES

ANDERSON, D. and ISLAM, M. A. (1979) Design of multi storey frames to sway deflection limitations. *Structural Engineer*, **57B**(1), 11–17.

BOWLES, R. E. and MERCHANT, W. (1958) Critical loads of tall building frames, part 4. *Structural Engineer*, **36,** 187.

CHAJES, A. (1974) *Principles of Structural Stability Theory*, Prentice-Hall, Inc., New Jersey, USA.

HORNE, M. R. (1975) An approximate method for calculating the elastic critical loads of multi-storey plane frames. *Structural Engineer*, **53**(6), 242–8.

HORNE, M. R. (1978) Continuous construction and special problems of single storey frames—Lecture 9. Design of multi storey frames—Lecture 10. *The Background to the New British Standard for Structural Steelwork*, Imperial College, London, July.

HORNE, M. R. and MERCHANT, W. (1965) *The Stability of Frames*, Pergamon Press Ltd, Oxford.

LIVESLEY, R. K. and CHANDLER, D. B. (1956) *Stability Functions for Structural Frameworks*, Manchester University Press, Manchester, UK.

MAJID, K. I. (1968) The evaluation of the failure load of plane frames. *Proc. Roy. Soc. A*, **306,** 297–311.

MAJID, K. I. (1972) *Non-linear Structures—Matrix Methods of Analysis and Design by Computers*, Butterworth, London.

MAJID, K. I. and OKDEH, S. (1982) Limit state design of sway frames. *Structural engineer*, **60B**(4), 76–82.

McMINN, S. J. (1962) *Matrices for Structural Analysis*, Spon, London.

MERCHANT, W. (1955) Critical loads of tall building frames. *Structural Engineer*, **33,** 85.

ROBERTS, T. M. (1970) Behaviour of Nonlinear Structures. PhD Thesis, Department of Civil and Structural Engineering, University College, Cardiff.

ROBERTS, T.M. (1981) Second order effects and elastic critical loads of plane, multi storey, unbraced frames. *Structural Engineer*, **59A**(4), 125–7.

ROBERTS, T. M. and ASHWELL, D. G. (1971) The use of finite element mid-increment stiffness matrices in the post-buckling analysis of imperfect structures. *Int. J. Solids Structures*, **7,** 805–23.

ROBERTS, T. M. and AZIZIAN, Z. G. (1983) Instability of thin walled bars. *Proc. ASCE, Journal of Engineering Mechanics*, **109**(3), 781–94.

SMITH, R. B. L. and MERCHANT, W. (1956) Critical loads of tall building frames, Part 2. *Structural Engineer*, **34,** 284.

TIMOSHENKO, S. P. and GERE, J. M. (1961) *Theory of Elastic Stability*, McGraw-Hill Book Co. Inc., New York.

VIRGIN, L. (1982) Second order effects and elastic critical loads of unbraced plane frames. MSc Dissertation, Department of Civil and Structural Engineering, University College, Cardiff.

WASHIZU, K. (1968) *Variational Methods in Elasticity and Plasticity*, Pergamon Press Ltd, Oxford.

WOOD, R. H. (1974) Effective lengths of columns in multi storey buildings. Parts 1, 2 and 3. *Structural Engineer*, **52**(7), 235–44; (8), 295–302; (9), 341–6.

ZIENKIEWICZ, O. C. (1971) *The Finite Element Method in Engineering Science*, Second Edition, McGraw-Hill Publishing Co. Ltd, New York.

Chapter 3

DESIGN OF MULTI-STOREY STEEL FRAMES TO SWAY DEFLECTION LIMITATIONS

D. ANDERSON

Department of Engineering, University of Warwick, Coventry, UK

SUMMARY

Methods are described for the design of multi-storey steel frames to specified limits on horizontal sway deflection. Approximate methods for rectangular frames require only simple calculations, and their use is illustrated by a worked example. More general approaches are also given. These necessitate iterative calculation and take the form of specialised computer programs. Accurate allowance can then be made for secondary effects which are of particular significance in the design of very slender unbraced structures.

NOTATION

a	Constant
B	Width of frame
E	Young's modulus of elasticity
F	Shear in columns due to wind
h	Storey height
I, I'	Moment of inertia of cross-section
K	Member stiffness
k	Distribution factors
L	Bay width
M	Parameter
M [with suffix]	Bending moment at end of member

m	Number of bays
N	Parameter
n	Power
O	Parameter
P	Axial forces in columns
p_y	Design strength
r	Ratio of bay width to storey height
S	Cladding stiffness
$\bar{s}$	Non-dimensional cladding stiffness
V	Shear force in beam
W	Parameter
X	Parameter
x	Horizontal deflection
Y	Vertical distance
Z	Cost
γ_f	Partial safety factor for loading
Δ	Sway deflection
θ	Joint rotation
ϕ	Sway angle
$\bar{\phi}$	Non-dimensional parameter

3.1 INTRODUCTION

Failure of a structure has been defined by Bate (1973) as 'unfitness for use', one possible cause being excessive deformation. This results in damage to the cladding or finishes of a building, hinders operations within, and may cause alarm or unpleasant sensations to the occupants. It follows that, when considering serviceability, the designer should calculate deflections under the working (unfactored) loads expected in normal use of the structure.

In multi-storey steel building frames, beam deflections can readily be determined by analysis of a limited frame consisting of the member under consideration and the adjacent beams and columns (Joint Committee, 1971). The main problem relating to deflection concerns horizontal sway in unbraced frames. This form of deflection arises mainly from wind, and its control may govern the member sections.

In the past, there has been no firmly established limit for the sway: height ratio ϕ to be used in design. One survey reported by the

Council on Tall Buildings (1979) showed that the limiting value had varied from 1/1000 to 1/200, whilst the British code BS449 (BSI, 1969) gave no recommendation for multi-storey structures. More recently, however, both British and European guides have settled on a value of 1/300 for calculations based on bare frames (BSI, 1977; ECCS, 1978). If the complete structure including cladding is considered, then the more restrictive 1/500 has been proposed (ECCS, 1978).

For adequate safety, ultimate strength is checked under enhanced values of loading, obtained by factoring the working loads. Under combined loading the factors range typically from 1·2 to 1·4. These values are sufficiently high to prevent significant plasticity at working load, and therefore deflection calculations are based usually on elastic behaviour.

A number of approximate methods are available for the calculation of sway, some enabling direct design to specified limits. These methods are suitable for hand calculation, and are sufficiently accurate for medium-rise frames. If a given frame is to be analysed for sway, the charts produced by Wood and Roberts (1975) are most convenient. An alternative procedure is due to Moy (1974), which has the advantage that it also provides guidance on what changes will be required in section properties if deflections in a trial design are found to be excessive. However, if control of sway is likely to govern member sizes then equations due to Anderson and Islam (1979*a*) enable a suitable design to be obtained directly, without the need for a trial set of sections.

For slender high-rise structures, it is most economical to provide a stiff core. However, if an unbraced frame is preferred for architectural or functional reasons, then secondary effects, particularly loss of stiffness due to compressive axial forces and sway due to differential axial shortening, should be considered. The former effect can be included in approximate methods without difficulty, but if axial shortening is significant it will be preferable to use a more accurate computer-based approach. If only analysis is required, many standard programs are available. Direct design by computer is also possible, as demonstrated by, amongst others, Anderson and Salter (1975) and Majid and Okdeh (1982).

In order to choose the most convenient procedure in design, it is helpful to know at any early stage whether ultimate strength or the serviceability limit on sway will dominate the choice of sections.

Guidance on this has been given recently by Anderson and Lok (1983), following a parametric study on medium-rise frames. Their work is described below, before proceeding to the design methods referred to above.

3.2 GOVERNING DESIGN CRITERION UNDER COMBINED LOADING

3.2.1 Design Studies

The frames examined were rectangular in elevation, of four, seven and ten storeys in height, and from two to four or five bays in width. Two ratios of bay width to storey height r were considered, namely 1·33 and 2·0, although within a particular frame these two dimensions were constant. All bases were fixed.

The unfactored loads are given in Table 3.1 together with the maximum and minimum basic wind speeds. For simplicity, the resulting horizontal wind pressure was taken as uniform over the height of the frame, although designers often use a reduced pressure on the lower storeys. On the other hand, no allowance was made for eccentricity of vertical loading arising from fabrication and erection tolerances, and no account was taken of any reduction in live loading permitted for the design of columns. In the studies, the maximum value of floor loading was combined with minimum values of wind loading, and vice-versa. A number of other load combinations were examined also.

The design strength of structural steel, p_y, was taken as 240 N/mm^2, corresponding to the grade commonly used in medium-rise unbraced frames. Sway due to unfactored horizontal wind load was to be restricted to 1/300 of each storey height for the bare frame, in accordance with recent recommendations (BSI, 1977; ECCS, 1978).

TABLE 3.1
LOADING VALUES

Loading	*Maximum*	*Minimum*
Dead on roof (kN/m^2)	3·75	3·75
Imposed on roof (kN/m^2)	1·50	1·50
Dead on floor (kN/m^2)	4·79	4·79
Imposed on floor (kN/m^2)	5·00	2·50
Basic wind speed (m/s)	50	38

Minimum sections were determined by designing against failure by beam-type plastic hinge mechanisms or by squashing, using partial safety factors γ_f of 1·4 and 1·6 on dead and imposed load, respectively (BSI, 1977). These sections were then increased, as appropriate, to satisfy the restriction on sway at working load. The method of Anderson and Islam (1979*a*) described below was used. Column sections were made continuous over at least two storeys, but beam sections were changed at each floor level if required. The designs were then subjected to a second-order elasto-plastic computer analysis (Majid and Anderson, 1968), with γ_f values of 1·4, 1·2 and 1·2 applied to dead, imposed and wind loads, respectively (BSI, 1977). If the factored load level was achieved before collapse occurred, then ultimate strength under combined loading was not the governing criterion for that particular frame.

3.2.2 Results

The results are summarised in Table 3.2. These are applicable to frames whose steel design strength is in the region of 240 N/mm^2, such as British Grade 43 and European Fe 360 material. The designer needs to determine the ratio of the sum of the column axial forces P to

TABLE 3.2

LIMITING VERTICAL: HORIZONTAL LOAD RATIO, P/F

Frame	*Bay width : storey height*	*Average P/F*
Four-storey	2·0	40
Seven-storey	2·0	40
Ten-storey	2·0	40
Four-storey	1·33	75
Seven-storey	1·33	65
Ten-storey	1·33	55

the corresponding total column wind shear F in each storey. P and F are calculated using the factored combined loads. The ratio is then averaged over all the storeys of the frame. Ultimate strength under combined load is not likely to be critical, provided the limits on P/F are not exceeded.

3.2.3 Example

Consider the six-storey two-bay frame shown in Fig. 3.1 which is subjected to the unfactored loadings shown in Table 3.3. The dynamic

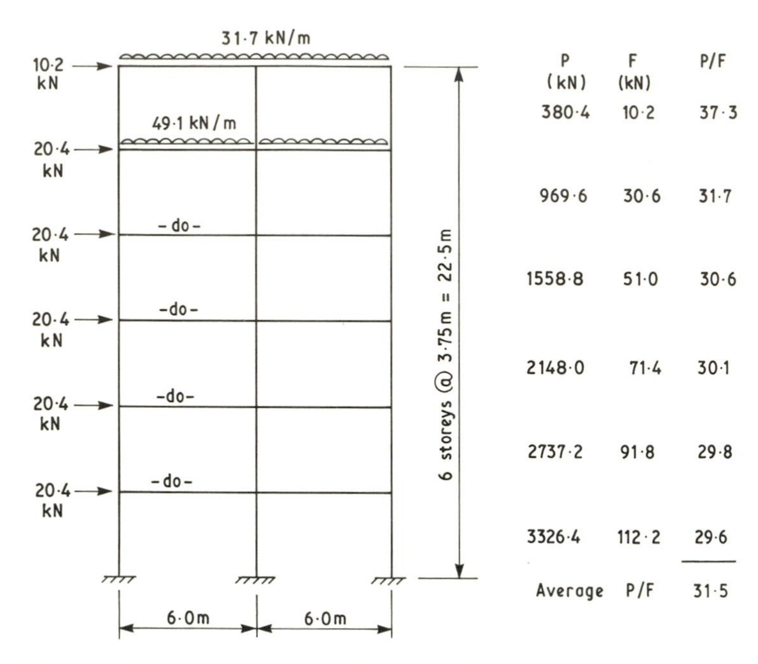

FIG. 3.1. Six-storey two-bay frame with factored combined loading.

wind pressure has been calculated from a basic wind speed of 44 m/s. With frames spaced longitudinally at 4·5 m centres, the factored combined loads are as given in Fig. 3.1, the corresponding values of P/F being stated alongside each storey. The average value is 31·5 and $r = L/h = 1{\cdot}6$.

It is clear from Table 3.2 that ultimate strength will not be critical for design. The appropriate procedure is therefore to calculate first the

TABLE 3.3
LOADING FOR SIX-STOREY TWO-BAY FRAME

Loading	*Unfactored value* (kN/m^2)	γ_f *combined*
Dead on roof	3·75	1·4
Imposed on roof	1·50	1·2
Dead on floor	4·80	1·4
Imposed on floor	3·50	1·2
Dynamic wind pressure	1·005	1·2

sections required to sustain the factored values applicable to dead plus imposed vertical load, and then to increase the sections as necessary in order to limit sway at working load. A final check analysis can then be undertaken to confirm adequate ultimate strength under combined loading.

3.3 APPROXIMATE METHOD FOR DIRECT DESIGN

When it is expected that the serviceability limit on sway will be dominant, the method due to Anderson and Islam (1979*a*, *b*) enables suitable section properties to be calculated directly for rectangular frames. The design equations are based on three assumptions:

(i) Vertical loads have a negligible effect on horizontal displacements.
(ii) A point of contraflexure exists at the mid-height of each column (except in the bottom storey) and at the mid-length of each beam.
(iii) The total horizontal shear is divided between the bays in proportion to their relative widths.

These assumptions render a frame statically determinate, except in the bottom storey, and enable each storey to be considered in isolation. Expressions relating the sway deflection over a storey height to the inertias of the corresponding columns and surrounding beams can then be derived.

3.3.1 Intermediate Storey

Figure 3.2 shows an intermediate storey of height h_2 subject to horizontal load. By treating each bay individually, it can be shown that assumptions (ii) and (iii) result in zero axial load in the internal columns. It is also implied by the assumptions that the column inertias are related as follows

$$\frac{I_{3,1}}{L_1}=\frac{I_{3,2}}{(L_1+L_2)}=\frac{I_{3,3}}{(L_2+L_3)}=\cdots=\frac{I_{3,m+1}}{L_m} \tag{3.1}$$

where the symbols are defined in Fig. 3.2. The frame is m bays in width. Furthermore, the stiffnesses of the beams must vary according

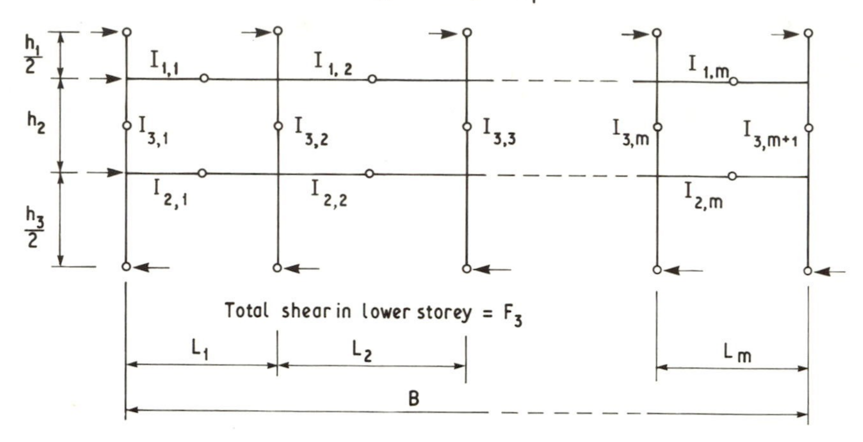

FIG. 3.2. Intermediate storey.

to bay width

$$\frac{I_{1,1}}{L_1^2}=\frac{I_{1,2}}{L_2^2}=\cdots=\frac{I_{1,m}}{L_m^2} \tag{3.2}$$

$$\frac{I_{2,1}}{L_1^2}=\frac{I_{2,2}}{L_2^2}=\cdots=\frac{I_{2,m}}{L_m^2} \tag{3.3}$$

where $I_{1,1}$, $I_{1,2}$, etc. are the inertias of the upper beams in the storey, and $I_{2,1}$, $I_{2,2}$, etc. are those of the lower beams. The design equations are derived in terms of $I_{1,2}$, $I_{2,2}$ and $I_{3,2}$ by analysing the subassemblage shown in Fig. 3.3.

As the moment is assumed to be zero at C, the slope–deflection equations give

$$M_{BC}=\frac{6EI_{1,2}}{L_2}\,\theta_B \tag{3.4}$$

where M_{BC} is the clockwise bending moment acting on BC at B, and θ_B is the corresponding rotation. Let the shear at C be V_c, and F_2 be the total horizontal shear in all the columns of the storey being designed; F_1 and F_3 are the total values in the columns of the storeys immediately above and below. The shear in the column at E is therefore $F_2(L_1+L_2)/(2B)$, where B is the total width of the frame.

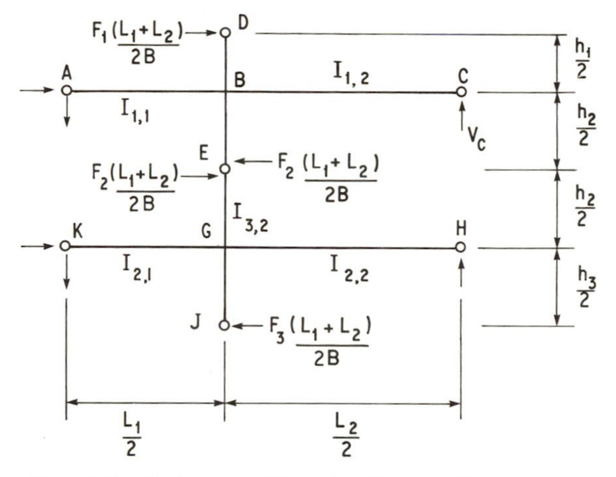

FIG. 3.3. Sub-assemblage for intermediate storey.

Taking moments about A for the region ABCDE it can be shown that

$$\frac{V_c L_2}{2} = \frac{(F_1 h_1 + F_2 h_2) L_2}{4B} \tag{3.5}$$

As the left-hand sides of eqns (3.4) and (3.5) are equal

$$\theta_B = \frac{(F_1 h_1 + F_2 h_2) L_2^2}{24 E B I_{1,2}} \tag{3.6}$$

Similarly, from the region EGJHK

$$\theta_G = \frac{(F_2 h_2 + F_3 h_3) L_2^2}{24 E B I_{2,2}} \tag{3.7}$$

where θ_G is the clockwise rotation at G. For a point of contraflexure at mid-height, θ_B and θ_G must be equal. Hence from eqns (3.6) and (3.7)

$$I_{1,2} = \frac{(F_1 h_1 + F_2 h_2) I_{2,2}}{(F_2 h_2 + F_3 h_3)} \tag{3.8}$$

Let M_{GB} be the clockwise end moment in BG at G. Equating the slope–deflection equation for M_{GB} to the moment of the shear at E about G

$$M_{GB} = \frac{2 E I_{3,2}}{h_2}\left(\theta_B + 2\theta_G - \frac{3\Delta}{h_2}\right) = -\frac{F_2 h_2 (L_1 + L_2)}{4B} \tag{3.9}$$

where, in design, Δ equals the allowable sway over the storey height h_2. Equations (3.6)–(3.9) now permit an expression for $I_{2,2}$ to be derived

$$I_{2,2}=\frac{(F_2h_2+F_3h_3)h_2L_2^2I_{3,2}}{24E\,\Delta BI_{3,2}-F_2h_2^3(L_1+L_2)} \tag{3.10}$$

Equations (3.10) and (3.8) enable $I_{2,2}$ and $I_{1,2}$ to be determined once $I_{3,2}$ is known. The need to choose trial values can be avoided by introducing an element of optimisation into the design.

The cost Z of an intermediate storey is represented by the product of inertia and length for each member. Thus

$$Z=0{\cdot}5(L_1I_{1,1}+L_2I_{1,2}+\cdots+L_mI_{1,m})+0{\cdot}5(L_1I_{2,1}+L_2I_{2,2}+\cdots+L_mI_{2,m})+(I_{3,1}+I_{3,2}+\cdots+I_{3,m+1})h_2 \tag{3.11}$$

The factor of 0·5 is introduced so that the cost of the complete frame is the sum of the individual storey costs. It is assumed that a beam is equally effective in restraining the sways immediately above and below its own level, and the factor is therefore 0·5.

Using eqns (3.1)–(3.3), (3.8) and (3.10), Z can be obtained in terms of $I_{3,2}$ only. For minimum cost

$$\frac{dZ}{dI_{3,2}}=0 \tag{3.12}$$

which gives

$$I_{3,2}=\frac{F_2h_2^3(L_1+L_2)+h_2L_2(L_1+L_2)\sqrt{\left(\dfrac{F_2h_2W(F_1h_1+2F_2h_2+F_3h_3)}{2B}\right)}}{24E\,\Delta B} \tag{3.13}$$

where

$$W=\frac{L_1^3+L_2^3+\cdots+L_m^3}{2L_2^2} \tag{3.14}$$

Once $I_{3,2}$ is determined, the other inertias follow from eqns (3.10), (3.8) and (3.1)–(3.3).

Usually, it will not be possible to use sections which correspond exactly to the required inertias. The beam inertias can be calculated using an effective value for $I_{3,2}$ based on the properties of the column sections actually adopted. From eqn (3.1), the effective value of $I_{3,2}$ is given by

$$(I_{3,2})_{\text{effective}} = \text{Least of } \left[\frac{(L_1+L_2)}{L_1} I_{3,1};\ I_{3,2};\ \frac{(L_1+L_2)}{(L_2+L_3)} I_{3,3};\ \cdots;\ \frac{(L_1+L_2)}{L_m} I_{3,m+1}\right] \quad (3.15)$$

3.3.2 Intermediate Storey of a Regular Frame

It is common for the intermediate storeys to have a constant storey height h and bay width L. In this case, eqn (3.13) simplifies to the following expression for the inertia I_3 of an internal column

$$I_3 = \frac{2F_2h^3L + h^2L^2\sqrt{(F_2(F_1+2F_2+F_3))}}{24E\,\Delta B} \quad (3.16)$$

The inertia I_3' for an external column is taken as $0{\cdot}5I_3$. Once column sections are chosen, the effective value of $I_{3,2}$ to be used in beam design must be the lesser of the value actually provided for the internal column and twice the actual value of I_3'.

3.3.3 Bottom Two Storeys of a Fixed Base Frame

A separate analysis is required for the bottom two storeys because it is grossly inaccurate to assume a point of contraflexure at mid-height of a ground storey column. Design equations have been derived for pinned base frames (Anderson and Islam, 1979*a*, *b*), but such bases result in very high inertias for the bottom storey members, in comparison with elsewhere in the frame. Fixed bases are preferable for multi-storey frames, unless the need to minimise stress on the soil is of over-riding importance. The equations given below apply to fixed base conditions.

The subassemblage is shown in Fig. 3.4. It is assumed that the fixity of the base and the avoidance of reverse column taper result in sections being governed by the sway Δ of the storey next to the bottom. The column section is therefore continuous over the bottom two storeys. Anderson and Islam also derived equations applicable to fixed base frames when sway of the bottom storey governed design. This case can arise when the height of the bottom storey is much greater than that of the storey above. These special equations are avoided by checking bottom storey sway using the analysis method described later, and modifying sections if necessary.

By comparing Figs. 3.3 and 3.4 it can be seen that eqn (3.6) still applies for the rotation θ_B. θ_G is obtained by deriving the sway

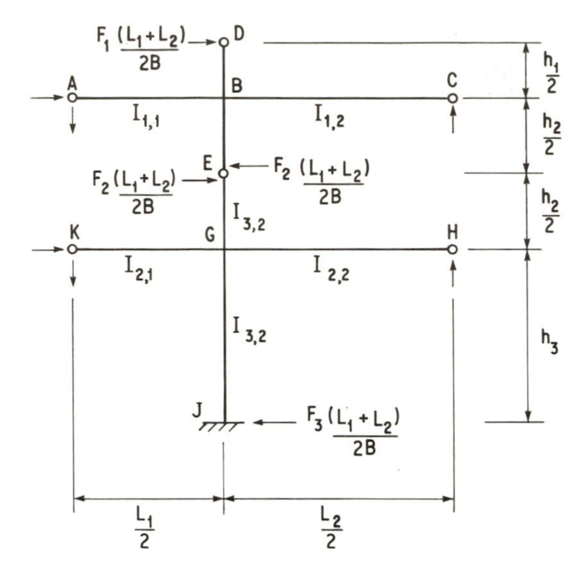

FIG. 3.4. Sub-assemblage for bottom two storeys.

equation for GJ and considering equilibrium of moments at G, giving

$$\theta_G = \frac{(F_2h_2 + F_3h_3)h_3(L_1 + L_2)L_2^2}{24EBh_3(L_1 + L_2)I_{2,2} + 4EBL_2^2I_{3,2}} \tag{3.17}$$

Expressions for $I_{1,2}$ and $I_{2,2}$ are obtained by replacing eqn (3.7) by (3.17)

$$I_{1,2} = \frac{(F_1h_1 + F_2h_2)h_2L_2^2I_{3,2}}{24E\,\Delta BI_{3,2} - F_2h_2^3(L_1 + L_2)} \tag{3.18}$$

$$I_{2,2} = \frac{(F_2h_2 + F_3h_3)h_2L_2^2I_{3,2}}{24E\,\Delta BI_{3,2} - F_2h_2^3(L_1 + L_2)} - \frac{L_2^2I_{3,2}}{6h_3(L_1 + L_2)} \tag{3.19}$$

The cost of the bottom two storeys is expressed as

$$Z = 0{\cdot}5\left(\sum_{i=1}^{i=m} L_iI_{1,i}\right) + 1{\cdot}0\left(\sum_{i=1}^{i=m} L_iI_{2,i}\right) + (h_2 + h_3)\sum_{j=1}^{j=m+1} I_{3,j} \tag{3.20}$$

Proceeding as before, differentiation leads to

$$I_{3,2} = \frac{(F_2h_2 + L_2X)h_2^2(L_1 + L_2)}{24E\,\Delta B} \tag{3.21}$$

with

$$X = \sqrt{\frac{3F_2h_3W(F_1h_1 + 3F_2h_2 + 2F_3h_3)}{6Bh_3(h_2 + h_3) - WL_2^2}} \tag{3.22}$$

Once column sections are chosen, an effective value of $I_{3,2}$ may again be used for beam design, provided eqn (3.15) is satisfied.

Anderson and Islam (1979*a*, *b*) also derived special equations to control sway of the top storey. However, experience has shown that strength under vertical loading will control the design of this region, and equations for sway are therefore unnecessary.

3.3.4 Example

The method is demonstrated by designing the six storey frame, discussed earlier, to a limiting sway index of 1/300 under unfactored wind loading. With a longitudinal spacing of 4·5 m, the resulting loads are as shown in Fig. 3.5(a). E is taken as 205 kN/mm². Column sections will only be changed every second storey, to reduce fabrication costs. As a result, it is only necessary to design three storeys from the frame.

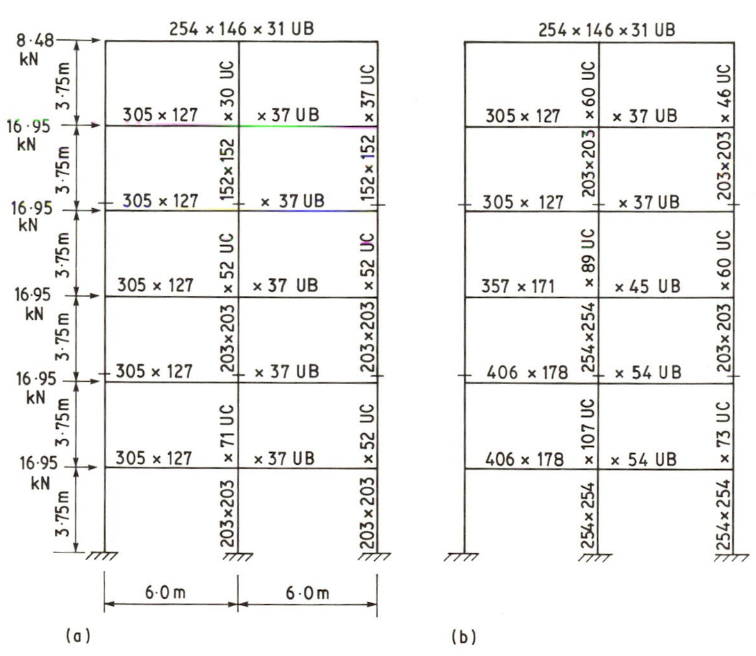

FIG. 3.5. Designs for six-storey two-bay frame: (a) initial design; (b) design.

TABLE 3.4

MOMENTS OF INERTIA IN CM^4 FOR SIX-STOREY TWO-BAY FRAME

Storey	*Internal column*		*External column*		*Lower beam*	*Upper beam*
	I_3 *required* (cm^4)	I_3 *provided* (cm^4)	I'_3 *required* (cm^4)	I'_3 *provided* (cm^4)	I_2 *required* (cm^4)	I_1 *required* (cm^4)
5	5 672	6 088	2 836	4 564	7 252	3 626
3	13 234	14 307	6 617	{7 647	14 447	10 835
				{6 088	15 993	11 995
2	16 314	17 510	8 157	11 360	16 243	14 862

The sections shown in Fig. 3.5(a) are the minimum ones which withstand dead plus imposed vertical loading only, using $\gamma_f = 1{\cdot}4$ and $1{\cdot}6$, respectively. They are obtained by simple plastic theory, taking $p_y = 240\ \text{N/mm}^2$, and selecting from the range of British Universal sections.

Design for sway is commenced at the fifth storey, using eqn (3.16) to calculate an inertia for the internal column of 5672 cm^4, as shown in Table 3.4. As can be seen from the sections in the Appendix, the nearest Universal Column (UC) is 203 × 203 × 60 kg/m ($I = 6088$ cm^4). This is adopted, as shown in Fig. 3.5(b). The required inertia for the external column is half that for the internal member. The lack of a UC with a suitable property necessitates the provision of a 203 × 203 × 46 kg/m section ($I = 4564$ cm^4). The beam inertias are now calculated from eqns (3.10) and (3.8), taking the effective value of $I_{3,2}$ as the value actually provided for the internal column (6088 cm^4). The required beam inertias are given in Table 3.4, and the chosen sections in Fig. 3.5(b). The minimum beam section 305 × 127 × 37 kg/m is retained for the lower beam as its inertia is 99% of that required.

A similar procedure is followed for the third storey, except that it is worth considering two alternatives for the external column. In the first case a 203 × 203 × 71 kg/m UC is chosen ($I = 7647$ cm^4). Beam design is then based on an effective $I_{3,2}$ of 14 307 cm^4 corresponding to the actual section of the internal column. For the alternative, a 203 × 203 × 60 kg/m UC ($I = 6088$ cm^4) is proposed for the external column. As this is less than twice the inertia of 14 307 cm^4 provided for the internal column, an effective value of $I_{3,2} = 2 \times 6088 = 12\,176$ cm^4 must be used for beam design. When the required beam inertias in Table 3.4 are compared with the list of available Universal Beams (UB), it is found that the same sections are required in both cases. Hence it is more economical to choose the lighter section for the external columns, as shown in Fig. 3.5(b).

The results for the second storey come from the use of eqns (3.22), (3.21), (3.19) and (3.18). After column sections are chosen, beam design is based on an effective $I_{3,2}$ of 17 510 cm^4.

3.4 APPROXIMATE METHOD FOR ANALYSIS

Provided secondary effects are not significant, the above method successfully provides section properties needed to satisfy limits on sway. Rarely, if ever, will these properties correspond exactly to available sections, but additional column stiffness can be offset by reduced beam stiffness, and vice-versa. The use of eqns (3.1)–(3.3), though, precludes a similar trade-off between, for example, internal and external columns. A designer may wish, therefore, to use an analysis to modify slightly the sections, to achieve greater economy. Suitable methods are due to Wood and Roberts (1975) and Moy (1974). The former method has been included in ECCS recommendations (1978) and is described below. Such analyses can also be used to check deflections in the top or bottom storeys if, exceptionally, these could be critical.

3.4.1 Derivation of the Analysis

The analysis is based on the frame shown in Fig. 3.6 which acts as the substitute for an individual storey of a multi-bay frame. The relation-

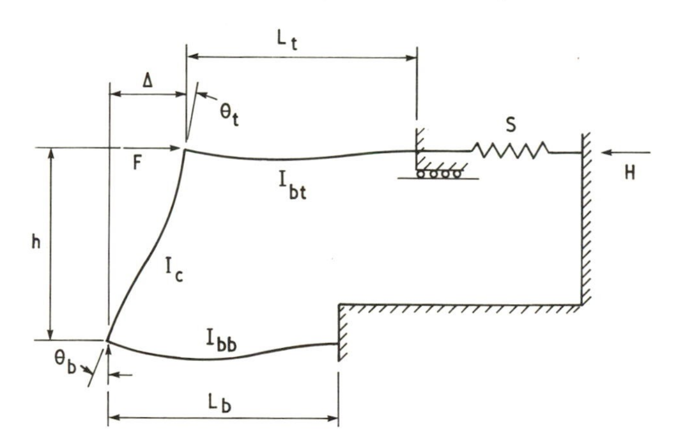

FIG. 3.6. Single-storey substitute frame.

ship between the substitute and real structures is discussed in a later section of this chapter. Cladding may be included by a spring of stiffness S. Wood and Roberts (1975) used stiffness distribution (Wood, 1974) in their paper, but the same results can be obtained in a more straightforward manner by considering member and joint equilibrium.

When the column sways by Δ, then for equilibrium

$$M_t + M_b + (F - S\Delta)h = 0 \tag{3.23}$$

where $S\Delta$ is the restoring force exerted by the spring, and M_t and M_b are the moments acting clockwise on the top and bottom ends of the member. The moments can be expressed in terms of Δ and the rotations θ_t and θ_b by slope–deflection

$$M_t = \frac{2EI_c}{h}\left(2\theta_t + \theta_b - \frac{3\Delta}{h}\right) \tag{3.24}$$

the expression for M_b being obtained by interchanging θ_t and θ_b. Substitution for the moments in eqn (3.23) then gives an expression for the required sway

$$\frac{\Delta}{h} = \frac{Fh^2 + 6EI_c(\theta_t + \theta_b)}{Sh^3 + 12EI_c} \tag{3.25}$$

The unknown rotations θ_t and θ_b are determined from joint equilibrium. To do this, a modified expression for column end moment is first obtained by substituting eqn (3.25) into (3.24). Hence

$$M_t = -\frac{FhM}{2} + EK_c(N\theta_t - O\theta_b) \tag{3.26}$$

where

$$K_c = I_c/h \tag{3.27}$$

$$M = 12/(12 + \bar{s}) \tag{3.28}$$

$$\bar{s} = Sh^2/(EK_c) \tag{3.29}$$

$$N = 4 - 3M \tag{3.30}$$

$$O = -2 + 3M \tag{3.31}$$

For joint equilibrium at the top of the column

$$M_t + M_{bt} = 0 \tag{3.32}$$

where M_{bt} is the moment in the top beam at the beam–column

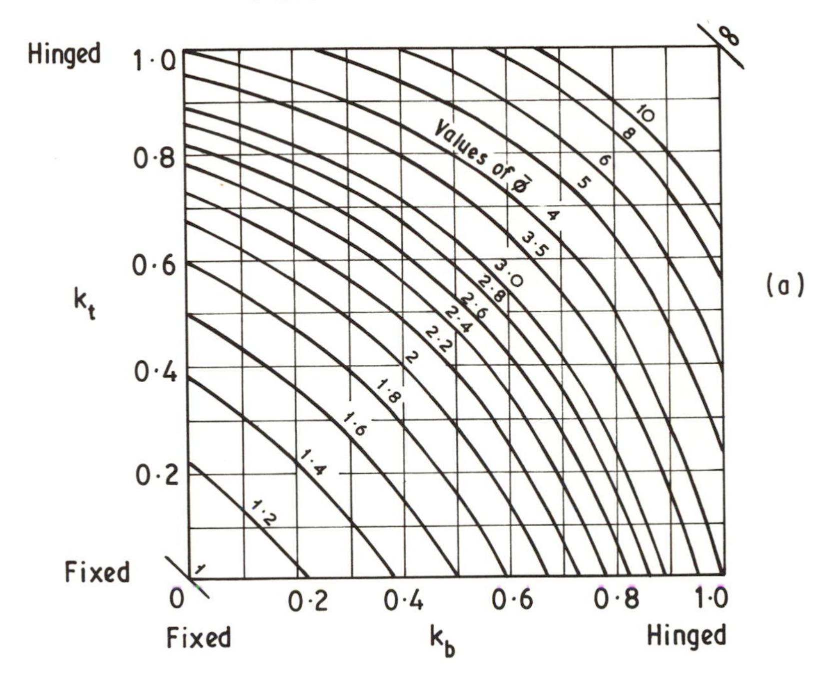

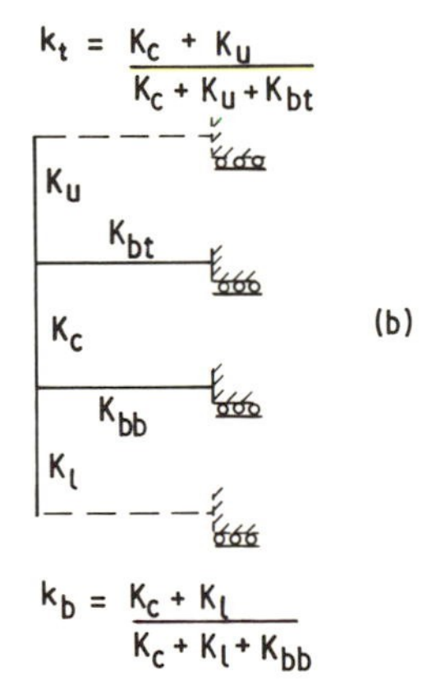

FIG. 3.7. Sidesway deflection for unbraced frame: (a) values of $\bar{\phi} = (\Delta/h)/(Fh/12EK_c)$; (b) distribution coefficients.

junction. M_{bt} is given by slope–deflection as

$$M_{bt} = \frac{4EI_{bt}}{L_t}\theta_t \tag{3.33}$$

On substitution of eqns (3.26) and (3.33) into (3.32), an expression in terms of θ_t and θ_b is obtained. A second expression is derived from equilibrium at the lower end of the column. The two expressions can then be solved for θ_t and θ_b to give

$$\theta_t = \frac{FhM(OK_c + NK_c + 4K_{bt})}{2E((NK_c + 4K_{bt})(NK_c + 4K_{bb}) - O^2K_c^2)} \tag{3.34}$$

where

$$K_{bt} = I_{bt}/L_t, \qquad K_{bb} = I_{bb}/L_b \tag{3.35}$$

The expression for θ_b is obtained from eqn (3.34) by interchanging K_{bt} and K_{bb}. The sway can now be determined by substituting for the rotations in eqn (3.25). After rearrangement, the following non-dimensional expression, given by Wood and Roberts (1975), is obtained

$$\bar{\phi} = \frac{\Delta/h}{Fh/(12EK_c)} = M\left[1 + \frac{3(k_b + k_t - k_bk_t)}{4 - 3k_b - 3k_t + 2k_bk_t + \bar{s}(1 - k_bk_t/4)/3}\right] \tag{3.36}$$

where k_t and k_b are the distribution factors

$$k_t = \frac{K_c}{K_c + K_{bt}}, \qquad k_b = \frac{K_c}{K_c + K_{bb}} \tag{3.37}$$

To assist designers, Wood and Roberts presented their analysis in the form of charts, such as Fig. 3.7(a). The charts are constructed by selecting values for $\bar{\phi}$, $\bar{s}$ and k_t, and solving eqn (3.36) for k_b.

3.4.2 Substitute Frame

To use the analysis, each storey of the actual frame must be replaced by an equivalent structure having the form of Fig. 3.6. This is done by first transforming the actual frame into a substitute beam–column structure, as shown in Fig. 3.8 for the six storey frame. The basis of the substitute frame (Wood, 1974) is that:

(i) for horizontal loading on the real frame, the rotations of all joints at any one level are approximately equal, and
(ii) each beam restrains a column at both ends.

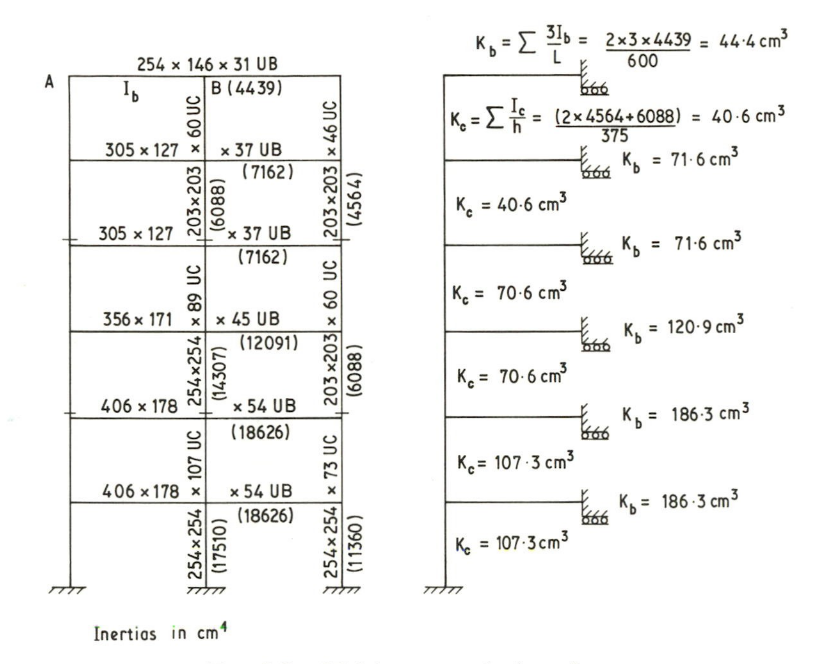

FIG. 3.8. Multi-storey substitute frame.

Hence for a typical beam AB in the actual frame (Fig. 3.8), with end rotations $\theta_A = \theta_B = \theta$, slope–deflection gives

$$\text{end moment} = \frac{2EI_b}{L}(2\theta_A + \theta_B) = \frac{6EI_b}{L}\,.\,\theta \tag{3.38}$$

This compares with eqn (3.33) which applies to Fig. 3.6 and also to the beams of the substitute frame in Fig. 3.8.

It can be seen, therefore, that to represent the equal end rotations of the real frame, a beam inertia in the substitute frame should be based on $(6/4)I_b, = 1{\cdot}5I_b$. It follows from (ii) that the total stiffness K_b of a beam in the substitute frame is

$$K_b = \sum\left(2 \times 1{\cdot}5\,\frac{I_b}{L}\right) = \sum\frac{3I_b}{L} \tag{3.39}$$

For the substitute column

$$K_c = \sum (I_c/h) \tag{3.40}$$

In both cases the summation is over all the beams or columns in the real frame at the level being considered.

To allow for continuity of columns in a multi-storey structure, it is recognised that each floor beam restrains column lengths above and below its own level. Hence the distribution coefficients are modified to

$$k_t = \frac{K_c + K_u}{K_c + K_u + K_{bt}}, \qquad k_b = \frac{K_c + K_l}{K_c + K_l + K_{bb}} \tag{3.41}$$

where K_u and K_l refer to the upper and lower columns (Fig. 3.7(b)). Note also that when calculating Δ in a multi-storey frame, F is the *total* shear in the storey being considered.

It remains to assign a value to the cladding stiffness, $\bar{s}$. Many designers will prefer to assign zero value, and then use the less restrictive limit of 1/300 recommended when calculations are made on the bare frame (ECCS, 1978). The assessment of $\bar{s}$ is a topic requiring further research, and is outside the scope of this chapter. However, some guidance is given by Wood and Roberts (1975) on this subject.

3.4.3 Example

The frame shown in Fig. 3.8 is that designed in Section 3.3.4, using the method of Anderson and Islam (1979*a*, *b*). The resulting sway deflections are determined using Fig. 3.7, the calculations being set out in Fig. 3.8 and Table 3.5. Comparison is also made in Table 3.5 with deflections given by a standard linear elastic computer program, including the effects of axial shortening. It will be observed that the values given by Wood and Roberts' analysis are reasonably accurate, and that the method of Anderson and Islam has given a design which satisfies the deflection limit of 12·5 mm. However, storeys 1/2 and 5/6 may be

TABLE 3.5
SWAY DEFLECTIONS FOR SIX-STOREY FRAME (FIG. 3.5(b))

Storey	F (*kN*)	K_c (*cm*^{3})	k_t	k_b	$\bar{\phi}$	Δ (*mm*)	Δ (*computer*) (*mm*)
6	8·48	40·6	0·48	0·53	2·5	3·0	4·7
5	25·43	40·6	0·53	0·61	3·0	10·7	9·8
4	42·38	70·6	0·61	0·54	3·0	10·3	11·0
3	59·33	70·6	0·54	0·49	2·6	12·5	11·8
2	76·28	107·3	0·49	0·54	2·6	10·6	10·8
1	93·23	107·3	0·54	0	1·7	8·5	8·5

overdesigned, because the maximum deflections are significantly below the permissible value. It is worthwhile, therefore, to consider some reduction in sections at these levels.

It will be recalled that for an equal bay frame, the design method of Anderson and Islam ideally requires the inertia of an external column to be half that of the corresponding internal member. The analysis of Wood and Roberts (1975) enables the designer to avoid any such restrictions. It is proposed, therefore, that the internal column for storeys 5/6 be reduced to the same section as that for the external

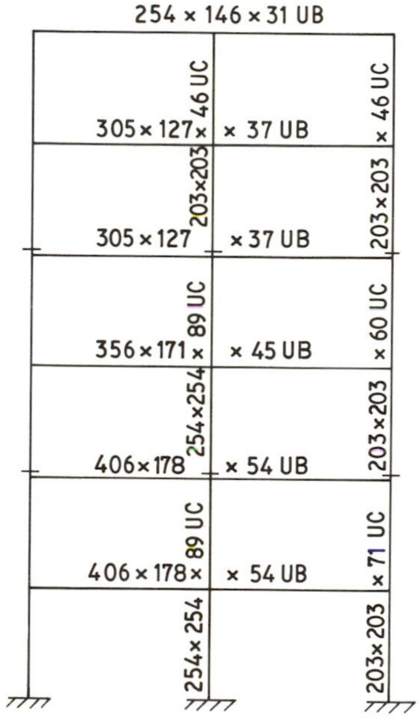

FIG. 3.9. Final design for six-storey two-bay frame.

members, namely 203×203×46 kg/m UC. For the bottom two-storeys, all columns will be reduced by one section, as shown in Fig. 3.9. The calculations for the resulting deflections follow the same procedure as those given earlier in Fig. 3.8 and Table 3.5. Using Fig. 3.7, it is predicted that the sway at storeys 2 and 5 will now be 12·4 mm and 11·2 mm, respectively, which are both acceptable. For comparison, the linear elastic computer program gives values of 12·3 mm and 10·3 mm.

3.4.4 Ultimate Strength

The calculation of ultimate strength is outside the scope of this chapter. However, it should be noted that when the design of Fig. 3.9 is subjected to the factored loading (Fig. 3.1), it is found to possess adequate strength. This confirms the prediction made in Section 3.2.3, that this criterion would not be critical for design.

If it is expected that for a particular frame ultimate strength will be dominant, then the structure should be designed first to this criterion. The method of Wood and Roberts then provides a useful check for the sway at working load. If in fact some deflections are found to be excessive, Moy (1974) has shown that the most economical procedure is usually to increase the stiffness of the beams.

3.5 SECONDARY EFFECTS AND COMPUTER METHODS

The simplified methods described above do not allow for the reduction in frame stiffness due to compressive forces, nor for the effects of axial shortening and unsymmetrical loading on sway.

The reduction in stiffness can easily be included by using additional horizontal shears (Vogel, 1983). If P_u denotes the total vertical loading carried by the columns at storey level u, then the total shear at this level should be increased by $1{\cdot}2P_u \,.\, \phi$.

Moy (1977) has given an estimate of the sway due to differential axial shortening. This was derived for frames subject to uniform horizontal loading, with the floors assumed to be rigid and column cross-sectional areas varying linearly from the top level to the bottom. Anderson and Islam (1979*b*) used this approach on a 15-storey frame and found the accuracy to be reasonable. However, when sway due to differential axial shortening or unsymmetrical vertical loading is likely to be significant, then simplified methods become appropriate only to the initial design stage. The final design should be obtained using a standard elastic analysis program to examine trial sections, or from specialised programs, such as those of Anderson and Salter (1975) or Majid and Okdeh (1982).

3.5.1 Use of Linear Programming

A flow diagram for the procedure due to Anderson and Salter is shown in Fig. 3.10. Due to the non-linear relationship between deflection and stiffness, a number of iterations will usually be necessary before the

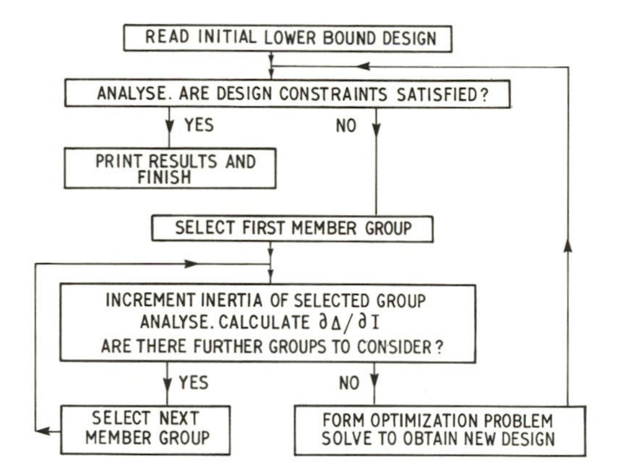

FIG. 3.10. Flow diagram (Anderson and Salter, 1975).

deflection limits are satisfied. The method is based on standard routines for elastic analysis and linear programming, and therefore can be developed easily. The procedure is applicable to a wide variety of frames, and secondary effects are readily included.

3.5.2 Minimum Cost Design

The previous method aims to minimise cost by generating a light design, but no account is taken of the variations in price per tonne that exist in the supplier's section catalogue, nor of the restricted number of sections available. The program developed by Majid and Okdeh (1982) overcomes these limitations by costing alternative selections of available sections using a supplier's price list. This method also adopts an iterative procedure to avoid direct solution of the structure's stiffness equations. The general procedure is explained with reference to the single storey fixed base frame of Fig. 3.11. For simplicity, axial deformations will be ignored, and the shear in each column will therefore be $F/2$, with equal rotations θ at B and C.

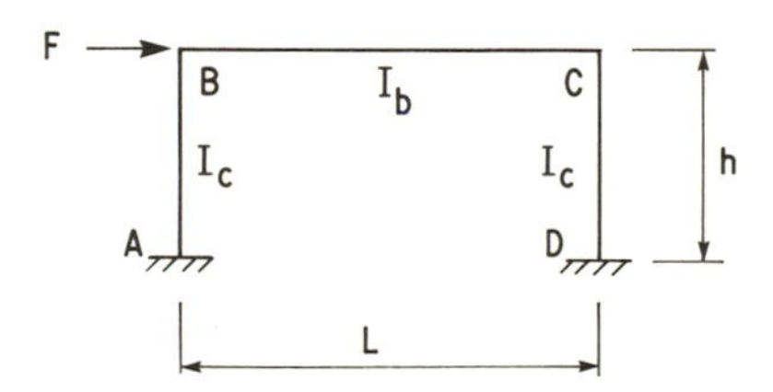

FIG. 3.11. Single-storey frame.

For equilibrium of AB

$$M_{AB}+M_{BA}+\frac{F}{2}\cdot h=0 \tag{3.42}$$

where M_{AB} and M_{BC} are the clockwise moments acting on AB at A and B, respectively. Slope–deflection enables these moments to be expressed in terms of θ and the specified sway Δ. Substituting for the moments in eqn (3.42) leads to

$$I_c=\frac{Fh^2/2}{12E\Delta/h-6E\theta} \tag{3.43}$$

For equilibrium at B, the sum of the column and beam end moments must be zero. Using the slope–deflection equations for these moments enables the following expression to be obtained

$$\theta=\frac{6EI_c\,\Delta/h^2}{4EI_c/h+6EI_b/L} \tag{3.44}$$

A value of I_b is selected to correspond to a beam section, and an initial value assigned to θ, say zero. I_c is then calculated from eqn (3.43), and θ from eqn (3.44). This enables I_c to be recalculated, using the value for θ just determined. Iteration continues until the required degree of accuracy is attained. A section is then selected for the columns. Other designs are initiated with different beam sections, and the various designs priced.

For multi-storey multi-bay frames, expressions are derived for column inertias and vertical and rotational displacements by applying slope–deflection and axial stiffness equations at each joint in turn. To treat the sway at each storey as a known quantity, the horizontal deflection x is given by

$$x=a_1+a_2Y+a_3Y^n \tag{3.45}$$

with the vertical distance Y measured from the roof downwards. The constants a_1–a_3 are determined from the following conditions:

(i) at the top of the frame ($Y=0$), $dx/dY=\phi$,
(ii) at ground level ($Y=H$), $x=0$,
(iii) at ground level, $dx/dY=0$ due to fixity at the bases.

Hence

$$x=\frac{\phi(YnH^{n-1}-Y^n-(n-1)H^n)}{nH^{n-1}} \tag{3.46}$$

As n increases the frame becomes more flexible and the profile approaches a linear profile of slope ϕ. The value of n used in the calculations is therefore the highest integer value that does not give exponential overflow in the computer.

It can be assumed that the column inertias satisfy eqn (3.1), but the economy of other relative values can also be investigated. Initial beam sections can be generated by rigid–plastic design under vertical loading, or by use of the design equations due to Anderson and Islam. Other beam sections can also be specified in the search for further economy.

Secondary effects are included in the method, which has been demonstrated on very large frames, up to 32 storeys in height with unequal bay widths. The results are generally satisfactory, although the form of the deflection profile may impose unnecessarily severe restrictions on sway of the bottom storeys.

3.6 CONCLUDING REMARKS

This chapter has described two approximate methods to design medium-rise unbraced rectangular frames in which sway deflections control the choice of sections. The method of Anderson and Islam has the advantage that it actually generates a design without the need for trial analyses. Provided secondary effects, particularly differential axial shortening, are not significant, the design will be satisfactory. The choice of sections can be refined by using the analysis due to Wood and Roberts. Both methods can be used by hand, or programmed for a micro-computer. The second method enables account to be taken of cladding stiffness.

If sway due to differential axial shortening is significant, as in very slender frames, then specialised computer methods are more appropriate. Two such methods have been described.

REFERENCES

ANDERSON, D. and ISLAM, M. A. (1979*a*) Design of multi-storey frames to sway deflection limitations. *Structural Engineer*, **57B**(1), 11–17.

ANDERSON, D. and ISLAM, M. A. (1979*b*) Design equations for multi-storey frames subject to sway deflection limitations. University of Warwick, Research Report CE3.

ANDERSON, D. and LOK, T. S. (1983) Design studies on unbraced, multistorey steel frames. *Structural Engineer*, **61B**(2), 29–34.

ANDERSON, D. and SALTER, J. B. (1975) Design of structural frames to deflexion limitations. *Structural Engineer*, **53**(8), 327–33.

BATE, S. C. C. (1973) Design philosophy and basic assumptions. *Concrete*, **7**(8), 43–4.

BRITISH STANDARDS INSTITUTION (1969) Specification for the use of structural steel in building, BS 449, Part 2.

BRITISH STANDARDS INSTITUTION (1977) Draft standard specification for the structural use of steelwork in building, 13908DC, Part 1, Simple construction and continuous construction.

COUNCIL ON TALL BUILDINGS, COMMITTEE 17 (1979) Stiffness. *Monograph on Planning and Design of Tall Buildings*, ASCE, New York, Vol. SB, Chapter SB-5, pp. 345–400.

ECCS (1978) *European Recommendations for Steel Construction*, The Construction Press, London.

JOINT COMMITTEE OF THE INSTITUTION OF STRUCTURAL ENGINEERS AND THE WELDING INSTITUTE (1971) Fully-rigid multi-storey welded steel frames, Second Report.

MAJID, K. I. and ANDERSON, D. (1968) The computer analysis of large multi-storey framed structures. *Structural Engineer*, **46**(11), 357–65.

MAJID, K. I. and OKDEH, S. (1982) Limit state design of sway frames. *Structural Engineer*, **60B**(4), 76–82.

MOY, F. C. S. (1974) Control of deflexions in unbraced steel frames. *Proc. ICE, Pt.* 2, **57,** 619–34.

MOY, F. C. S. (1977) Consideration of secondary effects in frame design. *Journal of the Structural Division, ASCE,* **103,** ST10, 2005–19.

VOGEL, U. (1983) Simplified second-order elastic and elastic–plastic analysis of sway frames. *Proceedings, Third International Colloquium, Stability of Metal Structures*, Toronto, Structural Stability Research Council, Bethlehem, USA, pp. 377–87.

WOOD, R. H. (1974) Effective lengths of columns in multistorey buildings. *Structural Engineer*, **52**(7), 235–44; (8), 295–302; (9), 341–6.

WOOD R. H. and ROBERTS, E. H. (1975) A graphical method of predicting side-sway in the design of multi-storey buildings. *Proc. ICE, Pt 2,* **59,** 353–72.

APPENDIX

Major axis moments of inertia for Universal sections in cm^4 for design of six-storey two-bay frame.

Universal beams		*Universal columns*	
254×146×31	4 439	203×203×46	4 564
305×127×37	7 162	203×203×52	5 263
356×171×45	12 091	203×203×60	6 088
406×178×54	18 626	203×203×71	7 647
		254×254×73	11 360
		254×254×89	14 307
		254×254×107	17 510

The complete range of Universal sections is given in the *Structural Steelwork Handbook* published by BCSA/Constrado, London, 1978.

Chapter 4

INTERBRACED COLUMNS AND BEAMS

I. C. MEDLAND and C. M. SEGEDIN

Department of Theoretical and Applied Mechanics,
University of Auckland, New Zealand

SUMMARY

Methods of determining the buckling load factor for regular interbraced sets of columns and beams are developed and detailed. The equations which allow the critical load factor for any specific case to be evaluated are set out and summary charts prepared using such values. Such charts link buckling load to brace stiffness and number of columns, also allowing for brace eccentricity in some cases.

Basic in-plane buckling with linear elastic support is detailed and expanded to cases of flexural torsional buckling involving eccentricity of linear braces and the addition of rotational and torsional support. Beam systems are detailed for the cases of uniform moment and for uniform spread load applied at flanges. The charts included only present a sample of what might be assembled. The appendices contain the detailed composition of the 2×2 and 4×4 determinants from which the charts are assembled. The evaluation of these determinants by desk computer is straightforward.

NOTATION

A^*	A constant
a	Warping radius
B, $B_{n,m}$, B_m	Nodal slope (at node n,m)
B^*	A general amplitude
b_j, b^*	Amplitude of jth component of initial deformity, adjusted value

C	Torsional constant
C^*	A general amplitude
c	Carry-over factor
D, $D_{n,m}$, D_m	Nodal displacement (at node n,m)
D^*	A general amplitude
d_p	Scale factor involving number of bays
E	Young's modulus
e, e^*	Eccentricity of brace, nondimensional form
F^*	A general amplitude
f	Nondimensional brace stiffness
I	Second moment of area
I_W	Torsional section constant
k	EI/l
k_L	Linear brace stiffness
k_R	Rotational brace stiffness
k_T	Torsional brace stiffness
l	Length of interbrace element
M	Number of columns in parallel, general beam moment
M_0	Basic critical moment of simply supported beam
M_{cr}	Actual critical moment of beam
M_L, M_R	End moments
m	Column identifier
N	Number of internal brace lines
n	Number of bays, brace line identifier
P, P_E	Axial force, Euler value
p	Angle, measure of axial force
Q	Shape function within beam–column
R	Shape function within beam–column
s	Fixed ended beam rotational stiffness factor
t	Angle, measure of axial force
u, u_I	General lateral displacement, initial value
V_L, V_R	End shear forces
v	Shear stiffness function
x	Lateral displacement
y	Lateral displacement
z	Longitudinal position measure
α	Factor involving axial force and torsional constants
β, β^*, $\bar{\beta}$	Nondimensional linear brace stiffness, modified values
γ	Nondimensional rotational brace stiffness

$\delta, \delta_L, \delta_R$ Linear nodal displacements
λ Factor involving axial force
η Nondimensional torsional brace stiffness
$\theta, \theta_L, \theta_R$ Nodal rotation angles
θ_j Angle $j\pi/(N+1)$
ρ, ρ_{cr} Nondimensional axial force, critical value, critical value of beam moment
ϕ An angle
ϕ^* A general phase angle
ψ Angle $i\pi/(M+1)$

4.1 INTRODUCTION

Most structural forms include members for which elastic lateral or flexural–torsional buckling is a possible means of failure. The determination of the load level which would cause such a failure is a problem for the designer because it is one of the limits to overall load capacity. An estimate of this load must be made by determining an 'effective length', or in some other rational manner. Few members can be treated in isolation. Each component of a structure both draws support from and supports other members against buckling action. Bracing members are deliberately placed to provide such support.

Light industrial construction in steel and aluminium provides many examples of compression members which are deliberately braced at intervals throughout their length. In many cases they are arranged as an interbraced set of parallel members. The skeletal structure illustrated in Fig. 4.1, for example, includes the set of interbraced compression chords of the roof trusses and also the sets of side columns, which are similarly linked. In such structures the purlins and girts which make up the brace lines are usually effectively anchored or founded at intervals by attachment to cross-braced bays.

In many cases, the bracing is attached eccentrically to the shear centre of the braced member and is often connected in a manner which allows rotational and torsional support to be provided to that member. However, it also causes the buckling mode to involve flexural–torsional displacements.

In the idealised mathematical critical load problem it is assumed that the compressed member (or structure) remains perfectly straight and elastic until a critical load level is reached. At that level the member is capable of sustaining the load both in its perfect condition and in one

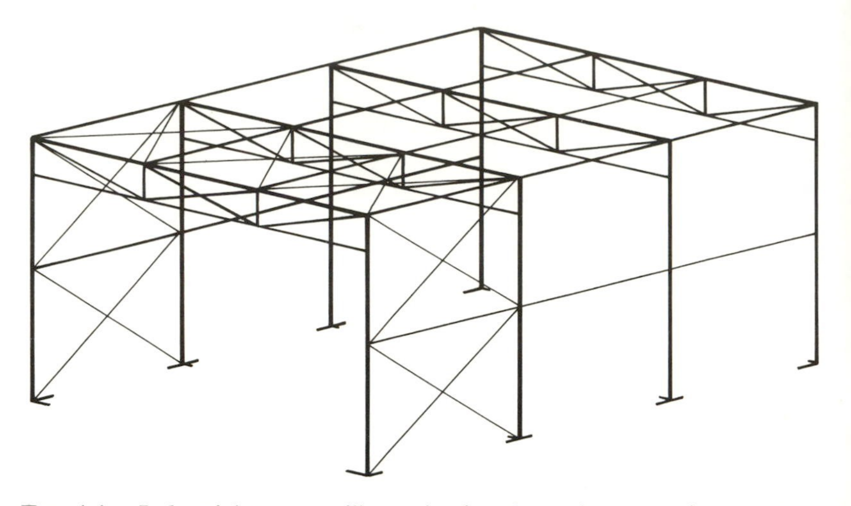

FIG. 4.1. Industrial structure illustrating interbraced compression members.

specific deformed shape, usually called the buckling or critical mode. This so-called 'bifurcation of equilibrium positions' typifies the phenomenon. It represents a neutral equilibrium state. In present day mathematical terms it is an eigenvalue problem. The result applies only to small displacements. Most structural texts cover this phenomenon at least as it applies to columns. The two classical stability texts, Bleich (1952) and Timoshenko and Gere (1961), provide clear physically-based coverage of buckling theory for basic structural components. The buckling of the simply supported uniform section column under uniform axial load, the Euler column, provides a base for this approach. Its lowest buckling load is $\pi^2 EI/l^2$ (known as the Euler load) and the associated buckling mode has the shape $\sin \pi z/l$. Simply supported beams under uniform moment and rectangular plates under uniform compression per unit width are comparable basic cases. All lend themselves to clean analytical solutions to their governing differential equations.

The determination of the critical load of a structure which is composed of several such members connected in series, or in a framework configuration, would basically involve the simultaneous solution of the governing differential equations under the appropriate boundary and continuity conditions. To expedite this process, an adaptation of normal matrix analysis of the structure is used. The basic rotational stiffness of a uniform beam, $4EI/l$, for example, is replaced by sEI/l. Similarly the moment carry-over factor $\frac{1}{2}$ is replaced by c. Both s and c are derived from the solution of the differential equation

$$EIy^{iv} + Py'' = 0 \tag{4.1}$$

which replaces the simple

$$EIy^{iv} = 0 \tag{4.2}$$

used to derive the factors 4 and $\frac{1}{2}$ for the non-axially loaded member. The basic beam element stiffness matrix equation

$$\begin{bmatrix} \frac{12k}{l^2} & \frac{6k}{l} & \frac{-12k}{l^2} & \frac{6k}{l} \\ \frac{6k}{l} & 4k & \frac{-6k}{l} & 2k \\ \frac{-12k}{l^2} & \frac{-6k}{l} & \frac{12k}{l^2} & \frac{-6k}{l} \\ \frac{6k}{l} & 2k & \frac{-6k}{l} & 4k \end{bmatrix} \begin{bmatrix} \delta_L \\ \theta_L \\ \delta_R \\ \theta_R \end{bmatrix} = \begin{bmatrix} V_L \\ M_L \\ V_R \\ M_R \end{bmatrix} \tag{4.3}$$

where

$$k = EI/l \tag{4.4}$$

applies when $P = 0$. For non-zero P, eqn (4.3) is replaced by one in which the numerical coefficients become functions of the axial force (in a nondimensional form) as developed by Livesley and Chandler (1956) and also discussed and tabulated in Horne and Merchant (1965). The other basic stiffnesses, such as shear stiffness factors, are combinations of s, c and ρ. The nondimensional axial force factor ρ is the actual compressive load P divided by the Euler load P_E $(=\pi^2 EI/l^2$ for a uniform column). In terms of these factors, the stiffness matrix (eqn (4.3)) is replaced, for the beam–column, by

$$\begin{bmatrix} \frac{vk}{l^2} & \frac{s(1+c)k}{l} & \frac{-vk}{l^2} & \frac{s(1+c)k}{l} \\ \frac{s(1+c)k}{l} & sk & \frac{-s(1+c)k}{l} & sck \\ \frac{-vk}{l^2} & \frac{-s(1+c)k}{l} & \frac{vk}{l^2} & \frac{-s(1+c)k}{l} \\ \frac{s(1+c)k}{l} & sck & \frac{-s(1+c)k}{l} & sk \end{bmatrix} \tag{4.5}$$

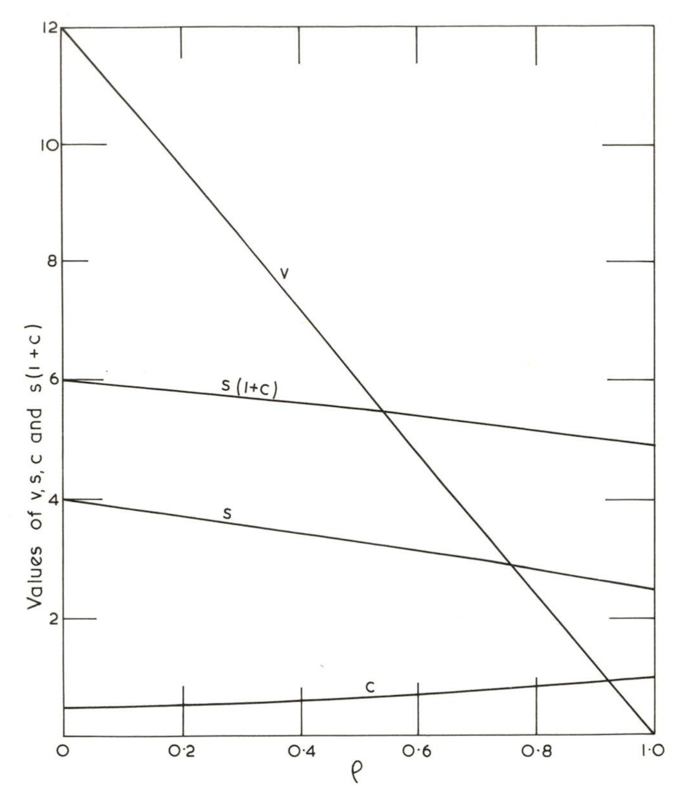

Fig. 4.2. Basic stiffness functions as functions of ρ.

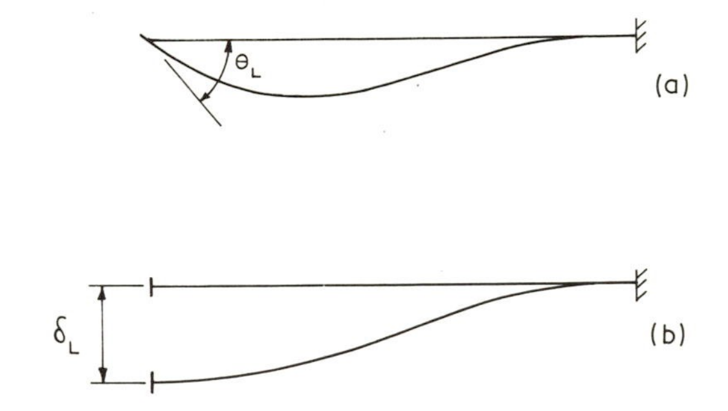

Fig. 4.3. Basic displacements defining stiffness functions.

in which the factor

$$v = 2s(1+c) - \pi^2\rho \qquad (4.6)$$

The variation of s, c, $s(1+c)$ and v with ρ is illustrated in Fig. 4.2, and the basic displacement cases used in their definition are shown in Fig. 4.3.

In the use of the matrix approach, the emphasis is changed from the solution of differential equations, using displacement compatibility, to the stiffness or resistance of the complete structure to a set of loads. The structure is stable if it can resist the applied loads. A measure of this corporate stiffness is a number known as the determinant of the structure matrix. The determinant decreases as the compressive axial loads increase. One reason for this decrease can be seen clearly in the definition of v (eqn (4.6)), where the term $\pi^2\rho$ is subtracted from a direct stiffness term on the matrix leading diagonal. Relative to this effect, the change in s and c with ρ is small. The function v measures the resistance to pure sway displacement within a column, as shown in Fig. 4.3(b). The shear force is $vk\delta/l^2$ and the second component of this when expanded using eqn (4.6) is

$$-\pi^2\rho EI\delta/l^3 = -\frac{P\delta}{l} \qquad (4.7)$$

This embodies the so-called P–δ effect, the overturning moment so destructive to the lateral stiffness of columns which are free to sway, and their dominant axial load effect. This dominance is further illustrated in the difference between the plots of v and $s(1+c)$ in Fig. 4.2.

The shapes into which the basic components (beams, columns) of a structure buckle are combinations of sinusoidal and linear forms. The single uniform column, for example, obeys the general differential equation (eqn (4.1)), for which the solutions have the basic form

$$y(z) = A^* \sin \lambda z + B^* \cos \lambda z + C^* z + D^* \qquad (4.8)$$

or

$$y(z) = F^* \sin(\lambda z + \phi^*) + C^* z + D^* \qquad (4.9)$$

where

$$\lambda = \sqrt{\left(\frac{P}{EI}\right)} = \frac{\pi}{l}\sqrt{\rho} \qquad (4.10)$$

Any sufficient set of end conditions allows the four constants F^*, ϕ^*, C^* and D^* in eqn (4.9) to be evaluated. It is also clear that the shape

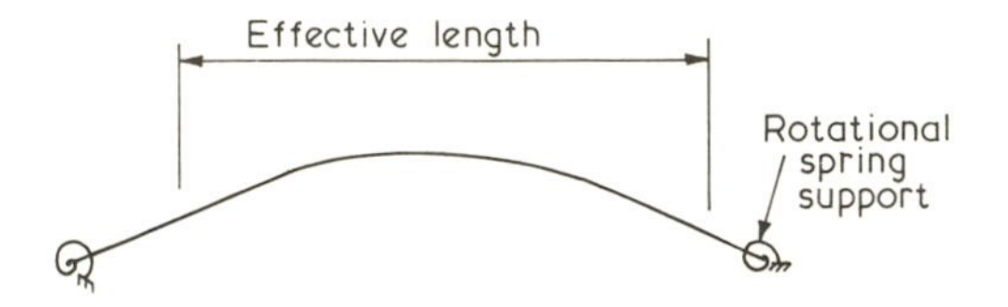

FIG. 4.4. Column supported by rotational springs at each end.

relative to the line joining the ends is sinusoidal with a half wave length $l/\sqrt{\rho}$, the 'effective length' of the column. This is illustrated in Fig. 4.4, where a section of a no-sway column having general elastic–rotational end restraint is shown relative to its prebuckling position. A sketch of the anticipated buckled form, incorporating its sinusoidal nature and the end support estimates, can often be used as a quick guide to the effective length. In sway cases, where the effective length is usually greater than l (i.e. $\rho_{cr} < 1{\cdot}0$), this procedure is likely to be less accurate but may still furnish a reasonable estimate to be used, say, for initial sizing. Charts and tables from which effective lengths may be estimated under a range of end support conditions are incorporated in many codes of practice.

4.2 SHAPE FUNCTIONS

The deformed shapes illustrated in Fig. 4.3 are two of the four basic shape functions which can be used to describe the deformed shape of a general uniform section beam–column interbrace length. As mentioned, the basic solution form of eqn (4.8) implies a linear combination of sine, cosine and linear terms in z, involving four constant scale factors for the components. In the case of a beam–column, it makes good physical sense to use the two end displacements y and the two end slopes y' (Fig. 4.3) as those parameters. This follows normal structural practice. Since the same forms arise for either x- or y-direction displacements in the case of the doubly symmetric member, the symbol $u(z)$ will be used to describe the general shape function forms within a member.

4.2.1 Basic Function Components

Two function forms are required to describe the type of shape function shown in Fig. 4.3:

$$Q(z) \text{ where } Q(l) = 1, \qquad Q(0) = Q'(0) = Q'(l) = 0$$

$$R(z) \text{ where } R'(l) = 1, \qquad R'(0) = R(0) = R(l) = 0$$

The functions

$$Q(z)=\frac{\dfrac{\lambda z-\sin\lambda z}{1-\cos t}-\dfrac{1-\cos\lambda z}{\sin t}}{\dfrac{t-\sin t}{1-\cos t}-\dfrac{1-\cos t}{\sin t}} \tag{4.11}$$

$$R(z)=\frac{\dfrac{\lambda z-\sin\lambda z}{t-\sin t}-\dfrac{1-\cos\lambda z}{1-\cos t}}{\dfrac{1-\cos t}{t-\sin t}-\dfrac{\sin t}{1-\cos t}} \tag{4.12}$$

in which

$$t=\lambda l \tag{4.13}$$

satisfy the above boundary conditions and a general solution to the original governing differential equation (eqn (4.1)) can be written

$$u(z)=Q(z)u(l)+R(z)u'(l)+Q(l-z)u(0)+R(l-z)u'(0) \tag{4.14}$$

Obviously Q and R are linear combinations of the components of the solution to eqn (4.1) as shown in eqn (4.8), the combinations being arranged so that their end values are zero or unity. The formulae for the basic s and c functions, for instance, are derived directly from these using the end curvatures and slopes of R.

Similar forms may be developed for the flexural–torsional buckling problems discussed in later sections.

4.3 MULTIPLY BRACED COLUMNS

4.3.1 Non-uniform Section, Bracing and Compression

Consider a column pinned at each end and braced at intervals along the length by elastic linear braces, as illustrated in Fig. 4.5. In such a case the buckling displacements are assumed to take place in the x–z plane. If the braces are irregularly spaced, the buckling load will continue to rise with the brace stiffness, though not necessarily steadily. The mode of buckling with very weak braces would be close to the half sine wave form into which the unbraced column would buckle. As the braces are stiffened the effective length decreases, until at some stage the mode associated with the lowest buckling load will involve two (distorted) half waves. The practical way to determine the buckling

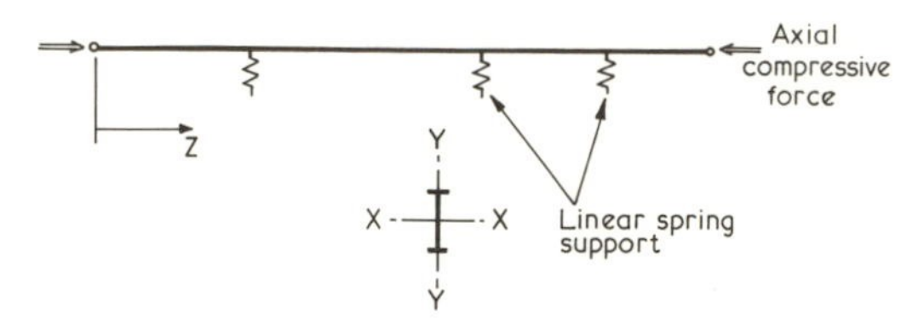

FIG. 4.5. Column supported by irregularly spaced elastic braces.

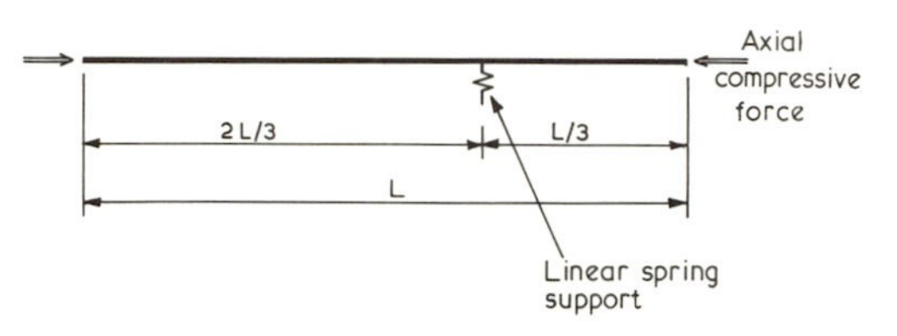

FIG. 4.6. Column supported by a single elastic brace at one third length.

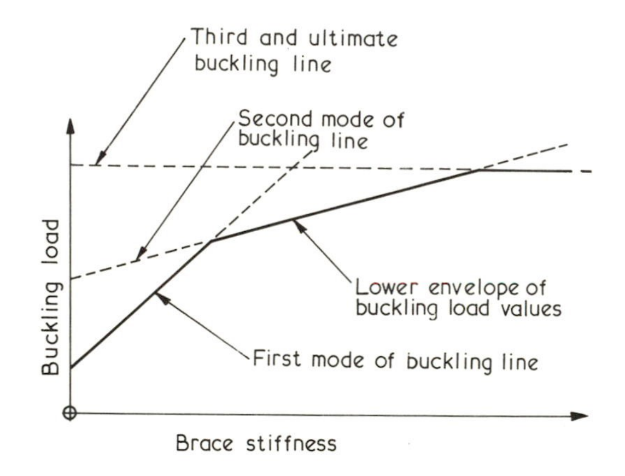

FIG. 4.7. Buckling load of column in Fig. 4.6 in three modes; lower bound to curves.

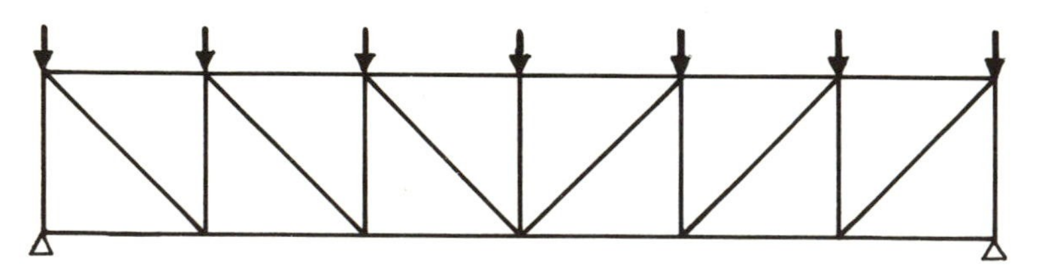

FIG. 4.8. Typical truss loaded at each panel point.

loads is to find the load factors at which the stiffness matrix of the structure, assembled using s and c functions, has a zero determinant. This is further illustrated by examining the simpler, but nonsymmetrical, case shown in Fig. 4.6. The initially straight uniform column would have a buckling load of $\pi^2 EI/l^2$, i.e. P_E, if unbraced and other natural modes of the unbraced column would occur at $4P_E$, involving two half sine waves with the length, $9P_E$ involving three half sine waves, etc. The brace shown would interfere with the pure first mode and would cause the first critical load to rise above P_E continually as the brace stiffness increases. The same would occur with the second mode, but the three half sine wave mode form naturally has a node at the brace point and will not cause the brace to be strained. This buckling load, $9P_E$, effectively represents an upper limit to the column axial compression capacity. A plot of buckling load versus brace stiffness would take the form shown in Fig. 4.7. The three regimes correspond to the first, second and third modes of buckling, the lower envelope being the effective critical value for the associated brace stiffness. Specific nondimensional design charts can be prepared and some are presented later in this chapter.

The compression chords of trusses, such as that shown in Fig. 4.8, constitute a particular case within this general class of irregular compression members. The panel lengths are usually equal, the braces equally spaced, but the compressive force in each chord section is different, varying in a stepped parabolic manner within a simply supported truss. Braces (purlins) are not always attached to each panel point of the chord. This class of structure is discussed in detail by Medland (1977), where the critical load levels are related to brace stiffness and spacing in a series of nondimensional charts, such as Fig. 4.9. This figure relates the maximum panel ρ value within the chord at buckling (ρ_{cr}) to the brace stiffness factor f. This nondimensional factor is the actual linear brace stiffness divided by $12EI/l^3$, the shear stiffness of a panel chord. In Fig. 4.9 the four curves relate to cases where only two brace lines are attached (at the 1/3 points) but the number of load points is varied. In this nondimensional form the three distinct segments of the plots corresponding to the single, double and finally triple 'half sine wave' modes of buckling are apparent and the ρ_{cr} value rises with the number of load points. This rise is a result of a lesser proportion of the chord length being at peak compression. It is also apparent that the critical load factor continues to rise if the number of load points is greater than two, in this two brace case. This is a

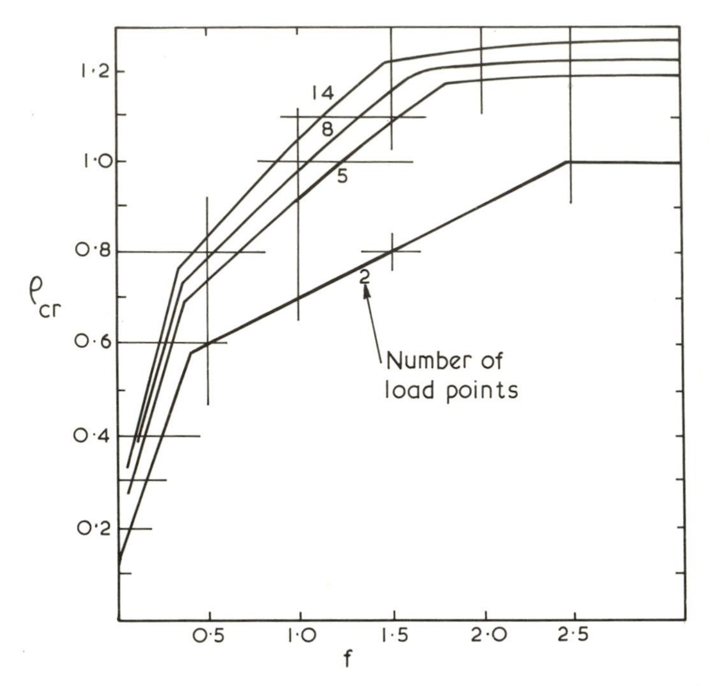

FIG. 4.9. Maximum panel compressive ρ: brace stiffness for parabolic compression chord load distribution.

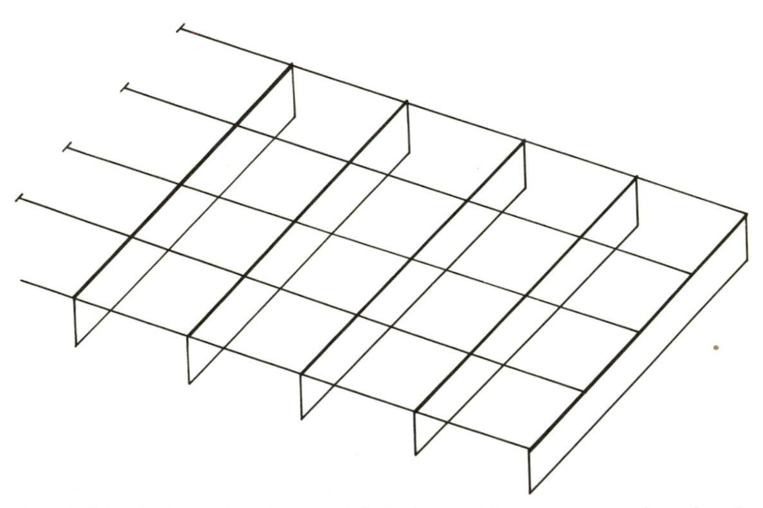

FIG. 4.10. Schematic of a set of interbraced truss compression chords.

consequence of the non-uniformity in the axial compression within the chord. No finite brace stiffness will force the chord into a mode which does not involve displacement at some brace.

The same general analysis applied to a structure comprising a number of trusses in parallel, their compression chords interbraced by equally spaced lines of braces as shown schematically in Fig. 4.10, indicates that the curves shown in Fig. 4.9 remain relevant if the nondimensional brace stiffness factor f is divided by the empirically derived factor

$$d_p = 0{\cdot}425n^2 + 1{\cdot}275n + 1{\cdot}0 \tag{4.15}$$

In eqn (4.15), n is the number of bays (number of columns -1). The one set of curves covers all n. A very similar factor applies to chords under uniform compression throughout their length.

Figure 4.11 further illustrates the type of design chart which can be prepared from such analyses. In this case it is assumed that every

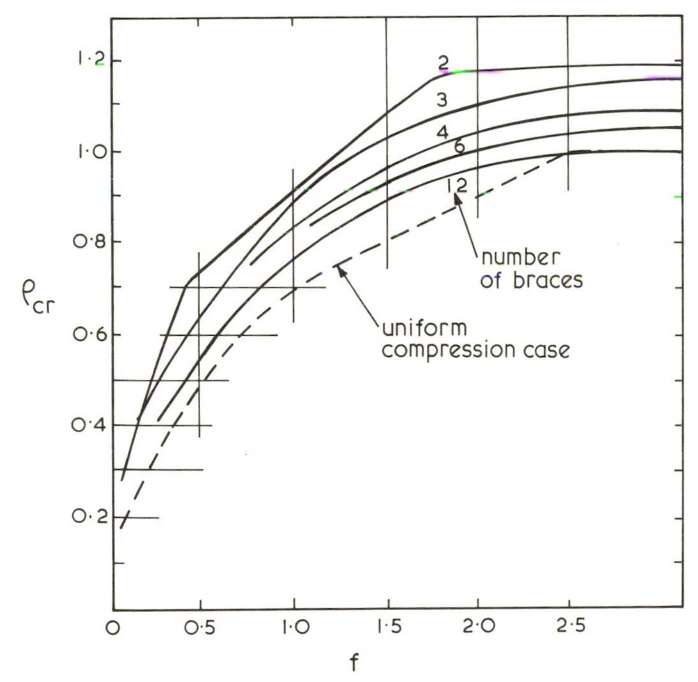

Fig. 4.11. Maximum panel compressive ρ: brace stiffness for every second panel point braced.

second load point of the chord is braced. As the number of braces (and panels) is increased the curves become smoother and lower, crowding towards a ρ_{cr} value of 1.0 for large brace stiffnesses. The broken line indicates the lower bound curve for the special case of column sets under uniform compression. This curve can be seen to be a not too conservative lower bound for the stepped parabolic cases, and the results of a more complete study allow such lower bound design curves to be calculated efficiently.

4.3.2 Uniform Section, Bracing and Compression

Figure 4.12 illustrates a highly regular set of parallel, pin-ended, uniformly compressed columns interbraced at equal intervals by equal stiffness braces. As stated above, this case provides a conservative estimate for some less regular cases. The interbraced column length (element) of column number m between brace lines n and $n+1$ is designated element n, m. There are N internal brace lines and $N+1$ column elements.

The displacements within column element n,m are governed by the basic differential equation (eqn (4.1)), arranged here in the general

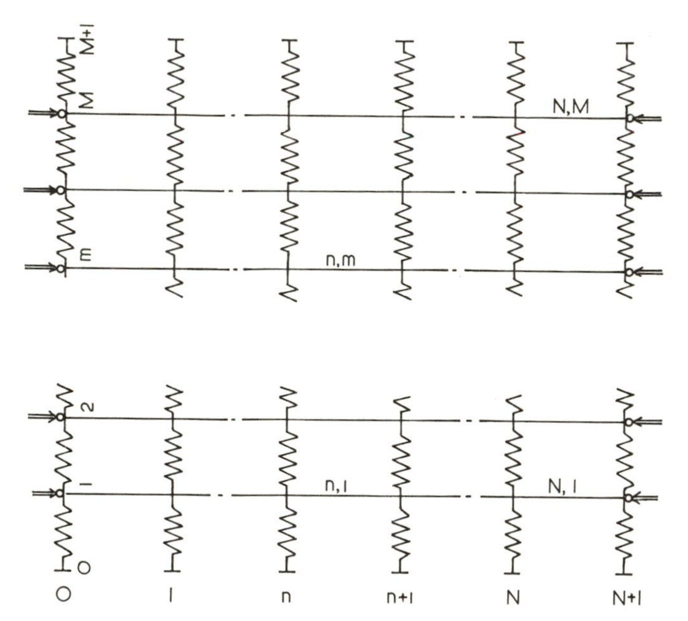

FIG. 4.12. Schematic of interbraced pin-ended columns.

form

$$u^{iv}_{n,m} + \lambda^2 u''_{n,m} = 0 \qquad (4.16)$$

for which a basic form of solution is given in eqn (4.14). It is reiterated that the functions Q and R are such that, at an end, one of them will have a value of 1.0, the other zero. The same applies to their derivatives.

4.3.3 Single Column Example

Consider initially the single column shown in Fig. 4.13. The linear springs brace each point to a solid foundation on each side. When buckling occurs under the uniform axial compression, the displacement, slope, moment and shear must be continuous at each brace point. Using the displacement form of eqn (4.14), and capitalising on the special forms of $Q(z)$ and $R(z)$, the continuities of one brace point provide a general relationship. At node n,m joining elements $n-1,m$ and n,m the four continuities in the above order are written

$$u_{n-1,m}(l) = u_{n,m}(0) = D_{n,m} \quad \text{(say)} \qquad (4.17)$$

$$u'_{n-1,m}(l) = u'_{n,m}(0) = B_{n,m} \quad \text{(say)} \qquad (4.18)$$

$$\left[\frac{1}{1-\cos t}\right](B_{n+1,m} + B_{n-1,m}) - \left[\frac{1}{\sin t}\right](D_{n+1,m} - D_{n-1,m}) + 2\left[\frac{\sin t}{t-\cos t} - \frac{\cos t}{1-\cos t}\right]B_{n,m} = 0 \qquad (4.19)$$

$$\left[\frac{1}{t-\sin t}\right](B_{n+1,m} - B_{n-1,m}) - \left[\frac{1}{1-\cos t}\right](D_{n+1,m} - 2D_{n,m} + D_{n-1,m}) - \frac{4\beta}{t^3}\left[\frac{t-\sin t}{1-\cos t} - \frac{1-\cos t}{\sin t}\right]D_{n,m} = 0 \qquad (4.20)$$

in which

$$\beta = k_L l^3/2EI \qquad (4.21)$$

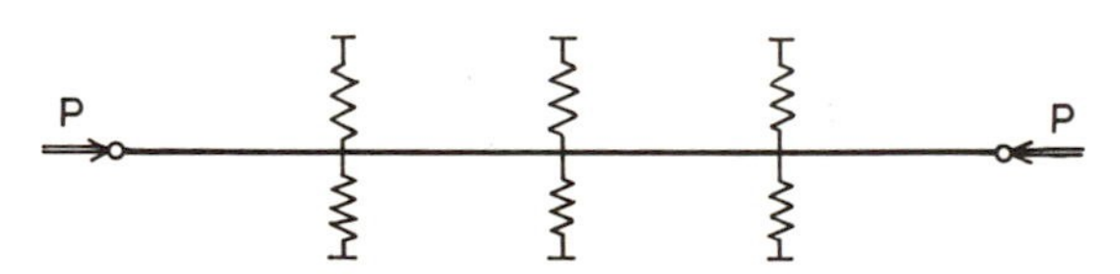

FIG. 4.13. Single multiply braced column.

Equations (4.19) and (4.20) provide two relationships between the B and D parameters. If the substitutions

$$\begin{aligned} B_{n,m} &= B_m \cos n\theta_j \\ D_{n,m} &= D_m \sin n\theta_j \\ \theta_j &= j\pi/(N+1) \end{aligned} \tag{4.22}$$

are made in eqns (4.19) and (4.20), those equations are condensed to

$$\left[\frac{\sin t}{t-\sin t}+\frac{\cos\theta_j-\cos t}{1-\cos t}\right]B_m-\left[\frac{\sin\theta_j}{\sin t}\right]D_m=0 \tag{4.23}$$

and

$$\left[\frac{-\sin\theta_j}{t-\sin t}\right]B_m+\left[\frac{1-\cos\theta_j}{1-\cos t}-\frac{2\beta}{t^3}\left(\frac{t-\sin t}{1-\cos t}-\frac{1-\cos t}{\sin t}\right)\right]D_m=0 \tag{4.24}$$

The form of eqns (4.22) simply recognises the recurrent form (physically and, hence, mathematically) of the structure and buckling modes. Equations (4.22) state that the distribution of brace point displacements D_m within the length of the column is sinusoidal, while the distribution of slopes B_m at those points is cosinusoidal. This implies pinned support at nodes 0 and $N+1$. The number of half sine waves within the length is obviously j. The detail of the approach is treated in Segedin and Medland (1978) and Medland (1979), and draws on a recurrence approach put forward for a vibration problem by Miles (1956).

Equations (4.23) and (4.24) constitute a pair of homogeneous linear equations in B_m and D_m, the slope and displacement amplitude factors. In the manner of all buckling problems, no unique solution is available. Apart from the trivial $B_m = D_m = 0$ solution, a range of proportions $B_m : D_m$ are able to be found and each corresponds to an axial force (represented in t through λ) which causes the 2×2 determinant of the coefficients in eqns (4.23) and (4.24) to become zero. While the form of the development of the two basic equations may appear complicated, the evaluation of the 2×2 determinant at a set β and θ_j and increasing trial values of t is almost trivial once programmed into a desk computer or calculator.

4.3.4 Multiple Column Case

The effect on the above development of there being M columns in parallel, as shown in Fig. 4.12, is confined to the fact that each spring at a node is extended by the relative displacements of the columns. As a result, the third component of eqn (4.20) becomes

$$\frac{2\beta}{t^3}\left[\frac{t-\sin t}{1-\cos t}-\frac{1-\cos t}{\sin t}\right](D_{n,m+1}-2D_{n,m}+D_{n,m-1}) \tag{4.25}$$

and eqn (4.24) becomes

$$\left[\frac{-\sin \theta_j}{t-\sin t}\right]B_m+\left[\frac{1-\cos \theta_j}{1-\cos t}\right]D_m + \frac{\beta}{t^3}\left[\frac{t-\sin t}{1-\cos t}-\frac{1-\cos t}{\sin t}\right](D_{m+1}-2D_m+D_{m-1})=0 \tag{4.26}$$

Equations (4.23) and (4.26) now govern the buckling behaviour of the system. The substitution of

$$\begin{aligned} B_m &= B \sin m\psi \\ D_m &= D \sin m\psi \end{aligned} \tag{4.27}$$

into the modified eqn (4.24) (using eqn (4.25)) and into eqn (4.26) results in the further compacted forms

$$\left[\frac{\sin t}{t-\sin t}+\frac{\cos \theta_j-\cos t}{1-\cos t}\right]B-\left[\frac{\sin \theta_j}{\sin t}\right]D=0 \tag{4.28}$$

$$\left[-\frac{\sin \theta_j}{t-\sin t}\right]B+\left[\frac{1-\cos \theta_j}{1-\cos t} - \left(\frac{2\beta(1-\cos \psi)}{t^3}\right)\left(\frac{t-\sin t}{1-\cos t}-\frac{1-\cos t}{\sin t}\right)\right]D=0 \tag{4.29}$$

As discussed with respect to eqns (4.23) and (4.24), the determinant of the coefficients of eqns (4.28) and (4.29) will be zero at a buckling load. Clearly the factor $2\beta(1-\cos \psi)$ is the effective linear brace stiffness for the multi-column case. The value of ψ will depend upon the outside foundation conditions for the brace lines ($m=0, M+1$). If both ends are fixed in position then

$$\psi = i\pi/(M+1) \tag{4.30}$$

For a given physical system, the critical load will be determined by finding the lowest load value for which the determinant of the coefficients of eqns (4.28) and (4.29) becomes zero having selected a specific j, the number of half sine waves within each column length, and i, the

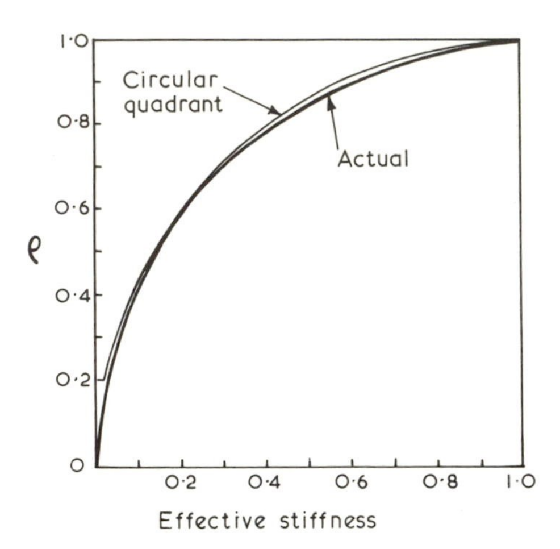

FIG. 4.14 Nondimensional summary of effective stiffness of a set of multiply interbraced columns.

number of zero displacement points within a brace line. The determinant of eqns (4.28) and (4.29) is shown in Appendix 1 with a summary of other details.

The lower envelope to the relationships between the effective brace stiffness and axial load can be very closely approximated by a quadrant of a circle if the nondimensional ρ is used instead of t and if the spring stiffness is represented using the factor

$$\beta^* = \beta(1 - \cos\psi)/\pi^2 \tag{4.31}$$

Figure 4.14 shows the relationship between true and approximate representations of this lower bound.

4.4 FLEXURAL–TORSIONAL BUCKLING OF COLUMNS

A uniform, doubly symmetric cross-section member under axial compression can buckle in three distinct forms. The lowest buckling load would normally involve bending about the minor axis in a simple Euler manner, at a load dictated by the section properties and end support conditions. If this mode is prevented, or considerably restrained by outside influence, both Euler buckling about the major axis and pure twist buckling about the longitudinal axis through the shear centre of the cross-section are possible. If the section has only one degree of

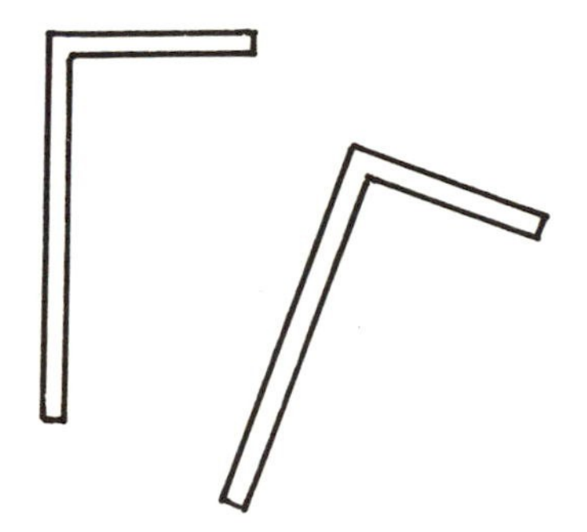

FIG. 4.15. Illustration of flexural–torsional buckling displacements.

symmetry (e.g. a channel section, an equal leg angle or an I-section with unequal flange widths) the shear centre will not coincide with the centroid and the buckling mode will be either pure displacement in the plane of the axis of symmetry or combined 'lateral' displacement and twist. If no symmetry exists, the buckling mode contains components of both lateral displacements as well as twist, as illustrated in Fig. 4.15. A doubly symmetric member on which the axial compression stress is not symmetrically disposed over the cross-section will behave in a similar manner. These latter cases are generally referred to as undergoing flexural–torsional buckling. Given simple support conditions at each end against each form of displacement, the buckling component forms will all vary sinusoidally within the length, each having its own amplitude dependent on its inherent stiffness in that mode.

In the following, such combined modes occur not as a result of the section properties (which are doubly symmetric) but because the elastic constraints (braces) are attached eccentrically to the compression member. As mentioned earlier, such eccentric attachments precipitate flexural torsional buckling but often provide some restraint to those movements in compensation.

In members which are subjected to torsion, warping occurs. The torsional shears cause the basic 'plane-sections-remain-plane' assumption to be violated in all but circular cross-section members. If this distortion is resisted by a stiffening attachment or, more subtly, by there being non-uniform torsion or torsional resistance within the member, additional local stresses are engendered. Resistance to warping considerably enhances the torsional stiffness of the member, particularly for I and channel shapes, but the resulting stresses can cause local problems.

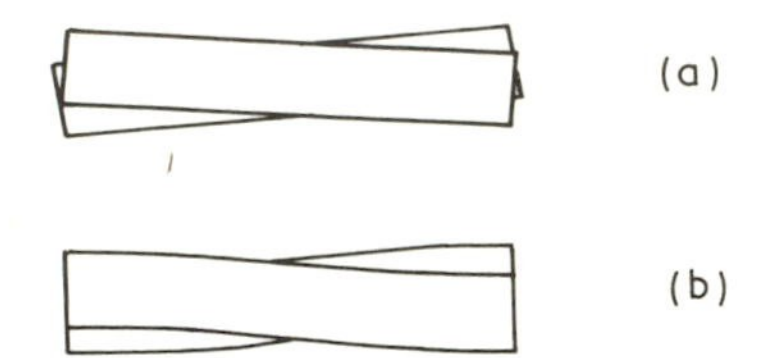

FIG. 4.16. Torsional displacement: (a) free-to-warp; (b) warp restrained.

Figure 4.16(a) illustrates the distortions involved in the free twisting of an I-section member, while Fig. 4.16(b) shows the extra distortion and consequently torsional resistance and energy absorption which accompanies the twist where warping is resisted. The figure illustrates how the warping effect in a flanged member is dominated by the flanges moving out of plane with one another.

4.4.1 Interbraced Columns in Flexural–Torsional Buckling

In most interbraced structures of the type discussed in Section 4.3.4, the bracing members are not simply tension–compression members but, like girts and purlins, have considerable bending resistance themselves. The connection with the braced member is often eccentric to the shear centre of that member and is also capable of transferring rotational forces in the plane of the structure. In such cases the brace may provide torsional, rotational and warping support as well as the basic lateral, and at the same time causes the buckling mode of the braced member to involve those four components. Figure 4.17 illustrates a typical connection within a grid and the associated support forms.

Since out-of-plane displacement is assumed to be zero in the buckling modes, the four further continuity equations of twist ϕ, rate of change of twist ϕ', torsional moment and warping moment must be satisfied. As an example, the torsional moment continuity across a

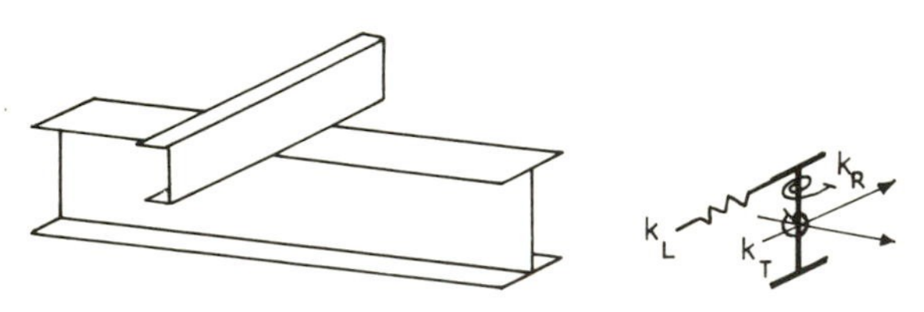

FIG. 4.17. Brace connection providing lateral, rotational and warping constraint.

brace point at which the effective stiffness resisting torsion is K_T (moment per unit twist), the linear in-plane spring stiffness is K_L and the eccentricity of that brace from the shear centre is e, can be written

$$T_{n-1,m}(l) - T_{n,m}(0) + K_T(2\phi_{n,m+1}(0) + 8\phi_{n,m}(0) + 2\phi_{n,m-1}(0)) + K_L e(u_{n,m+1}(0) - 2u_{n,m}(0) + u_{n,m-1}(0)) - K_L e^2(\phi_{n,m+1}(0) - 2\phi_{n,m}(0) + \phi_{n,m-1}(0)) = 0 \quad (4.32)$$

The coefficient of K_T contains the direct and carryover twist effects, while the other two involve the contributions due to the net extension of the linear spring, attached eccentrically, by the amount e. The sign convention for displacements and forces is shown in Fig. 4.18. The form selected for $\phi_{n,m}(z)$ is the same as that for $u_{n,m}$ in eqn (4.14) but with different coefficients and using α instead of λ. The full set of continuity equations across the general brace point can be written. These are presented in detail by Medland (1979). The factor α represents the axial load and is defined by

$$\alpha^2 = (PI_0/A - C)/C_1 \quad (4.32)$$

Upon substitution of eqns (4.22) and (4.27), the set of four equations which are the four degrees of freedom equivalent of the pair of equations (4.28) and (4.29) may be written and their 4×4 determinant evaluated at increasing levels of axial force until it becomes zero at the

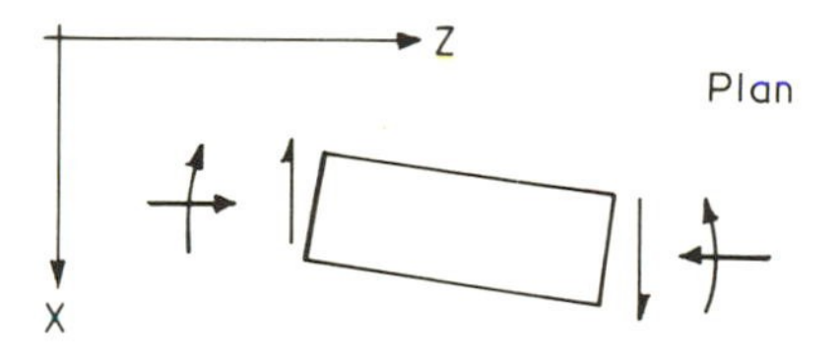

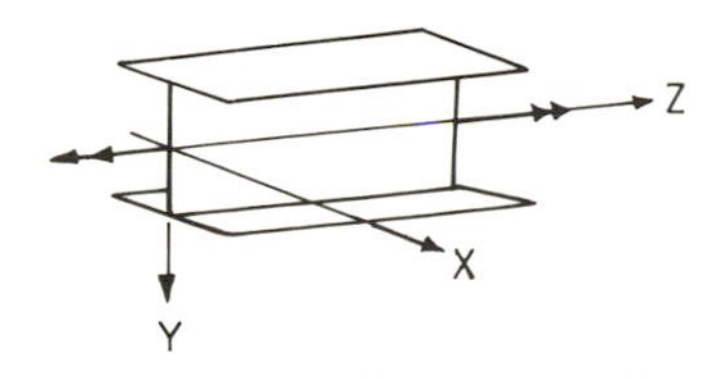

FIG. 4.18. Sign convention for forces: axial, shear, minor axis bending and warping.

critical level for the system. That determinant is shown in detail in Appendix 2.

4.4.2. Nondimensional Factors

While it is not possible to completely nondimensionalise the system, it is useful to employ the factors

$$\rho = P/(\pi^2 EI/l^2) \tag{4.34}$$

and

$$e^* = e\Big/\sqrt{\left(\frac{I_W}{I}\left(\pi^2+\frac{l^2}{a^2}\right)\right)} \tag{4.35}$$

$$\beta = K_L l^3(1-\cos\psi)/EI \tag{4.36}$$

$$\gamma = 2K_R l(2+\cos\psi)/EI \tag{4.37}$$

$$\eta = 2K_T l^3(2+\cos\psi)/EI_W \tag{4.38}$$

for a given column member (assumed doubly symmetric). Charts can be prepared to relate critical ρ levels to elastic support stiffnesses. The

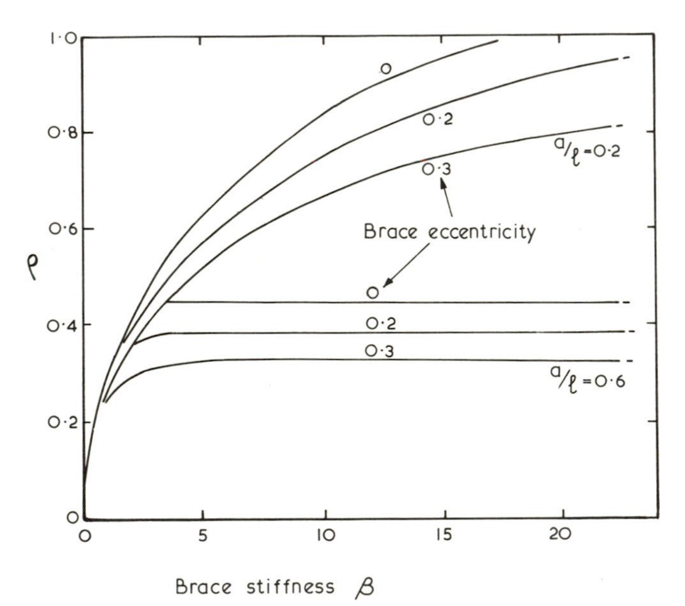

FIG. 4.19. Effect of linear brace eccentricity on flexural–torsional buckling of columns.

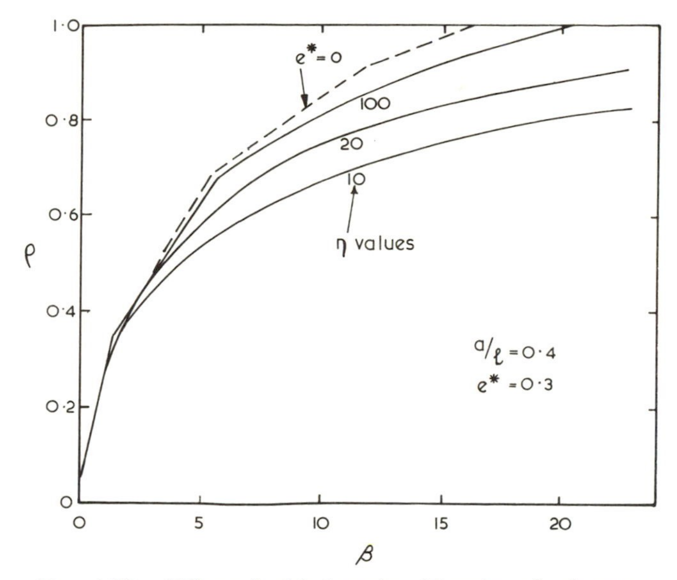

FIG. 4.20. Effect of added torsional bracing of columns.

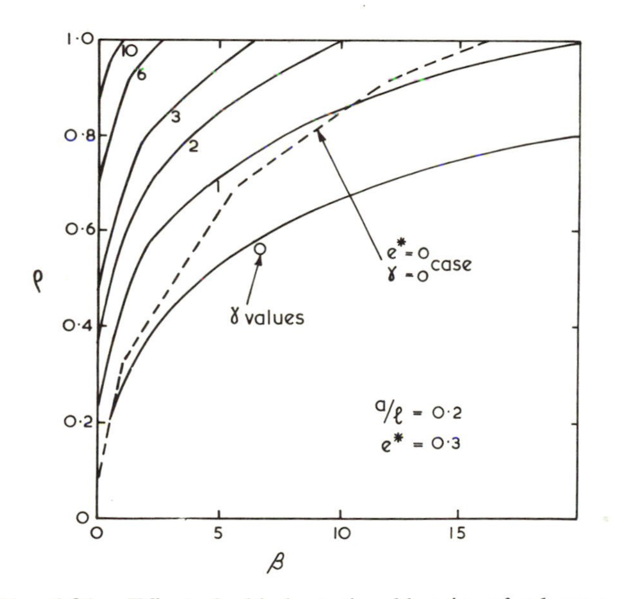

FIG. 4.21. Effect of added rotational bracing of columns.

fact that column members exist which have very similar areas and minor axis I-values but very different warping constants (for example) is the reason for the lack of further nondimensionalisation. Figure 4.19 illustrates the effects of linear brace eccentricity on critical ρ value for two columns of similar area and flange width but different torsional properties. Figures 4.20 and 4.21, respectively, indicate the effect of adding torsional and rotational bracing to a basic eccentric linear braced system. As values of the nondimensional torsional and rotational brace stiffnesses are raised, the critical ρ values associated with a given linear support stiffness rise, compensating in some measure for the eccentricity of the linear braces. The broken curve on each figure is the $\rho_{cr} : \beta$ plot for a zero eccentricity four linear brace datum case. For realistic cases, in-plane rotational braces are, not surprisingly, relatively inefficient in compensating for what is basically a torsional effect. The scale of the nondimensional rotational stiffness factor γ is small compared to β and η. Specific cases selected to illustrate the sensitivities are presented in Medland (1979).

4.5 BUCKLING OF INTERBRACED BEAMS

Flooring systems comprising parallel beams linked together by lateral members have the same general features as the column sets discussed previously. The lateral bracing members will normally be eccentric and may provide rotational, torsional and lateral bracing. The beams, however, will always buckle in a flexural–torsional mode. The governing differential equations for an interbrace beam length under uniform moment M are

$$EIu^{iv}_{n,m} + M\phi''_{n,m} = 0 \tag{4.39}$$

$$Mu''_{n,m} + C_1\phi^{iv}_{n,m} - C\phi''_{n,m} = 0 \tag{4.40}$$

and the critical uniform moment value of the simply supported element is

$$M_{cr} = \frac{\pi}{l}\sqrt{\left(EIC\left(1+\frac{\pi^2 a^2}{l^2}\right)\right)} \tag{4.41}$$

The nondimensional eccentricity factor e^* is again used for any attached bracing and plots of ρ_{cr} $(=M_{cr}/M_0)$ against brace stiffness and eccentricity can be assembled as in Fig. 4.22. The effect of adding a

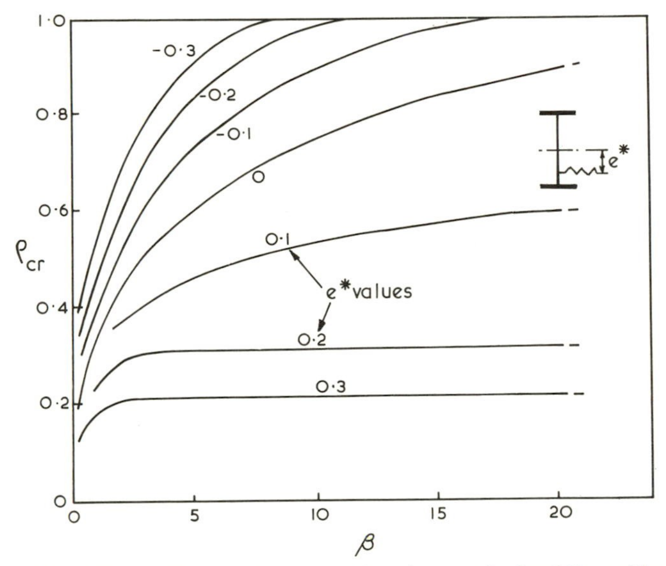

FIG. 4.22. Effect of eccentric lateral bracing on the buckling of beams.

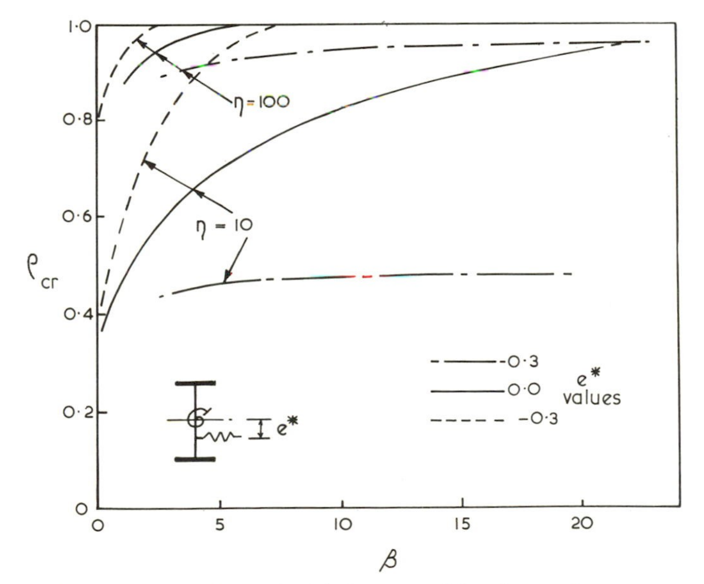

FIG. 4.23. Effect of added torsional bracing on beams.

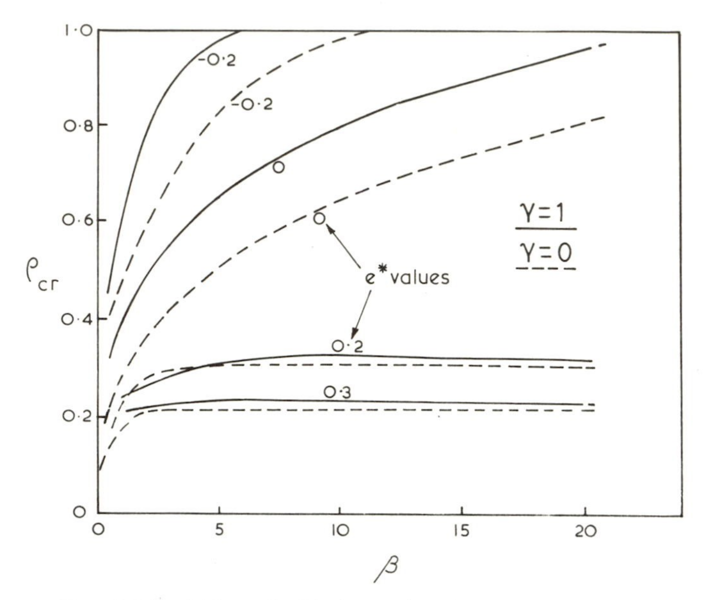

FIG. 4.24. Effect of added rotational bracing on beams.

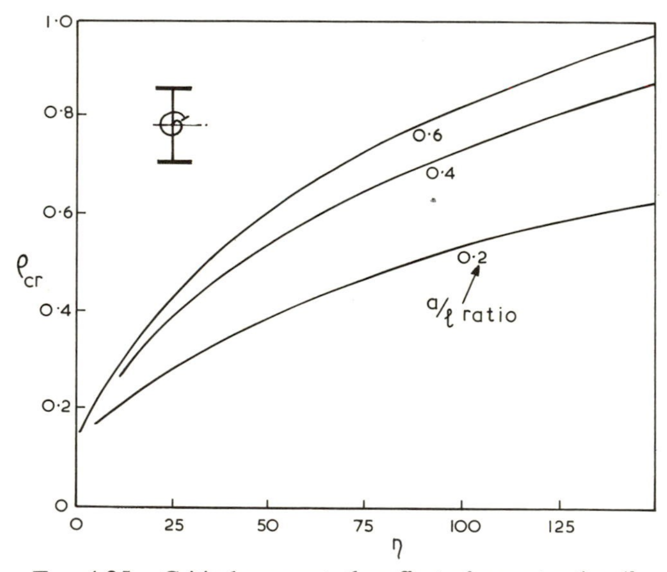

FIG. 4.25. Critical moment: the effect of aspect ratio a/l.

torsional brace at the shear centre of the section is illustrated in Fig. 4.23, while the effect of rotational bracing with eccentric linear is shown in Fig. 4.24. To determine any specific value on such charts involves the evaluation of the determinant of a 4×4 system of linear equations in the same way as the flexural–torsional braced column cases were handled. The detail of the component factors is contained in Medland (1980). The detailed set of equations whose determinant is zero at a buckling load is shown in Appendix 3, with explanatory notes.

For any buckling involving torsion, the geometry of the cross-section has a marked effect. The a/l ratio is a means of categorising this effect. Figure 4.25 indicates this sensitivity by comparing the critical uniform moment values of torsionally constrained members at different a/l ratios.

The case where the loading of the beam is due to a uniform spread load is more common than that of a uniform moment. Such loading is

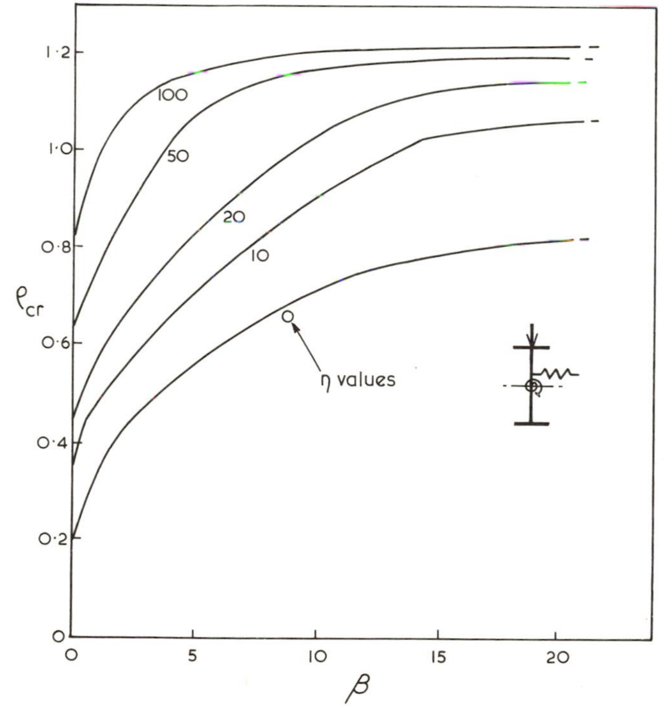

FIG. 4.26. Beam stability as affected by position of spread load application.

not often applied through the shear centre, normally being placed on or above the compression flange. This results in a further destabilisation effect. Figure 4.26 presents an example where top flange loading is applied. A finite element analysis (Barsoum and Gallagher, 1970; Davidson and Medland, 1974) was used to determine these values as the use of a recurrence technique is not suited to such variable moment cases.

4.6 BRACE STRENGTH CONSIDERATIONS

While the elastic critical loads, the determination of which has been the subject of Sections 4.1 to 4.5, are important parameters for the designer, a means of estimating the strength required of the braces must also be found. To determine the critical loads, only the brace stiffness needs to be considered. Strength is required if the braced members deform. Some bracing systems are primarily designed to carry loading to a foundation and will be designed accordingly. If a brace is placed basically to prevent buckling, it theoretically needs no strength until buckling occurs. In practice the compression element being braced (strut, compression flange, etc.) is not perfectly straight and is subjected to secondary loading which pushes it off line. This results in the braces being strained when the strut is loaded.

In this section the compression members are assumed to have a specific crookedness before axial loading is applied. In general, such an initial shape between the ends of the member can be expressed as a Fourier sine series which, in the case of a pin-ended column, can be regarded as a series containing the successive buckling mode shape functions, say

$$\mu_{\mathrm{I}}(\xi) = \sum b_j \sin \frac{j\pi\xi}{L} \tag{4.42}$$

Upon application of a compressive force P, the jth component of the series is magnified by division by a factor

$$\left(1 - \frac{p}{j^2 P_{\mathrm{E}}}\right) = (1 - \rho/j^2) \tag{4.43}$$

The differential equation governing the interbrace elements of the system is

$$EI(u^{iv} - u_{\mathrm{I}}^{iv}) + pu'' = 0 \tag{4.44}$$

which can be arranged in the form

$$u^{iv} + \lambda^2 u'' = u_{\mathrm{I}}^{iv} \tag{4.45}$$

The general solution of eqn (4.45) can be expressed as

$$u(z) = Q(z)u(l) + R(z)u'(l) + Q(l-z)u(0) + R(l-z)u'(0) + \sum b_j^* \sin(n + z/l)\theta_j \tag{4.46}$$

in which

$$\theta_j = j\pi/(N+1) \tag{4.47}$$

$$b_j^* = b_j/(1 - \rho/j^2) \tag{4.48}$$

Displacement, slope, moment and shear must still be continuous across a brace point. The first two are automatically so by the form of the Q and R functions and of $\sin(n + z/l)\theta_j$. Moment and shear continuities provide two further equations, the moment one being

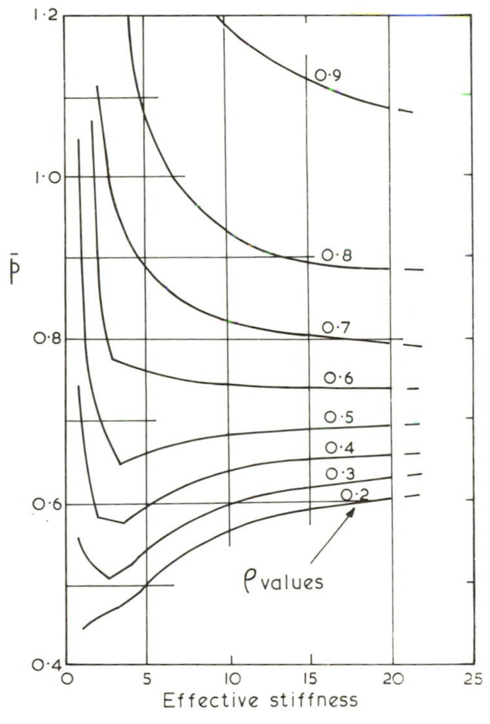

FIG. 4.27. Non-dimensional brace force related to brace stiffness.

homogeneous because no rotational springs are attached. The linear springs are stretched by the amount $(u-u_I)$ at each brace point and the u_I component makes that equation nonhomogeneous. A specific 'magnified' initial displacement set is defined by the solution of these two simultaneous equations under any applied axial force. If that force reaches any one of the basic critical loads of the original perfectly straight system the displacements become infinite. Obviously the applied axial force must remain below the lowest critical load of the system.

A detailed derivation of the relationship between the forces in the braces and the spring stiffnesses at a given axial force level for the type of structure shown in Fig. 4.23 is presented by Medland and Segedin (1979). Figure 4.27 illustrates the type of nondimensional chart which can be prepared from such an analysis. The factor $\bar{p}$ represents the brace force as a percentage of the axial column force P, divided by the number of columns in parallel M. The symbol $\bar{\beta}$ is the actual brace stiffness K, divided by the nondimensionalising factor $12EI/l^3$ and further divided by $2(1-\cos\theta_j)$ which incorporates the form of the initial displacements.

4.7 SUMMARY AND CONCLUSIONS

A coordinated approach to the problem of elastic lateral and flexural–torsional buckling in multiply braced column and beam members has been summarised. In very regular systems (e.g. uniform axial compression or moment) the calculation of the buckling load factor is reduced to the evaluation of 2×2 or 4×4 stiffness determinants at increasing load factors until the determinant becomes zero. This may have to be repeated for two or three trial modes of buckling.

Relationships between brace stiffness and column properties for any number of equally spaced brace lines and columns (or beams) are summarised in these determinants. In less regular cases (e.g. stepped parabolic axial compression) a larger, but sparse, matrix determinant is involved. For some of these cases it has been established that the corresponding uniform case provides a safe and not too conservative bound. For beams under uniform spread load a mixed analysis involving a beam–column finite element and some recurrence capitalisation is used. Eccentricity of load and bracing is covered in the analyses. By assuming initially deformed members, formulae and charts for brace strength requirements are put forward.

REFERENCES

BARSOUM, R. S. and GALLAGHER, R. H. (1970) Finite element analysis of torsional and torsional–flexural stability problems. *Int. J. Numerical Methods in Engineering*, **2,** 335–52.

BLEICH, F. (1952) *Buckling Strength of Metal Structures*, McGraw-Hill, New York.

DAVIDSON, B. J. and MEDLAND, I. C. (1974) Finite element approach to stability analysis in frames. *Proceedings, International Conference on Finite Elements in Engineering*, Sydney, 767–84.

HORNE, M. R. and MERCHANT, W. (1965) *The Stability of Frames*, Pergamon Press, Oxford.

LIVESLEY, R. K. and CHANDLER, D. B. (1956) *Stability Functions for Structural Frameworks*, Manchester University Press, Manchester.

MEDLAND, I. C. (1977) A basis for the design of column bracing. *Structural Engineer*, **55,** 301–7.

MEDLAND, I. C. (1979) Flexural–torsional buckling of interbraced columns. *Engineering Structures*, **1,** 131–8.

MEDLAND, I. C. (1980) Buckling of interbraced beam systems. *Engineering Structures*, **2,** 90–6.

MEDLAND, I. C. and SEGEDIN, C. M. (1979) Brace forces in interbraced column structures. *Proc. ASCE, J. Struct. Div.*, **105,** ST7, 1543–56.

MILES, J. W. (1956) Vibration of beams on many supports. *Proc. ASCE, J. Eng. Mech. Div.*, **82,** EM1, 1–9.

SEGEDIN, C. M. and MEDLAND, I. C. (1978) The buckling of interbraced columns. *International Journal of Solids and Structures*, **14,** 375–84.

TIMOSHENKO, S. P. and GERE, J. M. (1961) *Theory of Elastic Stability*, McGraw-Hill, New York.

APPENDIX 1

To determine the critical axial load of the interbraced set of columns shown in general form in Fig. 4.12, find the level of t at which the determinant

$$\begin{vmatrix} \dfrac{\sin t}{t-\sin t}+\dfrac{\cos\theta_j-\cos t}{1-\cos t} & -\dfrac{\sin\theta_i}{\sin t} & \\ -\dfrac{\sin\theta_j}{t-\sin t} & \dfrac{1-\cos\theta_j}{1-\cos t} & \\ & & -\left[\dfrac{2\beta(1-\cos\psi)}{t^3}\right]\left[\dfrac{t-\sin t}{1-\cos t}-\dfrac{1-\cos t}{\sin t}\right] \end{vmatrix}=0$$

where

$$t=\lambda l;\qquad \theta_j=\frac{j\pi}{N+1},\qquad j=1,N;\qquad \psi=\frac{i\pi}{M+1},\qquad i=1,M.$$

APPENDIX 2

The flexural–torsional cases involve the four basic displacements, those of lateral displacement and slope, twist and rate of change of twist. A 4×4 determinant governs the critical load. Let the elements be designated g_{ij}. Those not listed are zero.

$$g_{11}=\left[\frac{\sin t}{t-\sin t}+\frac{\cos\theta_j-\cos t}{1-\cos t}\right]-\frac{2\gamma(2+\cos\psi)}{t}\left[\frac{\sin t}{1-\cos t}-\frac{1-\cos t}{t-\sin t}\right]$$

$$g_{12}=-\frac{\sin\theta_j}{\sin t}$$

$$g_{13}=\frac{2\gamma(2+\cos\psi)pe}{t^2}\left[\frac{\sin p}{1-\cos p}-\frac{1-\cos p}{p-\sin p}\right]$$

$$g_{22}=\frac{2\beta(1-\cos\psi)EIe}{p^3C_1}\left[\frac{t-\sin t}{1-\cos t}-\frac{1-\cos t}{\sin t}\right]$$

$$g_{23}=-\frac{\sin\theta_j}{p-\sin p}$$

$$g_{24}=\left[\frac{1-\cos\theta_j}{1-\cos p}\right]-\left[\frac{2\eta(2+\cos\psi)}{p^3}+\frac{2\beta(1-\cos\psi)EIe^2}{p^3C_1}\right]\times\left[\frac{p-\sin p}{1-\cos p}-\frac{1-\cos p}{\sin p}\right]$$

$$g_{31}=-\frac{\sin\theta_j}{t-\sin t}$$

$$g_{32}=\left[\frac{1-\cos\theta_j}{1-\cos t}\right]-\frac{2\beta(1-\cos\psi)}{t^3}\left[\frac{t-\sin t}{1-\cos t}-\frac{1-\cos t}{\sin t}\right]$$

$$g_{34}=\frac{2\beta(1-\cos\psi)e}{t^3}\left[\frac{p-\sin p}{1-\cos p}-\frac{1-\cos p}{\sin p}\right]$$

$$g_{41}=-\frac{2\gamma(2+\cos\psi)tEIe}{p^2C_1}\left[\frac{\sin t}{1-\cos t}-\frac{1-\cos t}{t-\sin t}\right]$$

$$g_{43}=\left[\frac{\sin p}{p-\sin p}+\frac{\cos\theta_j-\cos p}{1-\cos p}\right]+\frac{2\gamma(2+\cos\psi)EIe^2}{pC_1}\left[\frac{\sin p}{1-\cos p}-\frac{1-\cos p}{p-\sin p}\right]$$

$$g_{44}=-\frac{\sin\theta_j}{\sin p}$$

where $\beta = K_L l^3/2EI$; $\gamma = K_R l/EI$; $\eta = K_T l^3/C_1$; $p = \alpha l$; $\alpha = \sqrt{((PI_0/A - C)/C_1)}$.

APPENDIX 3

The beam buckling also involves four continuities, those of minor axis bending moment and lateral shear force, torsional moment and warping bimoment. The determinant components are designated q_{ij} and any which are zero are not included.

$$q_{11} = -(1 - \cos\theta_j)$$

$$q_{12} = [(t_1 - \sinh t_1)\cos\theta_j + (\sinh t_1 - t_1\cosh t_1)]T_2$$

$$q_{14} = [(t_2 - \sin t_2)\cos\theta_j + (\sin t_2 - t_2\cos t_2)]T_1$$

$$q_{22} = q_{12}/T_2$$

$$q_{23} = q_{11}$$

$$q_{24} = q_{14}/T_1$$

$$q_{31} = 2(1 - \cos\psi_i)\beta$$

$$q_{32} = t_1^3(\cos\theta_j - \cosh t_1)T_2$$

$$q_{33} = -2e(1 - \cos\psi_i)\beta$$

$$q_{34} = -t_2^3(\cos\theta_j - \cos t_2)T_1$$

$$q_{41} = 2e(1 - \cos\psi_i)\beta EI/C_1$$

$$q_{42} = t_1^3(\cos\theta_j - \cosh t_1)$$

$$q_{43} = 2e^2(1 - \cos\psi_i)\beta EI/C_1 + 2(2 + \cos\psi_i)\eta$$

$$q_{44} = q_{34}/T_1$$

where $t_1 = \mu_1 l$, $t_2 = \mu_2 l$, $T_1 = (C - \mu_1^2 C_1)M$, $T_2 = (C + \mu_2^2 C_1)M$,

$$\mu_1, \mu_2 = \sqrt{\left(\left(\frac{C^2}{4C_1^2} + \frac{M^2}{EIC_1}\right)^{1/2} \pm \frac{C}{2C_1}\right)}.$$

In this beam under uniform moment case, the usual form of shape function (eqns (4.11) and (4.12)) is used with μ_2, while with μ_1 the transcendental functions are replaced by hyperbolic functions.

Chapter 5

ELASTIC STABILITY OF RIGIDLY AND SEMI-RIGIDLY CONNECTED UNBRACED FRAMES

G. J. SIMITSES and A. S. VLAHINOS

School of Engineering Science and Mechanics,
Georgia Institute of Technology, Atlanta, USA

SUMMARY

The nonlinear analysis of plane elastic and orthogonal frameworks is presented. The static loading consists of both eccentric concentrated loads (near the joints) and uniformly distributed loads on all or few members. The joints can be either rigid or flexible. The flexible joint connection is characterised by connecting one member to an adjoining one through a rotational spring (with linear or nonlinear stiffness). The supports are immovable but are also characterised with rotational restraint by employing linear rotational springs. The mathematical formulation is presented in detail and the solution methodology is outlined and demonstrated through several examples. These examples include two-bar frames, portal frames as well as multi-bay multi-storey frames. The emphasis is placed on obtaining sway buckling loads and prebuckling and postbuckling behaviours, whenever applicable. Finally, some general concluding remarks are presented on the basis of the generated results.

NOTATION

$\bar{A}$ Constant in the moment–relative rotation expression

A_i Cross-sectional area of bar i

A_{ij} Coefficient of general solution to equilibrium differential equation of bar i

A_{ij}^*	Coefficient of general solution to buckling differential equation o bar i
EI_i	Bending stiffness of bar i
e_i^0	Load eccentricity near $x_i = 0$
e_i^1	Load eccentricity neat $x_i = L_i$
$\bar{e}_i$	e_i/L_i
k_i	$\sqrt{(P_i L_i^2/EI_i)}$
$\bar{k}_i$	k_i on primary equilibrium path
L_i	Length of bar i
M_i	Bending moment in bar i
P_i	Axial force in bar i
P_i^*	Additional P_i corresponding to u^* and w^*
Q_i^0	Concentrated load applied on bar i at $x_i = e_i^0$
Q_i^1	Concentrated load applied on bar i at $x_i = L_i - e_i^1$
Q_{cl}	Critical load obtained by linear theory for special geometries
$\bar{Q}_i$	$Q_i L_1^2/EI_1$
q_i	Uniformly distributed load on bar i
$\bar{q}_i$	$q_i L_i^3/EI_i$
q_i^*	$q_i L_1^3/EI_1$
q_t	Total load carried by frame (q_i^* multiplied by number of contribu tions)
R_i	L_i/L_1
S_i	$EI_i L_1/EI_1 L_i$
U_i	u_i/L_i
U_i^*	Kinematically admissible variation of U_i
u_i	Axial displacement component along bar i
V_i	Shearing force of bar i
W_i	w_i/L_i
$\bar{W}_i$	W_i on primary equilibrium path
W_i^*	Kinematically admissible variation of W_i
w_i	In-plane normal displacement component along bar i
X_i	x_i/L_i
x_i	Axial coordinate of bar i
Z	Semi-rigid connection factor ($= 1/{}_0\beta$)
z_i	Normal coordinate of bar i
β_i^0	Rotational spring stiffness near $x = 0$
β_i^1	Rotational spring stiffness near $x = L_i$
${}_0\beta$	Initial slope of the moment–relative rotation curve

λ_c Q/Q_{cl}
λ_i $L_i/\sqrt{(I_i/A_i)}$
σ_i^* $P_i^* L_i^2/EI_i$

5.1 INTRODUCTION

Plane frameworks, composed of straight slender bars, have been widely used as primary structures in several configurations. These include one- or multi-storey buildings, storage racks, factory cranes and off-shore platforms. Depending on characteristics of geometry (symmetric or asymmetric and various support conditions) and loading (symmetric or asymmetric transverse and horizontal), plane frames may fail by general instability (in a sidesway mode or a symmetric mode) or they may fail by a mechanism or a criterion other than stability (excessive deformations and/or stresses, etc.). For example, a symmetric portal frame subjected to a uniformly distributed transverse load is subject to sway buckling. On the other hand if, in addition to the transverse load, a concentrated horizontal load is applied, excessive deformations and stresses will occur without the system being subject to instability (buckling).

The various frame responses, associated with the various geometries and loadings, have been the subject of many studies, both in analysis and in synthesis (design). A brief description and critique of these studies is presented in the ensuing chapter.

5.1.1 Rigid-Jointed Frames: Linear Analyses

The first stability analyses of rigid-jointed plane frameworks may be traced to Zimmermann (1909, 1910, 1925), Müller-Breslau (1908) and Bleich (1919). They only treated the problem for which a momentless primary state (membrane) exists and bifurcational buckling takes place through the existence of an adjacent bent equilibrium state (linear eigenvalue problem). Prager (1936) developed a method which utilises the stability condition of a column with elastic end restraints. The first investigation of a problem for which the primary state includes bending moments (primary moments) is due to Chwalla (1938). He studied the sway buckling of a rigid-jointed one-storey symmetric portal frame under symmetric concentrated transverse loads, not applied at the joints of the horizontal bar. In obtaining both the primary path and the bifurcation load, Chwalla employed linear equilibrium equations and assumed linearly elastic behaviour. In more recent years, similar problems have been studied by Baker *et al.* (1949), Merchant (1954, 1955), Chilver (1956), Livesley (1956), Goldberg (1960), Masur *et al.* (1961) and Horne (1962). The last two consider the effect of primary moments, which cause small deflections prior to instability, in their buckling analysis of portal frames. Many of the aforementioned analyses have been incorporated into textbooks, such as those of Bleich (1952), McMinn (1962), Horne and Merchant (1965) and Simitses (1976). Other investigations in this category include the studies of Halldorsson and Wang (1968) and Zweig and Kahn (1968). It is also worth mentioning the work of Switzky and Wang (1969), who outlined a simple procedure for designing rectangular rigid frames for stability. Their procedure employs linear theory and is applicable to load cases for which the primary state is a membrane state (free of primary moments).

5.1.2 Rigid-Jointed Frames: Nonlinear Analyses

The effects of finite displacements on the critical load and on the postbuckling behaviour of frameworks have only been investigated in the last 20 years or so. Saafan (1963) considered the effects of large deformations on the symmetric buckling of a gable frame. Similar effects were also considered by Britvec and Chilver (1963) in their studies of the buckling and postbuckling behaviour of triangulated frames and rigid-jointed trusses. The nonlinear behaviour of the two-bar frame was studied by Williams (1964), Roorda (1965), Koiter (1966), Huddleston (1967) and more recently by Kounadis *et al.* (1977) and by Simitses *et al.* (1977). Roorda's work contains experimental results, while Koiter's contribution employs his (1945) rigorous nonlinear theory for initial postbuckling behaviour, applicable to structures that exhibit bifurcational buckling. The studies of Kounadis and Simitses employ nonlinear kinematic relations (corresponding to moderate rotations) and assume linearly elastic material behaviour. Huddleston's nonlinear analysis is based on equations of the Elastica. A similar approach (Elastica-type equations) was outlined by Lee *et al.* (1968) for studying the large deflection buckling and postbuckling behaviour of rigid plane frameworks loaded by concentrated loads. They demonstrated their procedure by analysing a two-bar frame and a portal frame, and they used a modified Newton–Raphson procedure to solve the nonlinear equations. More recently, Elastica-type equations were employed by Qashu and DaDeppo (1983) for the analysis of elastic plane frames. They used numerical integration of the differential equations and their examples include one- and two-storey elastic rigid frames. Besides the inherent assumptions of Elastica-type equations, that make them applicable to very slender members, the difficulty of solving the highly nonlinear equations in a straightforward manner further limits the applicability of this approach to frames with a relatively small number of members. On the other hand, the nonlinear methodology, described herein, as developed by Simitses and his collaborators (Simitses *et al.*, 1977; Simitses and Kounadis, 1978; Simitses *et al.*, 1981; Simitses and Giri, 1982; Simitses and Vlahinos, 1982) employs first-order nonlinear kinematic relations (moderate rotations) but can be used, with relative ease, in analysing the large deformation behaviour (including buckling and postbuckling) of multistorey multi-bay elastic, rigid-jointed, orthogonal, plane frameworks, with a large number of members.

The interested reader is referred to the book by Britvec (1973), which presents some of the nonlinear analyses of frames. Moreover, those who are interested in the design of elastic frames are referred to the Design Guide of the Structural Stability Research Council (see Johnston, 1976).

5.1.3 Semi-Rigidly Connected Frames

All of the previously discussed analyses are based on the assumption that the bars are rigidly connected at the frame joints. This means that the angle between connected members, at the joints, remains unchanged during deformations.

Since the 1930s, there has been considerable interest and research into the behaviour of beam structural connections. A number of experimental and analytical studies have been carried out to measure the moment–relative rotation characteristics of various types of metal (primarily steel) framing connections. Various methods of analysis (moment distribution, slope-deflection, elastic line) have been employed, by Batho and Rowan (1934), Rathbun (1936) and Sourochnikoff (1949), in order to account for the flexibility of the connections. Moreover, some efforts have been made recently to account for the effect of flexible connections in frame design. DeFalco and Marino (1966) modified the effective column length used in frame design by obtaining and employing a modified beam stiffness, which is a function of the semi-rigid connection factor Z (slope of the relative rotation to moment curve at the origin) proposed by Lothers (1960). Frye and

Morris (1975) presented an iterative procedure which incorporates the effects of nonlinear connection characteristics. They assumed linearly elastic material behaviour and developed equations that depict moment–relative rotation relations for a wide range of frame connections. More recently Moncarz and Gerstle (1981) presented a matrix displacement method for analysing frames with flexible (nonlinear) connections. The effect of flexible joints on the response characteristic of simple two-bar frames which are subject to limit point instability (violent buckling) has been reported by Simitses and Vlahinos (1982). This subject will be further explored in a later section of this chapter. Finally, a brief summary of recent research on the effect of end restraints on column stability has been presented by Lui and Chen (1983).

In closing, it is worth mentioning that the analysis of plane frameworks, including stability studies, postbuckling behaviour and the study of the effect of flexible connections, has been the subject of several PhD theses, especially in the United States. Of particular interest, and related to the objective of the present chapter, are those of Ackroyd (1979) and Vlahinos (1983). Moreover, there exist a few reported investigations in which the frame has been used as an object of demonstration. In these studies, the real interest lies in some nonlinear numerical scheme, especially the use of finite elements. These works include, but are not limited to, those of Argyris and Dunne (1975), Olesen and Byskov (1982) and Obrecht *et al.* (1982).

5.2 MATHEMATICAL FORMULATION

5.2.1 Geometry and Basic Assumptions

Consider a plane orthogonal rigid-jointed frame composed of N straight slender bars of constant cross-sectional area. A typical ten-bar frame is shown in Fig. 5.1. Each bar, identified by the subscript i, is of length L_i, cross-sectional area A_i, cross-sectional second moment of area I_i, and subscribes to a local coordinate system, x, z, with displacement components u_i and w_i, as shown. The frame is subjected to eccentric concentrated loads Q_i^0 and Q_i^1 and/or uniformly distributed loadings q_i. For the concentrated loads, the superscript 0 implies that the load is near the origin of the ith bar ($x = 0$), while the superscript 1 implies that the load is near the other end of the ith bar ($x = L_i$). The

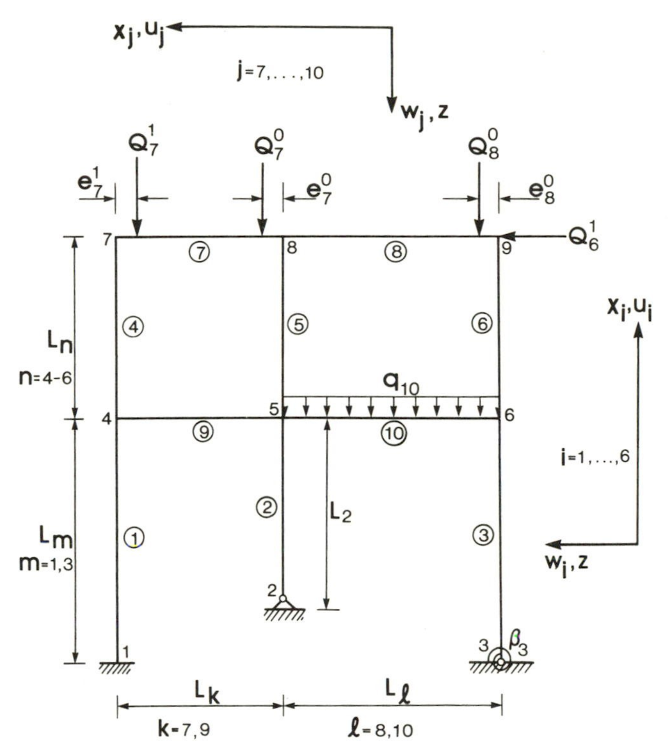

FIG. 5.1. Geometry and sign convention of a multi-bay multi-storey frame.

concentrated load eccentricities are also denoted in the same manner as the concentrated loads (e_i^0 and e_i^1). Moreover, these eccentricities are positive if the loads are inside the x-interval of the corresponding bar and negative if outside the interval. For example, in Fig. 5.1, e_7^0 is a positive number. But this same eccentricity (and therefore the corresponding load too) can be identified as e_8^1, in which case its value is negative. This is used primarily for corner overhangs (joint 7 or 9 with concentrated loads off the frame). The supports are such that translation is completely constrained, but rotation could be free. For this purpose, rotational linear springs are used at the supports (see Fig. 5.1, support 3). When the spring stiffness β is zero, we have an immovable simple support (pin). On the other hand, when β is a very large number ($\rightarrow\infty$) we have an immovable fixed support (clamped, built-in).

For clarity, all limitations of the mathematical formulation are compiled, in the form of assumptions. These are:

(1) The frame members are initially straight piecewise prismatic and joined together orthogonally and rigidly (this assumption can be, and is, relaxed later on).
(2) The material is homogeneous and isotropic and the material behaviour is linearly elastic with an invariant elastic constant, regardless of tension or compression.
(3) Normals remain 'normal' to the elastic member axis and inextensional (the usual Euler–Bernoulli assumptions).
(4) Deformations and loads are confined to the plane of the frame.
(5) The concentrated loads are applied near the joints (small eccentricities). This assumption can easily be relaxed, but it will lead to an increase in the number of bars. A concentrated load at the midpoint of a bar is treated by considering two bars and an additional joint at, or near, the location of the concentrated load.
(6) The effect of residual stresses on the system response (critical) load is neglected.
(7) The nonlinear kinematic relations correspond to small strains, but moderate rotations, for points on the elastic axes (first order nonlinearity).

On the basis of the above, the kinematic relations are

$$\varepsilon_{xx_i} = \varepsilon^0_{xx_i} + zk_{xx_i}$$

where (5.1)

$$\varepsilon^0_{xx} = u_{i,_x} + \tfrac{1}{2}w^2_{i,_x} \quad \text{and} \quad k_{xx_i} = -w_{i,_{xx}}$$

Furthermore, the axial force P_i and bending moment M_i in terms of the displacement gradients are

$$P_i = (EA_i)[u_{i,_x} + \tfrac{1}{2}w^2_{i,_x}]$$
$$M_i = EI_i w_{i,_{xx}} \tag{5.2}$$

where E is the Young's modulus of elasticity for the material. Similarly, the expression for the transverse shear force is

$$V_i(x) = -EIw_{i,_{xxx}} + P_i w_{i,_x} \tag{5.3}$$

5.2.2 Equilibrium Equations: Boundary and Joint Conditions

Before writing the equilibrium equations and the associated boundary and joint conditions, the following nondimensionalised parameters are introduced

$$\begin{gathered}
X = x_i/L_i, \qquad U_i = u_i/L_i, \qquad W_i = w_i/L_i \\
\bar{e}_i = e_i/L_i, \qquad \bar{q}_i = q_i L_i^3/EI_i, \qquad q_i^* = q_i L_1^3/EI_1 \\
\bar{Q}_i = Q_i L_1^2/EI_1, \qquad \bar{\beta} = \beta L_1/EI_1, \qquad \lambda_i = L_i/\sqrt{(I_i/A_i)} \\
k_i^2 = \mp P_i L_i^2/EI_i, \qquad S_i = EI_i L_1/EI_1 L_i \qquad R_i = L_i/L_1
\end{gathered} \tag{5.4}$$

The expression for the internal forces, in terms of the nondimensionalised parameters, are

$$\begin{gathered}
P_i = \mp k_i^2 (EI_i/L_i^2), \qquad M_i = W_{i,xx}(EI_i/L_i) \\
V_i = [\mp k_i^2 W_{i,x} - W_{i,xxx}](EI_i/L_i^2)
\end{gathered} \tag{5.5}$$

where the top sign holds for the case of compression in the bar and the lower for the case of tension (the axial force P_i is positive for tension and negative for compression; thus k_i^2 is always positive).

The equilibrium equations for the frame are (in terms of the nondimensionalised parameters)

$$\begin{gathered}
U_{i,x} + \tfrac{1}{2}(W_{i,x})^2 = \mp k_i^2/\lambda_i^2 \\
W_{i,xxxx} \pm k_i^2 W_{i,xx} = \bar{q}_i \qquad i = 1, 2, \ldots, N
\end{gathered} \tag{5.6}$$

where N is the number of bars, and the top sign holds for the compression case. The general solution to the equilibrium equations is given by

$$\begin{gathered}
U_i(X) = A_{i5} \pm \left(\frac{k_i}{\lambda_i}\right)^2 X - \frac{1}{2}\int_0^X [W_{i,x}(X)]^2 \, \mathrm{d}X \\
W_i(X) = A_{i1}\begin{pmatrix}\sin k_i X \\ \sinh k_i X\end{pmatrix} + A_{i2}\begin{pmatrix}\cos k_i X \\ \cosh k_i X\end{pmatrix} + A_{i3}X + A_{i4} \pm \frac{\bar{q}_i}{2k_i^2} X^2
\end{gathered} \tag{5.7}$$

where A_{ij} and k_i ($i = 1, 2, \ldots, N$; $j = 1, 2, \ldots, 5$) are constants (for a given level of the applied loads) to be determined from the boundary and joint conditions. For an N-member frame, the number of unknowns is $6N$. Therefore, $6N$ equations are needed for their evaluation.

These equations are provided by the boundary conditions and the joint conditions. At each boundary, three conditions must be satisfied (kinematic, natural or mixed: typical conditions are listed below). At each joint, three force and moment equations (equilibrium of a joint taken as a particle) and a number of kinematic continuity equations must be satisfied. This number depends on the number of members coming into a joint, and the equations represent continuity in displacement and continuity in rotation (typical conditions are listed below). For a two-member joint, we have three kinematic continuity conditions; two in displacement and one in rotation. For a three-member joint the number is six, and for a four-member joint (largest possible) the number is nine.

A quick accounting of equilibrium equations, and boundary and joint conditions, for the ten-bar frame, shown in Fig. 5.1, yields the following:

(i) The number of equilibrium equations is 60 (6×10).
(ii) The number of boundary conditions is *nine* (three at each of boundaries 1, 2 and 3).
(iii) The number of joint conditions is 51; of these, 18 are force and moment equilibrium conditions (three at each of the six joints 4, 5, 6, 7, 8 and 9) and 33 kinematic continuity conditions (three at each of joints 7 and 9, six at each of joints 4, 6, and 8 and nine at joint 5).

Therefore, the total number of available equations is 60. Here, it is implied that the loading is of known magnitude.

For clarity, typical boundary and joint conditions are shown below, with reference to the frame of Fig. 5.1 (in nondimensionalised form).

Boundary 3

$$\begin{gathered} U_3(0)=0, \qquad W_3(0)=0 \\ S_3 W_{3,xx}(0)-\bar{\beta}_3 W_{3,x}(0)=0 \end{gathered} \tag{5.8}$$

Joint 5

$$[\mp k_9^2 W_{9,x}(0)-W_{9,xxx}(0)]\frac{S_9}{R_9}+(\mp k_2^2)\frac{S_2}{R_2}-[\mp k_{10}^2 W_{10,x}(1) - W_{10,xxx}(1)]\frac{S_{10}}{R_{10}}-(\mp k_5^2)\frac{S_5}{R_5}=0 \tag{5.9a}$$

$$[\mp k_5^2 W_{5,x}(0)-W_{5,xxx}(0)]\frac{S_5}{R_5}+(\mp k_9^2)\frac{S_9}{R_9}-[\mp k_2^2 W_{2,x}(1) - W_{2,xxx}(1)]\frac{S_2}{R_2}-(\mp k_{10}^2)\frac{S_{10}}{R_{10}}=0 \tag{5.9b}$$

$$S_5 W_{5,xx}(0) + S_9 W_{9,xx}(0) - S_2 W_{2,xx}(1) - S_{10} W_{10,xx}(1) = 0 \quad (5.9c)$$

$$\begin{gathered} R_2 U_2(1) = -R_{10} W_{10}(1) = R_5 U_5(0) = -R_9 W_9(0) \\ R_2 W_2(1) = R_{10} U_{10}(1) = R_5 W_5(0) = R_9 U_9(0) \\ W_{2,x}(1) = W_{10,x}(1) = W_{5,x}(0) = W_{9,x}(0) \end{gathered} \quad (5.10)$$

Joint 7

$$\bar{Q}_7^1 + (\mp k_4^2)\frac{S_4}{R_4} - [\mp k_7^2 W_{7,x}(1) - W_{7,xxx}(1)]\frac{S_7}{R_7} = 0 \quad (5.11a)$$

$$[\mp k_4^2 W_{4,x}(1) - W_{4,xxx}(1)]\frac{S_4}{R_4} + (\mp k_7^2)\frac{S_7}{R_7} = 0 \quad (5.11b)$$

$$W_{4,xx}(1)S_4 + W_{7,xx}(1)S_7 + \bar{Q}_7^1 \bar{e}_7^1 = 0 \quad (5.11c)$$

$$\begin{gathered} R_4 U_4(1) = -R_7 W_7(1), \qquad R_4 W_4(1) = R_7 U_7(1) \\ W_{4,x}(1) = W_{7,x}(1) \end{gathered} \quad (5.12)$$

Note that, in these expressions as well, the top sign corresponds to the compression case and the bottom to the tension.

5.2.3 Buckling Equations

The buckling equations and the associated boundary and joint conditions are derived by employing a perturbation method (Bellman, 1969; Sewell, 1965). This derivation is based on the concept of the existence of an adjacent equilibrium position at either a bifurcation point or a limit point. In the derivation, the following steps are followed: (i) start with the equilibrium equations (eqns (5.6)) and related boundary and joint conditions, expressed in terms of the displacements, (ii) perturb them by allowing small kinematically admissible changes in the displacement functions and a small change in the bar axial force, (iii) make use of equilibrium at a point at which an adjacent equilibrium path is possible and retain first order terms in the admissible variations. The resulting inhomogeneous differential equations are linear in the small changes. Replace U_i and W_i in eqns (5.6) by $\bar{U}_i + U_i^*$ and $\bar{W}_i + W_i^*$, respectively. Moreover, replace $\mp k_i^2$ by $\pm \bar{k}_i^2 + \sigma_i^*$, where σ_i^* is the change in the nondimensionalised axial force ($= P_i^* L_i^2 / EI_i$) and it can be either positive or negative, regardless of tension or compression in the bar at an equilibrium position. The *bar* quantities denote parameters at a static primary equilibrium position and the *star* quantities denote the small changes.

The buckling equations are

$$
\begin{aligned}
U^*_{i,x} + \bar{W}_{i,x} W^*_{i,x} &= \sigma^*_i / \lambda^2_i \\
W^*_{i,xxxx} \mp \bar{k}^2_i W^*_{i,xx} &= \sigma^*_i \bar{W}_{i,xx}
\end{aligned}
\tag{5.13}
$$

The related boundary and joint conditions are presented, herein, only for the same boundaries and joints as those related to the equilibrium equations (eqns (5.9)–(5.12)).

Boundary 3

$$
\begin{aligned}
U^*_3(0) = W^*_3(0) &= 0 \\
S_3 W^*_{3,xx}(0) - \bar{\beta}_3 W^*_{3,x}(0) &= 0
\end{aligned}
\tag{5.14}
$$

Joint 5

$$
\begin{aligned}
&[\mp \bar{k}^2_9 W^*_{9,x}(0) - W^*_{9,xxx} + \sigma^*_9 \bar{W}_{9,x}] \frac{S_9}{R_9} + \sigma^*_2 \frac{S_2}{R_2} - [\mp \bar{k}^2_{10} W^*_{10,x}(1) \\
&\qquad - W^*_{10,xxx}(1) + \sigma^*_{10} \bar{W}_{10,x}(1)] \frac{S_{10}}{R_{10}} - \sigma^*_5 \frac{S_5}{R_5} = 0
\end{aligned}
\tag{5.15a}
$$

$$
\begin{aligned}
&[\mp \bar{k}^2_5 W^*_{5,x}(0) - W^*_{5,xxx}(0) + \sigma^*_5 \bar{W}_{5,x}(0)] \frac{S_5}{R_5} + \sigma^*_9 \frac{S_9}{R_9} \\
&\qquad - [\mp \bar{k}^2_2 W^*_{2,x}(1) - W^*_{2,xxx}(1) + \sigma^*_2 \bar{W}_{2,x}(1)] - \sigma^*_{10} \frac{S_{10}}{R_{10}} = 0
\end{aligned}
\tag{5.15b}
$$

$$
S_5 W^*_{5,xx}(0) + S_9 W^*_{9,xx}(0) - S_2 W^*_{2,xx}(1) - S_{10} W^*_{10,xx}(1) = 0
\tag{5.15c}
$$

$$
\begin{aligned}
R_2 U^*_2(1) &= -R_{10} W^*_{10}(1) = R_5 U^*_5(0) = -R_9 W^*_9(0) \\
R_2 W^*_2(1) &= R_{10} U^*_{10}(1) = R_5 W^*_5(0) = R_9 U^*_9(0) \\
W^*_{2,x}(1) &= W^*_{10,x}(1) = W^*_{5,x}(0) = W^*_{9,x}(0)
\end{aligned}
\tag{5.16}
$$

Joint 7

$$
\sigma^*_4 \frac{S_4}{R_4} - [\mp k^2_7 W^*_{7,x}(1) - W^*_{7,xxx}(1) + \sigma^*_7 \bar{W}_{7,x}(1)] \frac{S_7}{R_7} = 0
\tag{5.17a}
$$

$$
[\mp \bar{k}^2_4 W^*_{4,x}(1) - W^*_{4,xxx}(1) + \sigma^*_4 \bar{W}_{4,x}(1)] \frac{S_4}{R_4} + \sigma^*_7 \frac{S_7}{R_7} = 0
\tag{5.17b}
$$

$$
S_4 W^*_{4,xx}(1) + S_7 W^*_{7,xx}(1) = 0
\tag{5.17c}
$$

$$
\begin{aligned}
R_4 U^*_4(1) = -R_7 W^*_7(1) \qquad & R_4 W^*_4(1) = R_7 U^*_7(1) \\
W^*_{4x}(1) = W^*_{7,x}&(1)
\end{aligned}
\tag{5.18}
$$

The solution to the buckling equations is given by

$$U_i^*(X) = A_{i5}^* + \frac{\sigma_i^*}{\lambda_i^2} X - \int_0^X \bar{W}_{i,x} W_{i,x}^* \, dX$$

$$W_i^*(X) = A_{i1}^* \begin{pmatrix} \sin \bar{k}_i X \\ \sinh \bar{k}_i X \end{pmatrix} + A_{i2}^* \begin{pmatrix} \cos \bar{k}_i X \\ \cosh \bar{k}_i X \end{pmatrix} + A_{i3}^* X + A_{i4}^*$$

$$+ \frac{\sigma_i^* X}{2\bar{k}_i} \left[A_{i2} \begin{pmatrix} \sin \bar{k}_i X \\ \sinh \bar{k}_i X \end{pmatrix} + A_{i1} \begin{pmatrix} -\cos \bar{k}_i X \\ \cosh \bar{k}_i X \end{pmatrix} + \frac{\bar{q}_i X}{\bar{k}_i^3} \right] \quad (5.19)$$

Here also the top sign and expression correspond to the compression case (the ith bar is in compression at equilibrium) and the bottom to the tension case. Note that $\bar{k}_i$, A_{i1} and A_{i2} are the values of the constants (see eqns (5.7)) on the primary path (equilibrium). On the other hand, the *star* parameters are $6N$ in number (60 for the ten bar frame). Moreover, the boundary and joint conditions associated with the buckling equations are also $6N$ in number and they are linear, homogeneous, algebraic equations in the $6N$ *star* parameters. Thus, the characteristic equation, which leads to the estimation of the critical load condition, is obtained by requiring a non-trivial (all A_{ij}^* and σ_i^* are not equal to zero) solution of the buckling equations to exist.

5.2.4 Semi-rigid Joint Connections

The mathematical formulation presented so far is based on the assumption of rigid-jointed connections. In the case of semi-rigid connections, the only difference lies in some of the joint conditions. Two types of non-rigid connections are treated herein. Both come under the general but vague term of semi-rigid connections. The first corresponds to the case where a member, at a given joint, is connected to the remaining members through a linear rotational spring (type A). The second corresponds to the case of realistic flexible connections at frame joints (type B). In this latter case, especially for steel frame construction, the connections are usually bolted with the use of various connecting elements (top and bottom clip angles, end plates, web framing, etc.). In this case the bending moment–relative rotation curve (for a member connected to a group of members at a joint) is nonlinear. Initially, the slope is not infinite, as assumed in the case of rigid joints, but a very large number, which primarily depends on the beam depth and the type of connection (see Tables I–IV of DeFalco and Marino (1966)), but the slope decreases as the moment increases.

In this latter case we may still employ the idea of a rotational spring, but with nonlinear stiffness.

The required modification in the mathematical formulation is treated separately for each case (types A and B).

Type A

The only difference, from the case of rigid connections, is to modify the condition of kinematic continuity in rotation. For example, if member 7 is connected to member 4 through a rotational spring of linear stiffness β_7 (see Fig. 5.1), then the last of eqns (5.12) needs to be modified. Instead of

$$W_{4,x}(1) = W_{7,x}(1)$$

one must use

$$W_{4,xx}(1) + \bar{\beta}_4^1[W_{4,x}(1) - W_{7,x}(1)] = 0 \qquad (5.20)$$

where $\bar{\beta}_i^m$ is the stiffness of the rotational spring that connects member 4 to joint 7 (see Fig. 5.1) in a nondimensionalised form, or

$$\bar{\beta}_i^m = \beta_i^m \frac{L_i}{EI_i} \qquad i = 1, 2, \ldots, 10, \qquad m = 0, 1 \qquad (5.21)$$

Note that β_i^m is the rotational stiffness associated with member i. If $m = 1$ the spring is at $X = 1$ of the member, while if $m = 0$ the spring is at $X = 0$. Furthermore, note that eqn (5.20) relates the member 4 end moment to the relative rotation (of member 4 to member 7). Moreover, for a rigid-jointed frame $\bar{\beta}_i^m$ tends to infinity (for calculations a very large number is used). On the other hand, when $\bar{\beta}_i^m$ tends to zero (pin connection), eqn (5.20) implies that no moment is transferred through the pin.

Type B

For the case of realistic flexible connections, the member end moment M_i(1 or 0) is related to the relative rotation curve in a nonlinear fashion.

Again, if the same example is used as for type A, then

$$\bar{M}_4 = \frac{M_4 L_4}{EI_4} = f(\varphi_4) \qquad (5.22)$$

where $f(\varphi_4)$ is a nonlinear function of φ_4, and φ_4 is the relative rotation of member 4 to member 7 at their joint (for a multi-member joint, one member is considered immovable and φ_i is the relative rotation of the other members with respect to the immovable one)

$$\varphi_4 = W_{4,x}(1) - W_{7,x}(1) \tag{5.23}$$

One possible selection for the nonlinear function $f(\varphi_4)$ is a cubic relation, or

$$-\bar{M}_4 = {}_0\bar{\beta}_4^1 \varphi_4 - \bar{A}_4^1 \varphi_4^3 \tag{5.24}$$

where ${}_0\bar{\beta}_4^1$ denotes the slope of the member end moment to the relative rotation curve at the origin (or before the external loads are applied) and $\bar{A}_4^1$ is a constant, which can be obtained from experimental data.

In order to employ the same equations as for type A (linear spring) connections and therefore the same solution methodology (instead of increasing the nonlinearity of the problem) the following concept is introduced. First, solutions for the frame response are obtained by starting with small levels for the applied loads and by using small increments. Then, eqn (5.24) at load step $(m+1)$ can be written as

$$(-\bar{M}_4)_{m+1} = [{}_0\bar{\beta}_4^1 - \bar{A}_4^1 (\varphi_4)_m^2](\varphi_4)_{m+1} \tag{5.25}$$

This implies that for small steps in the load, the relative rotation experiences small changes. Thus, the required joint condition, eqn (5.24), becomes

$$W_{4,xx}(1) + [W_{4,x}(1) - W_{7,x}(1)](\bar{\beta}_4^1)_{m+1} = 0 \tag{5.26}$$

where $(\bar{\beta}_4^1)_{m+1}$ is evaluated at the previous load step by

$$(\bar{\beta}_4^1)_{m+1} = {}_0\bar{\beta}_4^1 - \bar{A}_4^1 [W_{4,x}(1) - W_{7,x}(1)]_m^2 \tag{5.27}$$

Clearly then, the solution scheme for type B connections is the same as the one for type A connections and the nonlinearity of the problem is not increased.

5.3 SOLUTION PROCEDURE

The complete response of an N-member frame is known, for a given geometry and level of the applied loads, if one can estimate the values of the $6N$ unknowns that characterise the two displacement functions $U(X)$ and $W(X)$ (eqns (5.7)). The $6N$ equations required are provided from satisfaction of the boundary and joint conditions. Furthermore, the estimation of the critical load condition requires the use of one more equation. This is provided by the solution to the buckling equations (eqns (5.13)). As already mentioned, satisfaction of the boundary and support conditions for the buckling solution leads to a system of $6N$ linear, homogeneous, algebraic equations in σ_i^* and A_{ij}^* ($i = 1, 2, \ldots, N$, $j = 1, 2, \ldots, 5$; these constants characterise the buckling modes). For a nontrivial solution to exist, the determinant of the coefficients must vanish. This step provides the necessary additional equation, which is one more equation in $\bar{k}_i$ and A_{ij}, and it holds true only at the critical equilibrium point (either bifurcation or limit point).

A solution methodology has been developed (including a computer algorithm) for estimating critical conditions, prebuckling response and postbuckling behaviour. The scheme makes use of the following steps.

(1) Through a simple and linear frame analysis program, the values of the internal axial load parameters $\bar{k}_i$ are estimated, for some low level of the applied loads. This can be used as an initial estimate for the nonlinear analysis, but most importantly it tells us which members are in tension and which in compression. Note that the solution expressions (eqns (5.7)) differ for the two cases (compression versus tension). Such a subroutine is outlined by Weaver and Gere (1980).

(2) Once the form of the solution has been established (from step 1 we know which members are in tension and which in compression), then through the use of the boundary and joint conditions one can establish the $6N$ equations that signify equilibrium states, for the load level of step 1.

In so doing, it is observed that $5N$, out of the $6N$, equations are linear in A_{ij} and nonlinear in $\bar{k}_i$. Two important consequences are directly related to this observation. First, through matrix algebra the $5N$ equations are used to express A_{ij} in terms of the $\bar{k}_i$, and the substitution into the remaining equations yields a system of N nonlinear equations in $\bar{k}_i$. Secondly, if the $\bar{k}_i$ values are (somehow) known, then the $5N$ equations (linear in A_{ij}) can be used to solve for A_{ij}.

(3) The N nonlinear equations are solved by employing one of several possible nonlinear solvers. There exist several candidates for this.

For the two-bar frame and for the portal frame (small number of nonlinear equations) the nonlinear equations $f_j = 0$ ($j \leq 3$) can be solved by first defining a new function:

$$F = \sum_{j=1}^{N} f_j^2 \tag{5.28}$$

Then, one recognises that the set of k_i that minimises F (note that the minimum value of F is zero) is the set that satisfies the nonlinear equations, $f_i = 0$. The mathematical search technique of Nelder and Mead (1964) can be used for finding this minimum. This nonlinear solver was employed by Simitses and co-workers (Simitses, 1976; Simitses *et al.*, 1977; Simitses and Kounadis, 1978; Simitses *et al.*, 1981; Simitses and Giri, 1982; Simitses and Vlahinos, 1982) for the two-bar and portal frame problems.

For multi-bay multi-storey frames ($N \geqslant 5$) the nonlinear equations $f_i = 0$ ($j \geqslant 5$), can be solved by Brown's (1969) method (see also Reinholdt (1974)). This method was employed by Vlahinos (1983) in generating results for all frames.

Regardless of the nonlinear solver, the $\bar{k}_i$ values obtained from step 1 are used as initial estimates.

Note that through steps 1–3, one obtains the complete nonlinear response of the system at the low level of the applied loads. Furthermore, note that *low* here means not necessarily small loads, but loads for which the linear analysis yields good estimates for $\bar{k}_i$, to be used as initial points in the nonlinear solver.

(4) The load level is step-increased and the solution procedure of steps 1–3 is repeated. Another possibility is to use small increments in the load and employ the values of $\bar{k}_i$ of the previous load level as initial points for the nonlinear solver. In this case, step 1 is used only once for a truly low level of the applied loads.

(5) At each load level, the stability determinant (see Section 5.2.3) is evaluated. If there is a sign change for two consecutive load levels, then a bifurcation point exists in this load interval. Note that the bifurcation point can be located, with any desired accuracy, by adjusting the size of the load increment. In the case of a limit point, the procedure is the same, but the establishment of the limit point requires special care. First, if the load level is higher than the limit point, the outlined solution steps either yield no solution or the solution does not belong to the primary path (usually this is a physically unacceptable solution for deadweight loading). If this is so, the load level is decreased until an acceptable solution is obtained. At the same time, as the load approaches the limit point the value of the determinant approaches zero. These two observations suffice to locate the limit point. Note that, when a non-primary path solution is obtained, the value of the buckling determinant does not tend to zero.

(6) Step 4 is employed to find postcritical point behaviour. The

establishment of equilibrium points on the postbuckling branch is numerically difficult. The difficulty exists in finding a point, which then can serve as an initial estimate for finding other neighbouring equilibrium points.

(7) The complete behaviour of the frame at each load level, regardless of whether the equilibrium point lies on the primary path or postbuckling branch, has been established if one has evaluated all A_{ij} and $\bar{k}_i$. Equilibrium positions can be presented, graphically, as plots of load or load parameter versus some characteristic displacement or rotation of the frame (of a chosen member at a chosen location).

Before closing this section, it should be noted that the procedure for the analysis of flexibly jointed frames is the same, with one small exception: the load increment must be small and the required spring stiffness at the $(m+1)$th load step is evaluated from the solution of the mth load step (see eqn (5.27)).

5.4 EXAMPLES AND DISCUSSION

The results for several geometries are presented and discussed in this section. The geometries include two-bar frames, which can be subject to limit point instability, as well as portal and multi-bay multi-storey frames, which for linearly elastic behaviour are subject to bifurcational (sway) buckling with stable postbuckling branch. The results are presented both in graphical and tabular form and they include certain important parametric studies. Each geometry is treated separately.

5.4.1 Two-Bar Frames

Consider the two-bar frame shown in Fig. 5.2. For simplicity, the two bars are of equal length and stiffness and the eccentric load is constant-directional (always vertical). Results are presented for both rigid and flexible connections. These results are presented and discussed separately.

5.4.1.1 Rigid Joint Connection

Results are discussed for the case of an immovable pin support at the right-hand end of the horizontal bar. For this geometry there are two important parameters that one must consider in generating results; first is the load eccentricity $\bar{e}$, and second the member slenderness ratio λ.

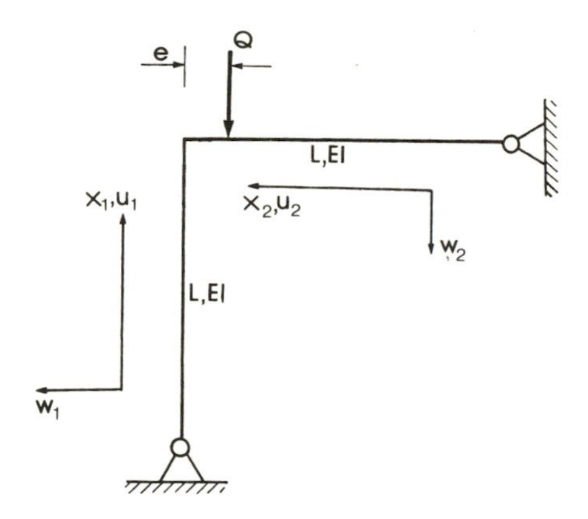

FIG. 5.2. Geometry of a two-bar frame.

Note that for this geometry $L_1 = L_2 = L$ and $\lambda_1 = \lambda_2 = \lambda$.

$$\begin{gathered} -0{\cdot}01 \leq \bar{e} \leq 0{\cdot}01 \\ \lambda = 40, 80, 120, \infty \end{gathered} \tag{5.29}$$

Note that the positive eccentricities correspond to loads applied to the right of the elastic axis of the vertical bar, while the negative ones correspond to the left (load applied, if needed, through a hypothetical rigid overhang).

For this configuration, it is clear from the physical system that, as the load increases (statically) from zero, with or without eccentricity, the response includes bending of both bars and a 'membrane state only' primary path does not exist. Therefore, there cannot exist a bifurcation point from a primary path that is free of bending. The classical (linear theory) approach, for this simple frame, assumes that the vertical bar experiences a contraction without bending in the primary state, while the horizontal bar remains unlôaded (zero eccentricity is assumed). Then a bifurcation exists and a bent state (buckling) is possible at the bifurcation load Q_{cl}, which is the critical load (see Simitses (1976) for analytical details)

$$Q_{cl} = 13{\cdot}89 \frac{EI}{L^2} \tag{5.30}$$

Results are presented graphically in Figs 5.3 and 5.4. In Fig. 5.3, the load parameter λ_c ($= Q/Q_{cl}$) is plotted versus the joint rotation $W_{1,x}(1)$ for several eccentricities and $\lambda = 80$ (slenderness ratio). The response

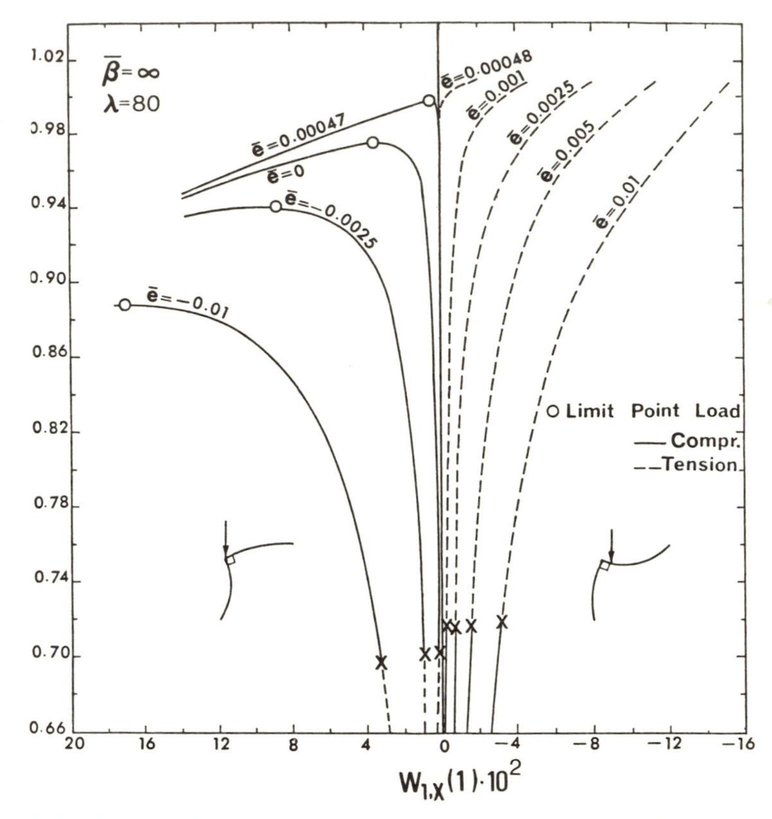

FIG. 5.3. Load–deflection curve; hinged two-bar frame with rigid joint connection.

for different values of λ is similar, and thus no other load–(characteristic) displacement curves are shown. It is seen from Fig. 5.3 that the response, regardless of whether it is stable (to the right) or subject to limit point instability (to the left), seems to be approaching asymptotically a line (almost straight) that makes an angle with the vertical and intersects it at $\lambda_c = 1{\cdot}00$. Moreover, the horizontal bar could be either in tension or in compression, regardless of the character of the response. Not shown in Fig. 5.3 are equilibrium points which belong to curves above the asymptote. These equilibrium paths cannot be attained physically under deadweight loading. In Fig. 5.4, limit point (critical) loads are plotted versus eccentricity for various λ values. Also, the experimental results of Roorda (1965), corresponding

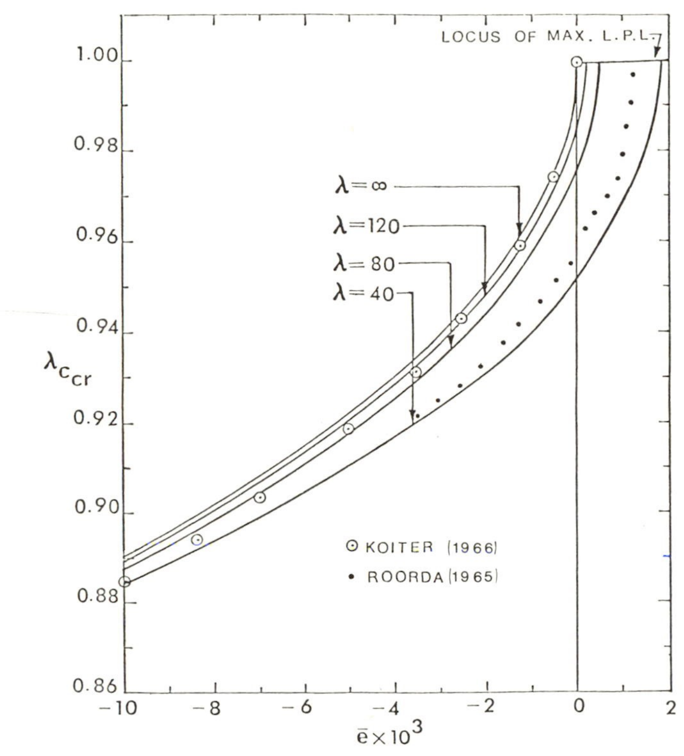

FIG. 5.4. Effects of eccentricity and slenderness ratio on critical loads (two-bar frame).

to $\lambda = 1275$, and the analytical results of Koiter (1966), based on his initial postbuckling theory, are shown for comparison. On the basis of the generated results, a few important observations and conclusions are offered. Depending on the value for the slenderness ratio, there exists a critical eccentricity which divides the response of the frame into two parts; on one side (see Fig. 5.3, on the right) the response is characterised by stable bent equilibrium positions for all loads (within the limitations of the theory), while on the other side the response exhibits limit point instability. The maximum limit point load, for each slenderness ratio value, corresponds to a specific eccentricity value (see Fig. 5.4) and is identical in value to that predicted by linear theory. The results also show that this two-bar frame is sensitive to load eccentricities (for $\bar{e} = -0{\cdot}01$, $\lambda_c \approx 0{\cdot}89$) and it might be sensitive to initial geometric imperfections. Details and more results (depicting the effect of the right hand support (movable along a vertical or a horizontal plane versus immovable) on the response) are found in Kounadis *et al.* (1977) and in Simitses *et al.* (1978).

5.4.1.2 Semi-rigid Joint Connection

Consider the two members connected at the joint through a rotational spring (Fig. 5.2). First, a linear spring is used at the joint and the nondimensionalised spring stiffness, $\bar{\beta}$, is varied from zero (pin connection) to 10^5 (rigid connection). Partial results are presented in graphical and tabular form, but the conclusions and observations are based on all

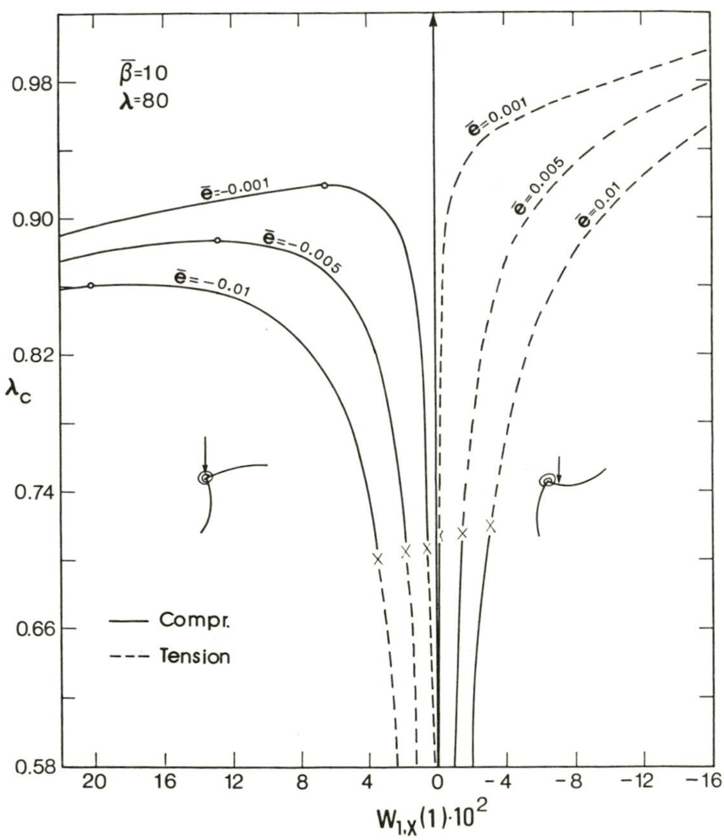

FIG. 5.5. Typical load–deflection curve; hinged two-bar frame with flexible joint connection.

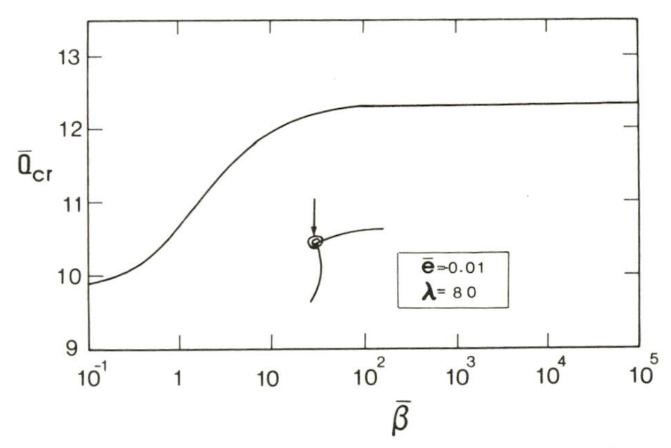

FIG. 5.6. Effect of joint spring constant on the critical load (hinged two-bar frame).

generated data (a wide range of eccentricities and slenderness ratios were used). Figure 5.5 depicts the response of the two-bar frame for $\bar{\beta} = 10$ and $\lambda = 80$. For the sake of economy and brevity, no attempt was made to find the critical eccentricity value for each $\bar{\beta}$ and λ. It is seen from Fig. 5.5 that the response for $\bar{\beta} = 10$ is similar to that for $\bar{\beta} = \infty$ (Fig. 5.3). Figure 5.6 is a plot of $\bar{Q}_{cr}$ (limit point load) versus $\bar{\beta}$ for $\bar{e} = -0{\cdot}01$. For very small values of $\bar{\beta}$, $\bar{Q}_{cr} \approx \pi^2$, which is the critical load of a column pinned at both ends (Euler load), while for very large values it approaches the value corresponding to $\lambda_{c_{cr}} = 0{\cdot}888$ (see Fig. 5.3, $\bar{Q}_{cr} = 0{\cdot}888(13{\cdot}89) = 12{\cdot}34$). Note that for $\bar{e} > -0.01$, similar curves can be obtained. For instance, for $\bar{e} = 0$ the curve would start from the value of π^2 for extremely small values of $\bar{\beta}$, and approach the value of 13·54 for $\bar{\beta} = 10^5 (\approx\infty)$. The influence of the slenderness ratio, for various $\bar{\beta}$ values, on the critical load is shown in Table 5.1.

For the case of realistic flexible connections, three depths of type II connections are considered (see Table 5.2). The required values are taken from DeFalco and Marino (1966) and the bars are assumed to be steel *I*-beams. The value of $\bar{A}$ (nonlinear flexible connection) is varied in accordance with the limitations presented in the mathematical formulation, and its effect, for all three cases, on the limit point loads for $\bar{e} = -0{\cdot}01$ and $\lambda = 100$ is shown in Table 5.3. An important conclusion here is that, for type II connections, the degree of nonlinearity of the rotational spring has negligibly small effect on limit point loads for a fixed eccentricity and bar slenderness ratio. For more details see Simitses and Vlahinos (1982).

TABLE 5.1
INFLUENCE OF SLENDERNESS RATIO ON THE CRITICAL LOADS OF THE TWO BAR FRAME
($\bar{e} = -0{\cdot}01$)

$\bar{\beta}$ \ λ	$\bar{Q}_{cr}$ 80	120	1000
0·1	9·902 8	9·904 5	9·905 1
1·0	10·681 7	10·686 8	10·690 8
10	11·950 4	11·963 8	11·974 4
100	12·293 1	12·308 9	12·321 6
∞	12·337 6	12·353 8	12·366 7

TABLE 5.2
DEPTH AND STIFFNESS OF FLEXIBLE CONNECTIONS (TYPE II)

Geometry	*Depth (in)*	$Z\times 10^5$ *(rad/kip-in)*	${}_0\beta\times 10^{-8}$ *(lb-in/rad)*	A_i *(in²)*	I_i *(in⁴)*	${}_0\bar{\beta}$	$\bar{A}$ *range*
1	8	0·046 0	21·739	6·71	64·20	361·17	$\bar{A}\leq(7{\cdot}5)\times 10^{10}$
2	18	0·015 0	66·667	20·46	917·70	167·79	$\bar{A}\leq(6)\times 10^{9}$
3	36	0·005 4	185·185	39·80	7 833·65	114·36	$\bar{A}\leq(2{\cdot}1)\times 10^{9}$

TABLE 5.3
EFFECT OF $\bar{A}$ (NON-LINEAR FLEXIBLE CONNECTION) ON THE CRITICAL LOADS ($e = -0{\cdot}01$, $\lambda = 100$)

Geometry 1 ${}_0\bar{\beta} = 361{\cdot}17$		*Geometry 2* ${}_0\bar{\beta} = 167{\cdot}79$		*Geometry 3* ${}_0\bar{\beta} = 114{\cdot}36$	
$\bar{A}$	$\bar{Q}_{cr}$	$\bar{A}$	$\bar{Q}_{cr}$	$\bar{A}$	$\bar{Q}_{cr}$
0	12·752 9	0	12·763 1	0	12·721 6
$1{\cdot}0\times 10^6$	12·752 9	$1{\cdot}0\times 10^5$	12·736 1	$1{\cdot}0\times 10^3$	12·721 6
$1{\cdot}0\times 10^7$	12·752 7	$5{\cdot}0\times 10^5$	12·735 9	$1{\cdot}0\times 10^4$	12·721 6
$5{\cdot}0\times 10^7$	12·751 5	$1{\cdot}0\times 10^6$	12·735 7	$1{\cdot}0\times 10^5$	12·721 4
$1{\cdot}0\times 10^8$	12·749 4	$1{\cdot}0\times 10^7$	12·729 8	$1{\cdot}0\times 10^6$	12·719 3
$1{\cdot}0\times 10^9$	12·745 6	$1{\cdot}0\times 10^8$	12·720 6	$1{\cdot}0\times 10^7$	12·699 1

5.4.2 Portal Frames

Consider the portal frame shown in Fig. 5.7. The loading consists of both eccentric concentrated loads near the joints and of a uniformly distributed load on bar 3.

When vertical concentrated loads are applied at joints 3 and 4 without eccentricity, and the geometry is symmetric ($EI_1 = EI_2 = EI$, $L_1 = L_2 = L$, $\beta_1 = \beta_2 = \beta$ but $\beta = 0$ or ∞), a primary state exists and beam–column theory can be employed to find critical loads for sway buckling, or for symmetric buckling (sideways prevented) and for antisymmetric buckling. Such analyses can be found in texts (see Bleich (1952) and Simitses (1976)).

For example, if the horizontal bar has the same structural geometry as the other two members ($EI_3 = EI$ and $L_3 = L$), then the critical load for sway buckling (referred to herein as classical) is given by

$$\text{simply supported, } (\beta = 0) \quad Q_{cl} = 1{\cdot}82 \frac{EI}{L^2} \tag{5.31}$$

$$\text{clamped } (\beta \rightarrow \infty) \quad Q_{cl} = 7{\cdot}38 \frac{EI}{L^2} \tag{5.32}$$

Results for loading that induces primary bending, and parametric studies associated with the effect of various structural parameters on the frame response, are presented below for rigidly connected portal frames. Moreover, some results corresponding to semi-rigidly connected portal frames are also presented.

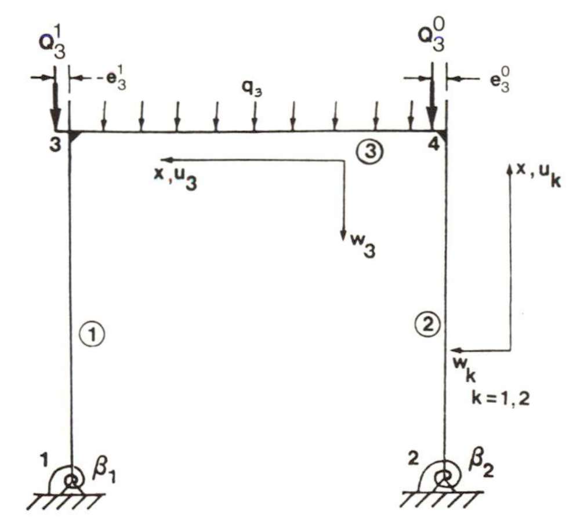

FIG. 5.7. Portal frames: geometry and loading.

5.4.2.1 Rigid Joint Connection
Partial results are presented both in graphical and in tabular form, but the conclusions are based on all available results.

Figures 5.8 and 5.9 deal with the effect of load eccentricity on the response characteristics of a square (structurally; $EI_i = EI$, $L_i = L$) symmetric ($\bar{\beta}_1 = \bar{\beta}_2 = 0$), rigid-jointed frame. Figure 5.8 shows primary path and postbuckling equilibrium positions for two symmetric eccentricities ($\bar{e}_3^1 = \bar{e}_3^0 = \bar{e}$). The value of the slenderness ratio ($\lambda_i = \lambda$) is taken as 1000, but the effect of slenderness ratio on the nondimensionalised response characteristics is negligibly small. The rotation of bar 1 at joint 3 is chosen as the characteristic displacement for characterising equilibrium states in this Figure. As seen from Fig. 5.8, bar 3 is in compression in the postbuckled branches and initially in the primary paths. As the eccentricity increases the sway buckling load decreases, but only slightly. This observation is in agreement with Chwalla's (1938) (see also Bleich (1952)) result, which was that the critical load when the eccentricity is one-third ($\bar{e} = 0{\cdot}333$) is equal to

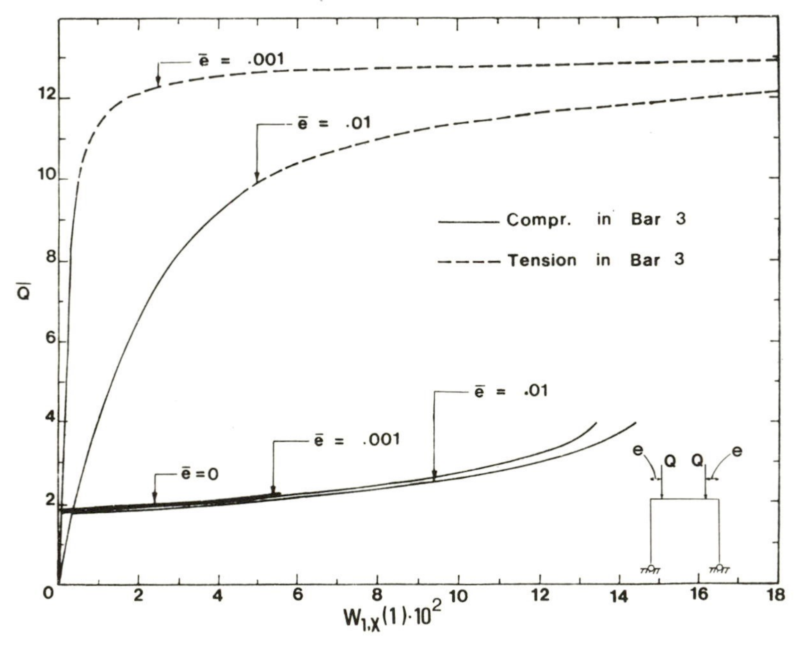

FIG. 5.8. Symmetrically and eccentrically loaded symmetric hinged portal frames ($S_i = R_i = 1$).

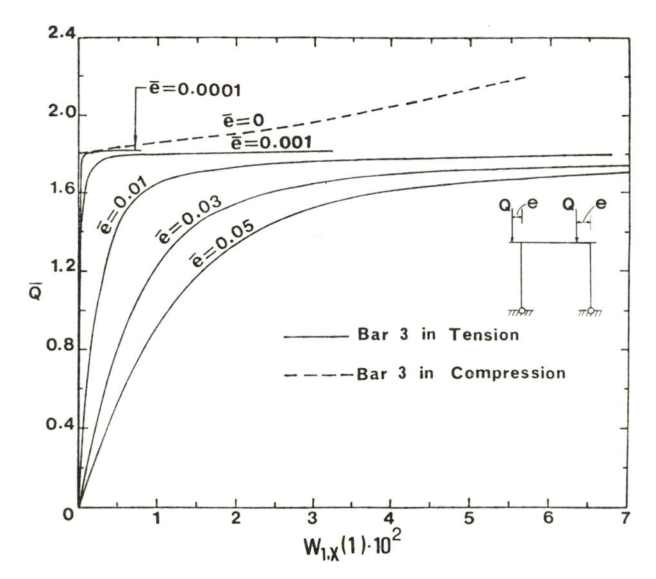

FIG. 5.9. Asymmetrically and eccentrically loaded symmetric hinged portal frames ($S_i = R_i = 1$).

$1{\cdot}78EI/L^2$. It is also observed that the primary path curves approach asymptotically the value of $\bar{Q}_{cr}$ corresponding to symmetric buckling of the portal frame (see Simitses (1976), Chapter 4, eqn (66)). This value, as computed from the said reference, is equal to $12{\cdot}91EI/L^2$. Figure 5.9 shows similar results but with antisymmetric eccentricity ($-\bar{e}_3^1 = \bar{e}_3^0 = \bar{e}$). Clearly for this case ($\bar{e} \neq 0$), there is a stable response that includes bending from the onset of loading. Moreover, this response approaches asymptotically a horizontal line corresponding to $\bar{Q} = \bar{Q}_{cl}$ (eqn (5.30)) and not the postbuckling branch ($\bar{e} = 0$). Furthermore, for asymmetric eccentricity bar 3 is in tension.

Table 5.4 presents sway buckling loads of a symmetric simply supported portal frame loaded by a uniformly distributed load on bar 3, for a wide range of horizontal bar (3) geometries. The value of $\lambda_1 = \lambda_2$ is taken to be 1000 and the value of λ_3 varies according to the changes in I_3 and L_3 by keeping the cross-sectional area, A_3, constant. This results in $50 \leqslant \lambda_3 \leqslant 4242$. Note that q^* is given in Table 5.4, instead of $\bar{q}$. This is done because L_3 is a variable. Moreover, if one is interested in comparing total load, q^* must be multiplied by L_3/L_1. Thus, the first row becomes $3{\cdot}52$ ($L_3/L_1 = 0{\cdot}5$), $2{\cdot}77$, $2{\cdot}27$, $1{\cdot}92$, $1{\cdot}65$

TABLE 5.4

EFFECT OF HORIZONTAL BAR GEOMETRY ON CRITICAL LOADS (HINGED PORTAL FRAMES)

$L_1 = L_2, EI_1 = EI_2$

EI_3/EI_1 \ L_3/L_1		0·5	1·0	1·5	2·0	2·5	3·0
0·5	q^*_{cr}	7·035	2·772	1·518	0·960 0	0·659 8	0·480 93
	$\bar{k}_1$	1·326 144	1·177 312	1·066 970	0·979 798	0·908 143	0·849 350
	$\bar{k}_3$	0·204 301	0·586 181	1·040 636	1·548 835	2·128 032	2·922 292
1·0	q^*_{cr}	8·142	3·522	2·075	1·394	1·011	0·776 9
	$\bar{k}_1$	1·426 682	1·327 027	1·247 465	1·180 678	1·123 931	1·079 532
	$\bar{k}_3$	0·128 840	0·410 684	0·778 291	1·212 997	1·725 474	2·412 621
2·0	q^*_{cr}	8·879	4·075	2·523	1·772	1·338	1·064
	$\bar{k}_1$	1·489 896	1·427 337	1·375 482	1·331 290	1·293 368	1·263 309
	$\bar{k}_3$	0·074 499	0·258 365	0·517 961	0·840 184	1·227 140	1·709 590
3·0	q^*_{cr}	9·166	4·309	2·721	1·945	1·491	1·200
	$\bar{k}_1$	1·513 758	1·467 748	1·428 456	1·394 528	1·365 357	1·341 829
	$\bar{k}_3$	0·052 459	0·189 001	0·389 313	0·643 696	0·951 338	1·324 863
10·0	q^*_{cr}	9·640	4·714	3·079	2·266	1·782	1·462
	$\bar{k}_1$	1·552 238	1·535 271	1·519 604	1·505 210	1·492 379	1·481 047
	$\bar{k}_3$	0·017 124	0·066 005	0·143 481	0·247 198	0·375 752	0·528 868
100·0	q^*_{cr}	9·865	4·909	3·266	2·444	1·951	1·622
	$\bar{k}_1$	1·570 430	1·566 634	1·564 618	1·563 342	1·561 408	1·559 621
	$\bar{k}_3$	0·002 500	0·007 062	0·015 835	0·028 044	0·043 648	0·062 619

and finally 1·44. Note also that the last row becomes 4·93, 4·91, 4·90, 4·89, 4·88 and 4·87, or all of them approximately equal to 2 ($\pi^2/4$). This load is the buckling load of the two vertical bars, which are pinned at the bottom and clamped at the top to a very rigid bar that can move horizontally. Finally, $\bar{k}_1$ and $\bar{k}_3$ are measures of the axial compressive force in the vertical bars ($\bar{k}_1 = \bar{k}_2$) and the horizontal bar, respectively.

The final result is shown in Fig. 5.10. This Figure shows the effect of small variations in the length of bar 2 on the response characteristics of a uniformly loaded frame. Clearly, the change in L_2 provides a geometric imperfection and the response, accordingly, approaches asymptotically the 'perfect geometry' response. The same can be said if an imperfection in bending stiffness exists such that the resulting

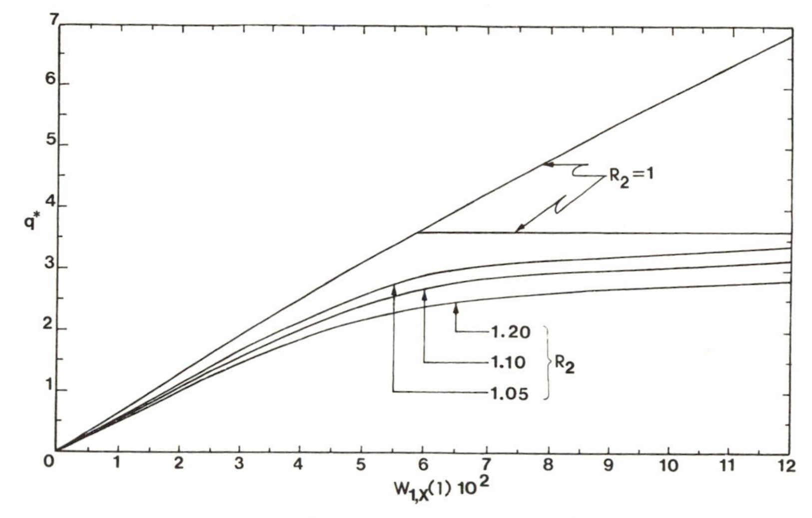

FIG. 5.10. Effect of variable vertical column length on the portal frame response ($R_3 = 1$; $S_i = 1$).

geometry becomes asymmetric. Details and more results can be found in Simitses *et al.* (1981) and Simitses and Giri (1982).

5.4.2.2 Semi-rigid Joint Connection

As in the case of the two-bar frame (Section 5.4.1.2), the horizontal bar is connected to the vertical bars through rotational springs. First, a linear spring is used, and its stiffness, $\bar{\beta}$, is varied from zero (10^{-1}) to infinity (10^5). Results are presented in tabular and graphical form for symmetric eccentric loading. Table 5.5 shows the effect of slenderness ratio for a square symmetric portal frame on the sway buckling load

TABLE 5.5

EFFECT OF SLENDERNESS RATIO λ ON SWAY-BUCKLING LOAD (SYMMETRIC LOADS, $\bar{e} = 0{\cdot}001$)

$\bar{\beta}$ \ λ	*40*	*100*	*1000*
1	0·659	0·659	0·660
5	1·355	1·355	1·360
100	1·781	1·787	1·790
1000	1·807	1·813	1·814

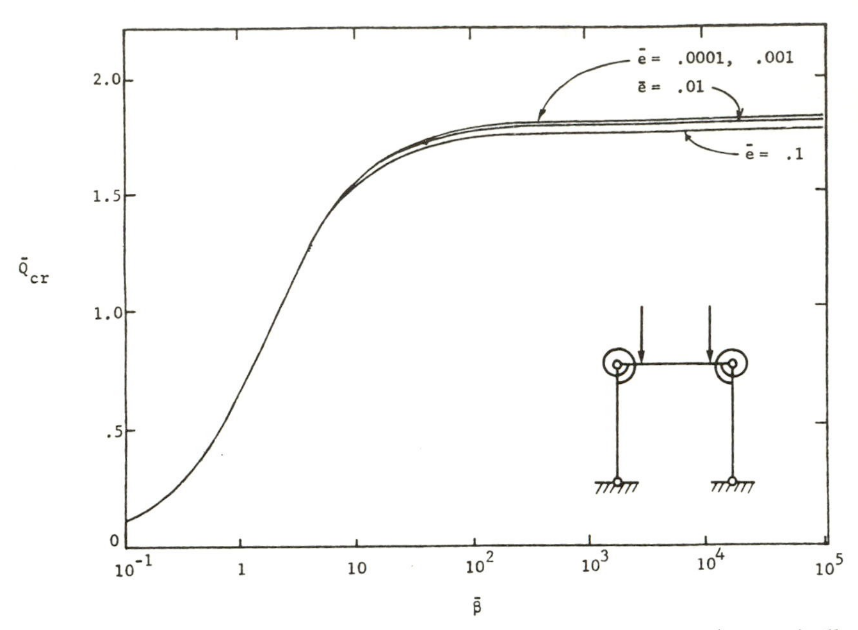

FIG. 5.11. Effect of joint rotational stiffness on critical loads (eccentrically loaded symmetric portal frame).

($\bar{e}=0{\cdot}001$) for various values of rotational spring stiffness (same at both joints). It is seen from Table 5.5 that this effect is negligibly small, as is the case for rigid connections. Figure 5.11 shows the effect of spring stiffness on the sway buckling load for various load eccentricities. For very small $\bar{\beta}$ values, the frame becomes unstable at very low load levels. Note that for $\bar{\beta}=0$ the frame becomes a mechanism. As the rotational stiffness increases, the critical load approaches that of a rigid-jointed portal frame ($\bar{Q}_{cr}=1{\cdot}82EI/L^2$).

Next, results are presented for flexibly connected portal frames using the same type II connections as for the two-bar frame (see Table 5.2). For the portal frame also it is concluded that the degree of nonlinearity of the rotational springs has a negligibly small effect on sway-buckling loads, for each specified geometry (see Table 5.6). From these and other studies (Vlahinos, 1983), it is concluded that nonlinearity in the rotational spring stiffness (variations in $\bar{A}$) has a negligibly small effect on the response characteristics of portal frames. In all generated results, it is required that the slope to the moment–relative rotation curve, for the flexible connection, be positive. This

TABLE 5.6
EFFECT OF $\bar{A}$ (NONLINEAR FLEXIBLE CONNECTIONS) ON CRITICAL LOADS $\bar{Q}_{cr}$ (SYMMETRIC CASE; $\bar{e} = 0{\cdot}01$)

Geometry 1 ($_0\bar{\beta} = 361{\cdot}17$)		*Geometry 2* ($_0\bar{\beta} = 67{\cdot}79$)		*Geometry 3* ($_0\bar{\beta} = 114{\cdot}36$)	
$\bar{A}$	$\bar{Q}_{cr}$	$\bar{A}$	$\bar{Q}_{cr}$	$\bar{A}$	$\bar{Q}_{cr}$
0	1·807	0	1·798	0	1·790
1×10^5	1·807	1×10^5	1·798	1×10^5	1·790
1×10^8	1·807	1×10^8	1·798	1×10^8	1·790
3×10^{10}	1·807	1×10^9	1·798	1×10^9	1·788
5×10^{10}	1·806	3×10^9	1·797	$1{\cdot}75\times10^9$	1·785
7×10^{10}	1·803	5×10^9	1·795	2×10^9	1·782
$7{\cdot}5\times10^{10}$	1·801	6×10^9	1·793	$2{\cdot}1\times10^9$	1·781

requirement is not only reasonable, it is also necessary for a good and efficient connection. Because of the above observations and those associated with the two-bar frame (Section 5.4.1.2), no further results are generated for flexibly connected frames.

5.4.3 Multi-bay Multi-storey Rigid–Jointed Frames

Several results are presented and discussed here.

First, results are presented for symmetric two-bay frames loaded transversely by uniformly distributed loads (Table 5.7). In this table, the length of the horizontal bars is varied ($L_4 = L_5 = L_h$; $L_1 = L_2 = L_3 = L_v$) as well as the stiffness. Here also, as in the case of portal frames, the slenderness ratio for the vertical bars is taken as 1000 ($\lambda_1 = \lambda_2 = \lambda_3 = 1000$) and the value of λ_h ($= \lambda_4 = \lambda_5$) is varied accordingly, as I_h and L_h vary, but the cross-sectional area is kept constant. The critical loads q^*_{cr} represent sway-buckling loads. The total load for the two-bay frame is obtained by multiplying q^* by $2L_h/L_v$. The factor of *two* is needed because of the two bays. In comparing the results of this table with those for the portal frame (Table 5.4) one observes that, by adding one bay (two bars; bars 5 and 3), the total sway-buckling load is increased by 50% or more, depending on the two ratios. The increase is larger with larger values for L_h/L_v and smaller values for EI_h/EI_v. The values for $\bar{k}_1$ ($\bar{k}_1 = \bar{k}_3$), $\bar{k}_2$ and $\bar{k}_4$ ($\bar{k}_4 = \bar{k}_5$) are measures of the axial loads (compressive for this case) in the five bars. Because of the distribution, the middle vertical bar carries more load than the other two, as expected. In spite of this, as the bending stiffness of the

TABLE 5.7

EFFECT OF HORIZONTAL BAR GEOMETRY ON CRITICAL LOADS (HINGED SYMMETRIC ONE-STOREY TWO-BAY FRAMES)

EI_h/EI_v \ L_h/L_v		0·5	1	2	3
		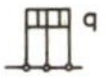	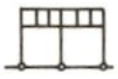		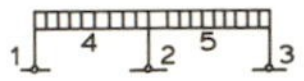
	q^*_{cr}	5·474	2·243	0·822	0·425
0·5	$\bar{k}_1$	1·079 615	1·001 068	0·872 701	0·768 667
	$\bar{k}_2$	1·773 156	1·575 093	1·328 265	1·169 728
	$\bar{k}_4$	0·152 456	0·467 452	1·302 599	2·342 564
	q^*_{cr}	6·190	2·739	1·124	0·635
1	$\bar{k}_1$	1·121 867	1·072 490	0·999 675	0·926 334
	$\bar{k}_2$	1·916 465	1·774 246	1·580 063	1·447 954
	$\bar{k}_4$	0·090 564	0·304 997	0·944 516	1·787 473
	q^*_{cr}	6·887	3·258	1·487	0·921
3	$\bar{k}_1$	1·155 458	1·138 553	1·108 115	1·079 077
	$\bar{k}_2$	2·049 744	1·980 715	1·868 985	1·787 660
	$\bar{k}_4$	0·034 894	0·129 310	0·455 144	0·925 647
	q^*_{cr}	7·221	3·530	1·703	1·101
10	$\bar{k}_1$	1·167 638	1·162 379	1·151 948	1·142 422
	$\bar{k}_2$	2·120 247	2·087 248	2·039 065	1·998 682
	$\bar{k}_4$	0·011 611	0·043 027	0·164 391	0·353 992

$L_h = L_4 = L_5$, $L_v = L_1 = L_2 = L_3$; $q^*_{cr} = q^*_{4cr} = q^*_{5cr}$; $EI_h = EI_4 = EI_5$, $EI_v = EI_1 = EI_2 = EI_3$.

horizontal bars approaches infinity the total sway-buckling load approaches $3(\pi^2/4)$. Note that for the portal frame the total load is $2(\pi^2/4)$. Thus, for this particular case ($EI_h \rightarrow \infty$) the increase in buckling load from a single bay to a two-bay frame is 50%, regardless of the ratio of L_h/L_v.

Limited results are also presented for a single-bay multi-storey frame and a two-bay two-storey frame. These results are generated only for special geometries. All lengths and all stiffnesses are taken to be equal and the loading is a uniformly distributed load of the same magnitude on every horizontal bar. The boundaries are simple supports and the bar slenderness ratio is taken to be 1000. Note, that for portal frames the effect of slenderness ratio on the nondimensionalised response is found to be negligibly small. This is found to be also true for two-bay one-storey, and one-bay multi-storey frames that were checked randomly. The value of λ_i was changed for a few geometries

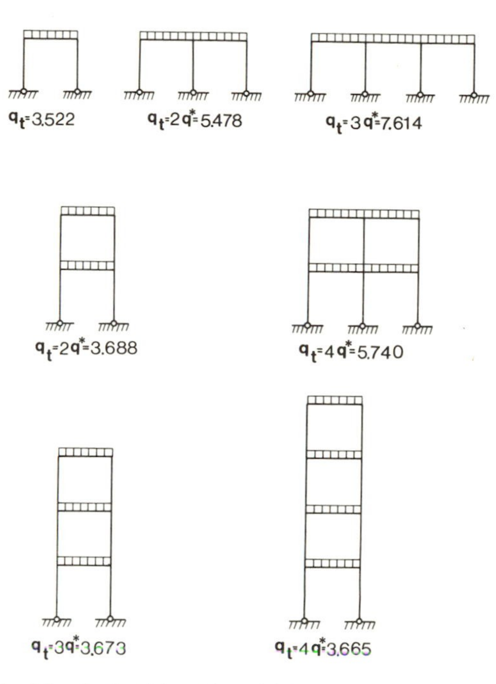

FIG. 5.12. Critical loads for hinged multi-bay, multi-storey frames ($R_i = S_i = 1$).

and this change did not affect the response appreciably. The results for the additional geometries are presented schematically in Fig. 5.12, by giving the total sway-buckling load next to a sketch of the frame. From this figure it is clearly seen that the sway-buckling load is increased appreciably by adding bays but the change is insignificant when storeys are added.

Another important result is related to the following study. A two-storey one-bay frame, with $L_i = L$ and $EI_i = EI$ (for all i), is loaded with uniformly distributed loads on the horizontal bars. The uniform loading is distributed in various amounts over the two horizontal bars. It is found that the total sway-buckling load does not change appreciably with this variation. When only the top horizontal bar is loaded (top 100%, bottom 0%) the total sway-buckling load is 3·677. When the top and bottom are loaded by the same amount, the total sway-buckling load is 3·688 (see Fig. 5.11). Finally, when the top is loaded by an amount which is much smaller than the bottom (top 5%, bottom 95%), the total sway-buckling load is 3·696.

When designing two-bay (or multi-bay) frames to carry uniformly distributed loads, inside columns must carry more load than outside columns. Because of this, inside columns are usually made stiffer. One possible design is to make the inside column(s) twice as stiff (in bending) as the outside one(s). Sway-buckling results for such a two-bay geometry are presented in Table 5.8. The lengths of all five members are the same, but the bending stiffness of the horizontal bars is varied. Axial load coefficients for all five bars are also reported in Table 5.8 ($\bar{k}_3 = \bar{k}_1$ and $\bar{k}_5 = \bar{k}_4$). Moreover, the total (nondimensionalised) sway-buckling load is given for each case. It is seen from Table 5.8 that as the stiffness of the horizontal bars increases, the total load increases. Moreover, a comparison with the results of Table 5.7,

TABLE 5.8
EFFECT OF HORIZONTAL BAR STIFFNESS ON CRITICAL LOADS FOR HINGED ONE-STOREY TWO-BAY FRAMES (WITH MIDDLE COLUMN STIFFNESS DOUBLED)

EI_h/EI_v	1	2	3	10
q^*_{cr}	3·599900	4·164400	4·391500	4·655000
$\bar{k}_1$	1·235737	1·299518	1·320376	1·334522
$\bar{k}_2$	1·439725	1·573468	1·627115	1·695136
$\bar{k}_4$	0·346890	0·207330	0·147837	0·048834
$q_t = 2q^*_{cr}$	7·199800	8·329880	8·783000	9·310000

corresponding to $L_h/L_v = 1$, reveals that by doubling the bending stiffness of the middle column the total sway-buckling load is increased by approximately 33%, regardless of the relative stiffness of the horizontal bars. Another important observation is that the ratio of axial forces (inside to outside, P_2/P_1; $P_i = \bar{k}_i^2 EI_i/L_i^2$) is not affected appreciably by the doubling of the bending stiffness of the middle column. This ratio varies (increases) with increasing bending stiffness of the horizontal bars.

All of the above observations indicate that there exists an optimum distribution of bending stiffness, in multi-bay multi-storey orthogonal frames which are subject to sway buckling, for maximising their load carrying capacity.

5.5 CONCLUDING REMARKS

From the several studies performed on elastic orthogonal plane frameworks, some of which are reported herein, one may draw the following general conclusions.

(1) The effect of flexible joint connections (bolted, riveted and welded connections are flexible rather than rigid) on the frame response characteristics is negligibly small. Hence, assuming rigid connections in analysing elastic plane frameworks will lead to accurate predictions.

(2) Eccentrically loaded two-bar frames lose stability through the existence of a limit point and do not experience bifurcational buckling. For these frames, the slenderness ratio of the bars has a small but finite effect on the critical load. Moreover, depending on the value for the slenderness ratio, there exists a critical eccentricity which divides the response of the frame into two parts. On one side the response is characterised by stable equilibrium positions and on the other side it exhibits limit point instability (within the limitations of the theory, $w_{i,x}^2 \ll 1$).

(3) Unbraced multi-bay multi-storey frames (including portal frames) are subject to bifurcational (sway) buckling with stable post-buckling behaviour. Sway buckling takes place when the frame is structurally symmetric and the load is symmetric. Because of this, the frame is insensitive to geometric imperfections regardless of the type (load eccentricity, variation in geometry: length, stiffness, etc.). In many respects, the behaviour of these frames is similar to the behaviour of columns, especially cantilever columns.

(4) The effect of slenderness ratio on the nondimensionalised response characteristics of plane frameworks (except the two-bar frame) is negligibly small.

(5) Starting with a portal frame, addition of bays increases appreciably the total sway-buckling load, while addition of storeys has a very small effect.

(6) For multi-storey frames, distributing the load in various amounts among the different floors does not alter appreciably the total sway-buckling load. In all cases, the first storey vertical bars (columns) carry the total load.

REFERENCES

Ackroyd, M. H. (1979) Nonlinear inelastic stability of flexibly-connected plane steel frames. PhD Thesis, University of Colorado.

Argyris, J. H. and Dunne, P. C. (1975) On the application of the natural mode technique to small strain large displacement problems. *Proceedings of World Congress on Finite Element Methods in Structural Mechanics*, Bournemouth, UK.

Baker, J. F., Horne, M. R. and Roderick, J. W. (1949) The behavior of continuous stanchions. *Proc. Roy. Soc. A.*, **198,** 493.

Batho, C. and Rowan, H. C. (1934) Investigation on beam and stanchion connections. Second report of the Steel Structure Research Committee, HMSO, London.

Bellman, R. (1969) *Perturbation Techniques in Mathematics, Physics and Engineering*, Holt, Rinehart and Winston, New York.

Bleich, F. (1919) Die Knickfestigkeit elasticher Stabverbindungen. *Der Eisenbau*, **10,** 27.

Bleich, F. (1952) *Buckling Strength of Metal Structures*, McGraw-Hill, New York.

Britvec, S. J. (1973) *The Stability of Elastic Systems*, Pergamon Press, New York.

Britvec, S. J. and Chilver, H. A. (1963) Elastic buckling of rigidly jointed braced frames. *J. Eng. Mech. Div., ASCE*, **89,** EM6, 217.

Brown, K. M. (1969) A quadratically convergent newton-like method based upon gaussian elimination. *SIAM Journal on Numerical Analysis*, **6**(4)**,** 560.

Chilver, A. H. (1956) Buckling of a simple portal frame. *Journal of Mechanics and Physics of Solids*, **5,** 18.

Chwalla, E. (1938) Die Stabilitat lotrecht belasteter Rechteckrahmen. *Der Bauingenieur*, **19,** 69.

DeFalco, F. and Marino, F. J. (1966) Column stability in type 2 construction. *AISC Engineering Journal*, **3**(2)**,** 67.

Frye, J. M. and Morris, G. A. (1975) Analysis of flexibly connected steel frames. *Canadian Journal of Civil Engng*, **2,** 280.

Goldberg, J. E. (1960) Buckling of one-storey frames and buildings. *Journal of the Structural Division, ASCE*, **86,** ST10, 53.

Halldorsson, O. P. and Wang, G. K. (1968) Stability analysis of frameworks by matrix methods. *Journal of the Structural Division, ASCE*, **94,** ST7, 1745.

Horne, M. R. (1962) The effect of finite deformations in elastic stability of plane frames. *Proc. Roy. Soc. A.*, **266,** 47.

Horne, M. R. and Merchant, W. (1965) *The Stability of Frames*, Pergamon Press, London.

Huddleston, J. V. (1967) Nonlinear buckling and snap-over of a two-member frame. *International Journal of Solids and Structures*, **3**(6)**,** 1023.

Johnston, B. G. (1976) *Guide to Stability Design Criteria for Metal Compression Members*, (3rd edition) John Wiley and Sons, Inc., New York.

Koiter, W. T. (1945) On the stability of elastic equilibrium. (English Translation NASA TT-2 F-10833), Thesis, Polytechnic Institute, Delft, The Netherlands.

Koiter, W. T. (1966) Postbuckling analysis of a simple two-bar frame. *Recent Progress in Applied Mechanics*, Almquist and Wiksell, Stockholm, p. 337.

Kounadis, A. N., Giri, J. and Simitses, G. J. (1977) Nonlinear stability analysis of an eccentrically loaded two-bar frame. *Journal of Applied Mechanics*, **44**(4), 701.

LEE, S.-L., MANUEL, F. S. and ROSSOW, E. C. (1968) Large deflections and stability of elastic frames. *Journal of the Engineering Mechanics Division, ASCE*, **94,** EM2, 521.

LIVESLEY, R. K. (1956) Application of electronic digital computers to some problems of structural analysis, *Structural Engineer*, **34**(6), 161.

LOTHERS, J. E. (1960) *Advanced Design in Structural Steel*, Prentice-Hall, Inc., New Jersey.

LUI, M. E. and CHEN, W. F. (1983) End restraint and column design using LRFD. *Engineering Journal, AISC*, First Quarter, 29.

MASUR, E. F., CHANG, I. C. and DONNELL, L. H. (1961) Stability of frames in the presence of primary bending moments. *Journal of the Engineering Mechanics Division, ASCE*, **87,** EM4, 19.

MCMINN, S. J. (1962) *Matrices for Structural Analysis*, John Wiley and Sons, Inc., New York.

MERCHANT, W. (1954) The failure load of rigid-jointed frameworks as influenced by stability. *Structural Engineer*, **32**(7), 185.

MERCHANT, W. (1955) Critical loads of tall building frames. *Structural Engineer*, **33**(3), 84.

MONCARZ, P. D. and GERSTLE, K. H. (1981) Steel frames with nonlinear connections. *Journal of the Structural Division, ASCE*, **107,** ST8, 1427.

MÜLLER-BRESLAU, H. (1908) *Die Graphische Statik der Bau-Konstruktionen* (Vol. II), Kroner, Berlin.

NELDER, J. A. and MEAD, R. (1964) A simplex method of function minimization. *Computer Journal*, **7,** 308.

OBRECHT, H., WUNDERLICH, W. and SCHRODTER, V. (1982) Large deflections and stability of thin-walled beam structures. *Stability in the Mechanics of Continua* (Ed. by F. H. Schroeder), Springer-Verlag, Berlin, 165.

OLESEN, J. F. and BYSKOV, E. (1982) Accurate determination of asymptotic postbuckling stresses by the finite element method. *Computers and Structures*, **15**(2), 157.

PRAGER, W. (1936) Elastic stability of plane frameworks. *Journal Aeronaut. Sci.*, **3,** 388.

QASHU, R. K. and DADEPPO, D. A. (1983) Large deflection and stability of rigid frames. *Journal of the Engineering Mechanics Division, ASCE*, **109,** EM3, 765.

RATHBUN, J. E. (1936) Elastic properties of riveted connections. *Transactions of American Society of Civil Engineers*, **101,** 524.

REINHOLDT, W. C. (1974) *Methods for Solving Nonlinear Equations*, Heyden, London.

ROORDA, J. (1965) Stability of structures with small imperfections. *Journal of the Engineering Mechanics Division, ASCE*, **91,** EM1, 87.

SAAFAN, S. A. (1963) Nonlinear behavior of structural plane frames. *Journal of the Structural Division, ASCE*, **89,** ST4, 557.

SEWELL, M. J. (1965) The static perturbation technique in buckling problems. *J. Mech. Phys. Solids*, **13,** 247.

SIMITSES, G. J. (1976) *An Introduction to the Elastic Stability of Structures*, Prentice-Hall, Inc., New Jersey.

SIMITSES, G. J. and GIRI, J. (1982) Non-linear analysis of unbraced frames of variable geometry. *International Journal of Non-Linear Mechanics*, **17**(1), 47.

SIMITSES, G. J., and KOUNADIS, A. N. (1978) Buckling of imperfect rigid-jointed frames. *Journal of the Engineering Mechanics Division, ASCE*, **104,** EM3, 569.

SIMITSES, G. J. and VLAHINOS, A. S. (1982) Stability analysis of a semi-rigidly connected simple frame. *Journal of Constructional Steel Research*, **2**(3), 29.

SIMITSES, G. J., KOUNADIS, A. N. and GIRI, J. (1977) Nonlinear buckling analysis of imperfection sensitive simple frames. *Proceedings, International Colloquium on Structural Stability under Static and Dynamic Loads*, Washington, DC, May 17–19, 1977. Published by ASCE, 1978, p. 158.

SIMITSES, G. J., GIRI, J. and KOUNADIS, A. N. (1981) Nonlinear analysis of portal frames. *International Journal of Numerical Methods in Engineering*, **17,** 123.

SOUROCHNIKOFF, B. (1949) Wind stresses in semi-rigid connections of steel framework. *Transactions of American Society of Civil Engineers*, **114,** 382.

SWITZKY, H. and WANG, C. (1969) Design and analysis of frames for stability. *Journal of the Structural Division, ASCE*, **95,** ST4, 695.

VLAHINOS, A. S. (1983) Nonlinear stability analysis of elastic frames, PhD Thesis, Georgia Institute of Technology, Atlanta.

WEAVER, W., JR. and GERE, J. M. (1980) *Matrix Analysis of Framed Structures*, D. Van Nostrand Co., London, 417.

WILLIAMS, F. W. (1964) An approach to the non-linear behavior of the members of a rigid jointed plane framework with finite deflections. *Quarterly Journal of Mechanics and Applied Mathematics*, **17,** 451.

ZIMMERMANN, H. (1909) Die Knickfestigkeit des geraden Stabes mit mehreren Feldern. *Sitzungsberichte der preussischen Akademie der Wissenschaften*, 180.

ZIMMERMANN, H. (1910) *Die Knickfestigkeit der Druckgurte offener Brucken*, W. Ernst und Sohn, Berlin.

ZIMMERMANN, H. (1925) *Die Knickfestigkeit der Stabverbindungen*, W. Ernst und Sohn, Berlin.

ZWEIG, A. and KAHN, A. (1968) Buckling analysis of one-storey frames. *Journal of the Structural Division, ASCE*, **94,** ST9, 2107.

Chapter 6

BEAM-TO-COLUMN MOMENT-RESISTING CONNECTIONS

W. F. CHEN and E. M. LUI
School of Civil Engineering, Purdue University, West Lafayette, Indiana, USA

SUMMARY

In a moment-resisting frame, the lateral stiffness is provided by the beams and columns connected together by moment-resisting connections. Experimental investigations on the behaviour of these connections under static and seismic conditions are presented. Current design practices are reviewed and improvements are suggested in the light of the experimental data.

NOTATION

A_f Area of one flange of beam
A_w Area of beam web
b_c Width of the column flange
C_1 Ratio of beam flange yield stress to column web yield stress
d_b Depth of beam
d_{bl} Depth of the left beam
d_{br} Depth of the right beam
d_c Depth of the column
d'_c Depth of column web clear of fillets
F_{yb} Beam flange yield stress
F_{yc} Column web yield stress
G Shear modulus
k_c Distance from outer face of flange to web-toe of fillet

L	Length of beam
M	Moment
M_l	Moment of the left beam
M_p	Plastic moment
M_{pl}	Plastic moment of the left beam
M_{pr}	Plastic moment of the right beam
P	Axial load
P_a	Axial load from top column
P_b	Axial load from bottom column
P_y	Yield load
T	Beam flange force
t_b	Thickness of the beam flange
t_c	Thickness of the column flange
V	Shear force
V_a	Shear force from upper column
V_b	Shear force from lower column
V_{col}	Average column shear
V_{mp}	Shear yielding force at critical section
V_p	Shear yielding force
w_c	Thickness of the column web
Δ_b	Beam deflection due to bending deformation
Δ_s	Beam deflection due to shear deformation
Δ_ϕ	Beam deflection due to rigid body rotation of the beam induced by the connection panel zone deformation
σ_y	Yield stress
τ_y	Shear yield stress

6.1 INTRODUCTION

Two of the most commonly used steel framing systems to resist lateral forces are the moment-resisting frames and braced frames. In a moment-resisting frame, the lateral stiffness is provided by the flexural rigidities of the beams and columns, which are connected by moment-resisting connections. These beam-to-column moment connections must have sufficient strength and ductility to ensure satisfactory performance of such frames.

Braced frames are those in which the lateral forces are resisted by braces. According to how the braces are connected to the frames, two types of braced frames can be identified:

(1) Concentrically braced frames—frames in which the centrelines of the braces, beams and columns all coincide at the joints. Lateral integrity of this type of frame is provided by the axial stiffness of the braces.
(2) Eccentrically braced frames—frames in which the line of action in a brace is offset from the joint or the intersection point of the centrelines of the beams and columns. In this type of braced frame, the vertical component of the axial force in the brace is transmitted through shear in the beam.

The obvious advantage of a moment-resisting frame over a braced frame is that there is more flexibility in architectural planning. However, the detailing of a moment-resisting connection in such a frame may be expensive. Furthermore, since the lateral stiffness depends on the flexural rigidities of the beams and columns, larger beam and column sections may be needed, resulting in a larger panel zone size. Distress of this panel zone may lead to early buckling and yielding. In addition, fracture has to be avoided.

The objective of this chapter is to investigate the behaviour of these beam-to-column moment-resisting connections. In particular, the behaviour of three types of moment connections—static flange moment connections, static web moment connections and seismic moment connections—will be discussed. Design practice and recent experiments on these types of connections will be presented. Improved methods for analysis and design will also be given.

6.2 DESIGN CRITERIA FOR MOMENT-RESISTING CONNECTIONS

The principal design criteria for these connections are:

(1) sufficient strength,
(2) adequate deformation capacity,
(3) adequate overall stiffness in the working load level, and
(4) economical fabrication and ease of erection.

A moment-resisting connection must have sufficient strength so that the full plastic moment of the adjoining members can be developed. The connection must possess ductility or deformation capacity to allow moment redistribution. Moment redistribution is essential for a mechanism to develop in plastically designed steel frames. Ductility is also important in seismically designed steel frames to absorb and dissipate energy from cyclic loadings. Adequate overall elastic stiffness of a moment connection under working load is essential to maintain the relative positions of all structural members and to prevent excessive drift. Since the cost of a moment-resisting framing system is highly influenced by the cost of the connections, economical fabrication and ease of erection of these connections are essential in reducing the cost of the construction.

A properly designed static moment-resisting connection should exhibit the behaviour of Curve A in Fig. 6.1. It not only has sufficient stiffness and enough strength to carry the plastic moment of the connecting members, but also possesses the adequate rotation capacity required for moment redistribution. Connections that exhibit the behaviour of Curve C are called unstiffened connections: they possess adequate deformation capacity but fail to develop the necessary stiffness and the full strength, or plastic moment capacity, of the adjoining members. As a result, plastic hinges with reduced moment capacity may form in these connections. The cost of these unstiffened connections is less than their stiffened counterparts. However, the use of larger beam sections may be necessary. This gives the designer the opportunity to balance the material cost of the beams against the labour cost of the connections in arriving at an optimum design. A more thorough discussion of these unstiffened connections is given elsewhere (Witteveen *et al.*, 1982). Connections that exhibit the

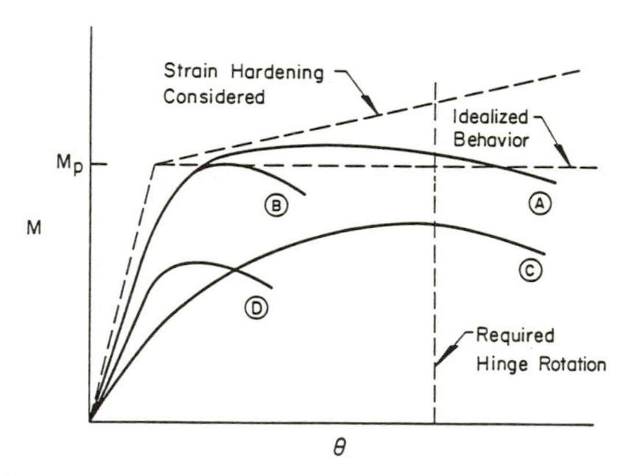

FIG. 6.1. Moment–rotation curves for static moment connections.

behaviour of Curves B and D are unacceptable as they lack the required deformation capacity. The discussion of static moment connections in the following two sections will be focused on stiffened connections, i.e. moment-resisting connections that exhibit the behaviour of Curve A in Fig. 6.1.

6.3 STATIC FLANGE MOMENT CONNECTIONS

A flange moment connection is a connection which joins a beam to the flange of the column. Thus, bending about the strong axis of the column will result, upon application of the beam loads. Another feature of a flange moment connection is the presence of a panel zone. The behaviour of this panel zone has a strong effect on the behaviour of the connection. In other words, the performance of such a connection is controlled to a large extent by the strength and stiffness of the panel zone.

The behaviour of a flange moment connection is also influenced by the types of connecting media. Fully-welded connections are regarded as 'ideal' connections, as they provide full continuity between the adjoining members. However, this type of connection is quite expensive. In order to reduce the cost of field welding, flange-welded web-bolted connections are often used to replace fully-welded connections in plastically and seismically designed structures. In most cases, it has been found that these flange-welded web-bolted connections give satisfactory performance in terms of strength and ductility. Connections which utilise moment plates and high strength bolts to connect the flanges of the beam to the flanges of the column are called fully-bolted connections. Fully-bolted connections usually exhibit sufficient moment and rotational capacities. However, bolt slip at working load for a moment plate designed as a bearing-type joint reduces the elastic stiffness of the connection, which may not be desirable. The reduction in stiffness can be alleviated by designing the moment plate as a friction-type joint. This, however, requires the use of larger moment plates and more bolts.

6.3.1 Behaviour of Flange Moment Connections

The study of the behaviour of this type of connection is exemplified by an experimental setup shown in Fig. 6.2. This setup resembles an interior joint of a moment-resisting frame. The performance of this

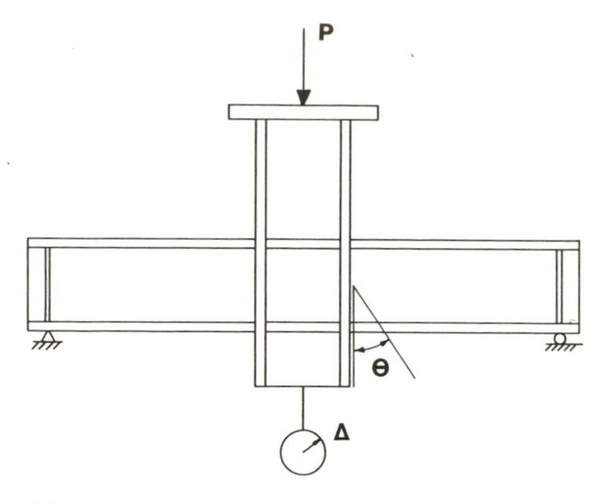

FIG. 6.2. Test setup for static flange moment connections.

joint is represented by the load–deformation behaviour of the connection. The load–deformation behaviour can be predicted satisfactorily by simple plastic analysis if the connection is subjected to a shear force of less than 60% of the shear force at yield $V_p = A_w \tau_y$, where A_w is the area of the beam web and τ_y is the shear stress at yield. τ_y is taken to be $\sigma_y/\sqrt{3}$ if the Von Mises yield criterion is used, in which σ_y is the yield stress of the beam web. The procedures for using this simple plastic analysis to predict the load–deformation behaviour of the connection are shown schematically in Fig. 6.3. Figures 6.3(a) and (b) show a loading condition and the resulting moment distribution of the subassemblage, respectively. By using the stress–strain relationship in Fig. 6.3(c), the curvature diagram can be constructed taking into consideration the property of the member cross-section. Finally, the moment–rotation curve and load–deflection curve can be obtained by the moment area theorems I and II, respectively. Two predicted curves are shown in Fig. 6.3(e). Curve A corresponds to elastic/perfectly plastic stress–strain behaviour, whereas Curve B corresponds to an elastic–plastic strain-hardening stress–strain relationship.

A comparison between the predictions using simple plastic analysis and test results of two connections, designated as C1 and C10, is shown in Fig. 6.4. These connections represent two of the twelve full-sized specimens tested at Lehigh University in the early seventies (Huang *et al.*, 1971). In this test series, A572 Grade 55 steel having a

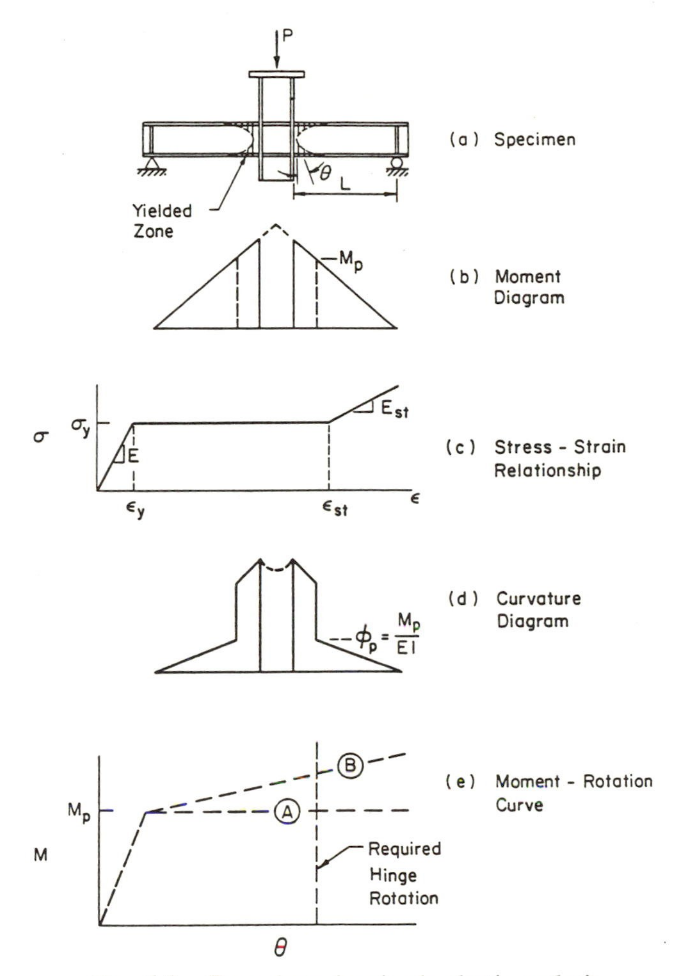

FIG. 6.3. Procedures for simple plastic analysis.

nominal yield stress of 55 ksi (379 MPa) was used. C1 is a flange-welded web-bolted connection and C10 is a fully-welded connection. Joint details for these two connections are shown in Figs 6.5 and 6.6, respectively. Both connections are experiencing a shear force less than 60% of the shear force at yield during the tests. It can be seen that there is a good correlation between the predictions and test results for the stiffness and moment capacities of these connections.

For connections that experience a shear force higher than 60% of

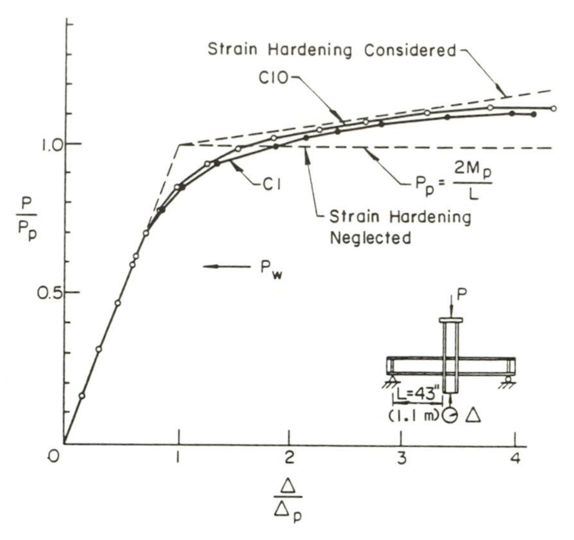

FIG. 6.4. Load–deflection curves of tests C1 and C10 using A572 Grade 55 steel (W14×74 beam, W10×60 column).

V_p, a more elaborate method of analysis should be used (Huang and Chen, 1973; Huang *et al.*, 1973). By using the assumption that, at the plastic limit load, the bending moment is carried only by the beam flanges while the shear force is solely resisted by the beam web, the total deflection Δ at the end of the beam can be expressed as the sum

$$\Delta = \Delta_b + \Delta_s + \Delta_\phi \tag{6.1}$$

where (see Fig. 6.7) Δ_b = deflection of beam due to bending deformation, Δ_s = deflection of beam due to shear deformation, Δ_ϕ = deflection of beam due to rigid body rotation of the beam induced by the connection panel zone deformation.

The bending deflection Δ_b is obtained by first assuming that the entire section is effective in the elastic range up to the initial yield moment M_y, and then that when the bending moment exceeds M_y, the total bending moment is carried only by the two flanges but the material in the flanges is assumed to be in the strain-hardening range. This assumption is shown schematically in Fig. 6.8, in which the non-dimensional moment–curvature relationship is plotted. With this moment–curvature relationship, the bending deflection Δ_b can be obtained by integration.

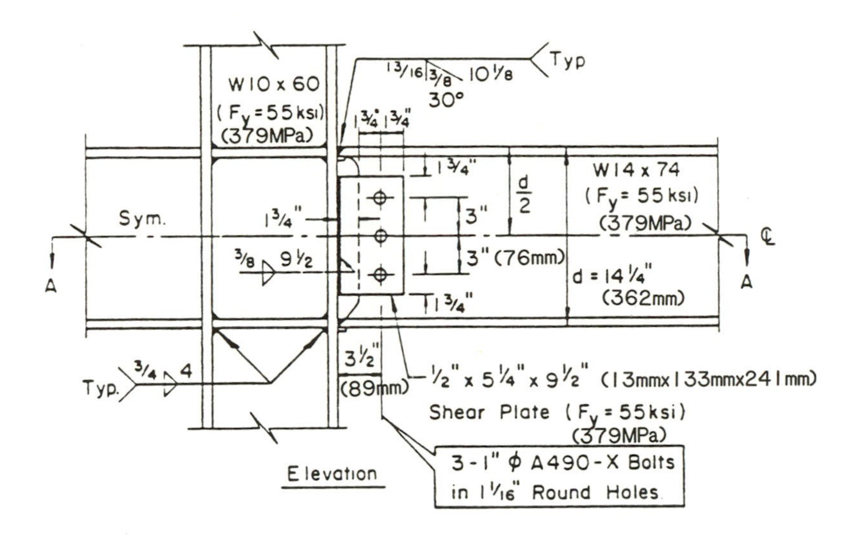

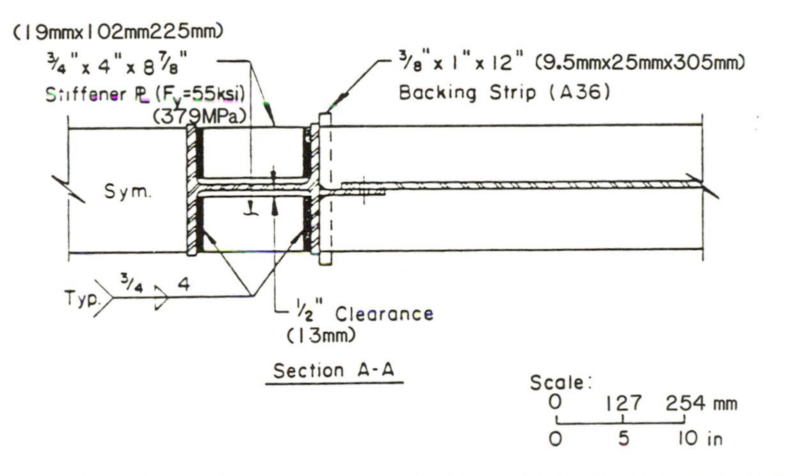

FIG. 6.5. Joint details of connection C1 (1 in. = 25·4 mm, 1 ksi = 6·895 MPa).

The deflection of the beam due to shear deformation Δ_s can be evaluated by

$$\Delta_s = \frac{VL}{A_w G} \tag{6.2}$$

where V = shear force in the beam, L = length of the beam, A_w = area of the beam web, G = shear modulus.

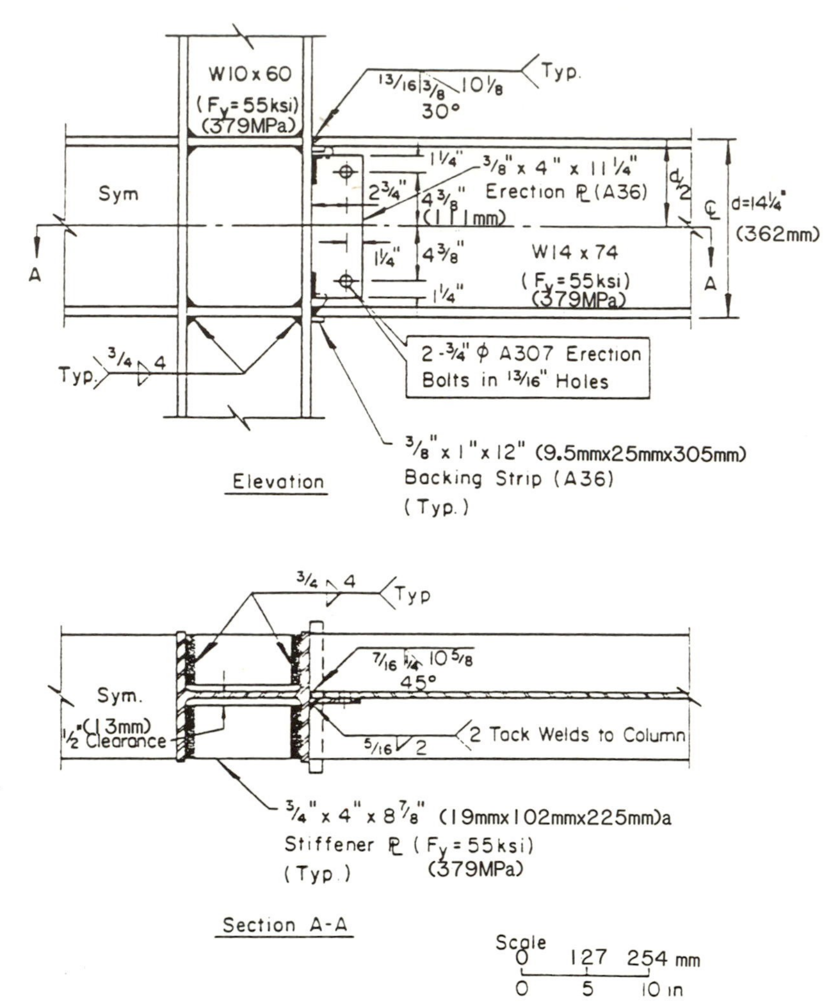

FIG. 6.6. Joint details of connection C10 (1 in. = 25·4 mm, 1 ksi = 6·895 MPa).

Equation (6.2) was written under the assumptions that (a) only the beam web is effective in resisting the shear force, (b) the shear stress distribution is uniform in the beam web, and (c) beam web buckling is precluded.

The deflection of the beam due to deformation of the panel zone Δ_ϕ can be calculated by considering the four stages illustrated in Fig. 6.9.

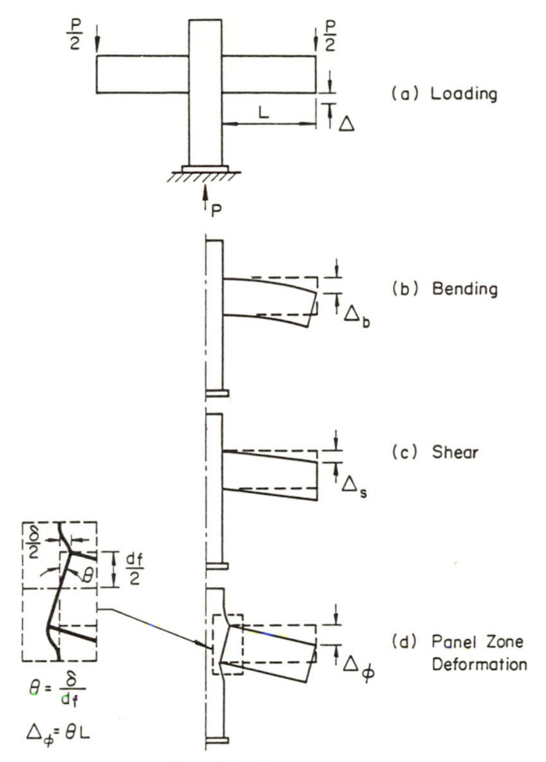

Fig. 6.7. Deflection components.

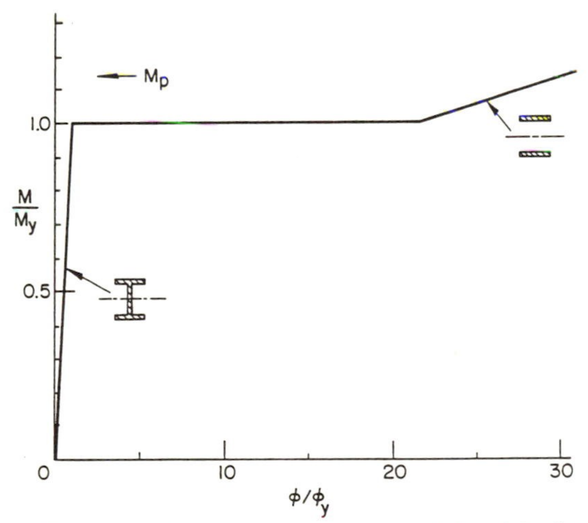

Fig. 6.8. Non-dimensional moment–curvature relationship (A572 Grade 55 steel, W27×94).

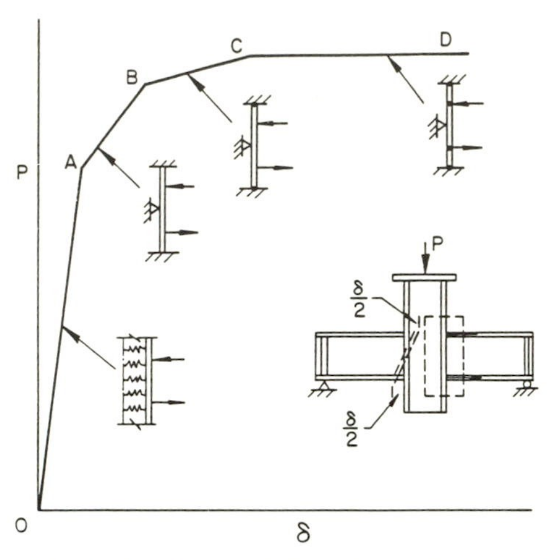

FIG. 6.9. Model for the prediction of panel zone deformation.

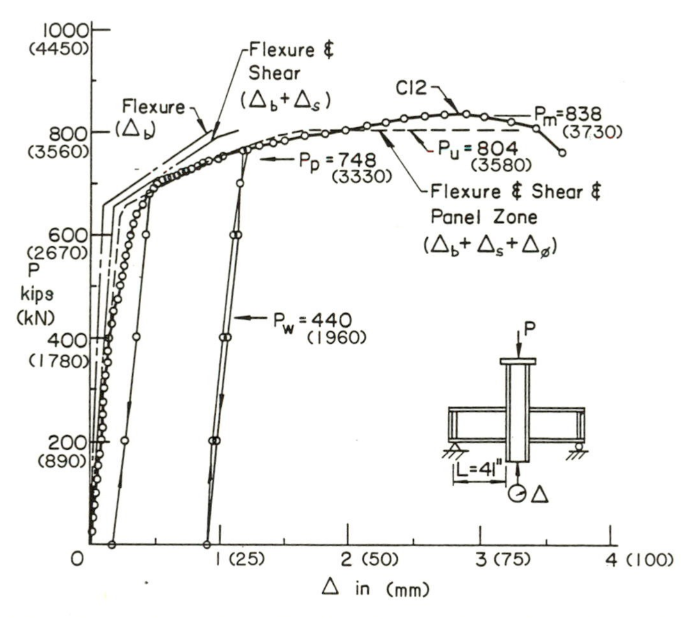

FIG. 6.10. Comparison of predicted and actual load–deflection behaviour of test C12 (1 kip = 4·448 kN, 1 in. = 25·4 mm; A572 Grade 55 steel, W27×94 beam, W14×176 column).

In the elastic range OA, the panel zone is modelled as a system of elastic springs supporting the column flanges. The deformation of the panel zone is thus predicted by the analysis of the bending of the continuous beam (column flange) on an elastic foundation (column web). When yielding in the column web spreads to a width of $(t_b + 5k_c)$, where t_b = thickness of the beam flange and k_c = distance from outer surface of column flange to web-toe of fillet, the column flange is modeled by a two-span continuous beam, with ends fixed at a distance of $(t_b + 7k_c)/2$ away from the application of load. The load–deformation behaviour of this model is shown as stage AB in the figure. The subsequent load–deformation behaviour is obtained according to simple plastic analysis by first considering the formation of plastic hinges at the two fixed-ended supports (stage BC) and then at the load points (stage CD). The ultimate load is said to have been reached when a mechanism is developed in the model.

The three computed deflection components are plotted in Fig. 6.10 together with the test result of a full-sized specimen C12. This connection is a fully-welded joint with no stiffening in the panel zone (Fig. 6.11). High shear force (>60% V_p) was experienced by the connection at the plastic limit load. It can be seen that there is a good correlation between the predicted load–deflection curve and the test results.

The above procedure is also applicable to flange-welded web-bolted connections. Figure 6.12 shows the load–deflection curves of two flange-welded web-bolted connections C2 and C3, together with the prediction curve (labelled 'Theory') and their control specimen C12, which is a fully-welded joint. The difference between connection C2 and C3 is that the shear plate which connects the beam web to the column flange has round holes for C2 and slotted holes for C3 (Figs 6.13 and 6.14). Both C2 and C3, like their control C12, were under high shear force at failure.

Several observations can be made from Fig. 6.12:

(1) The behaviour of flange-welded web-bolted connections is satisfactory, compared to that of fully-welded connections.
(2) The use of slotted holes in the shear plate is justified, as the performance of connection C3 is comparable to that of connection C2.
(3) The procedure outlined above, to predict the load–deflection behaviour for connections under high shear force, can be used for flange-welded web-bolted connections as well as fully-welded connections.

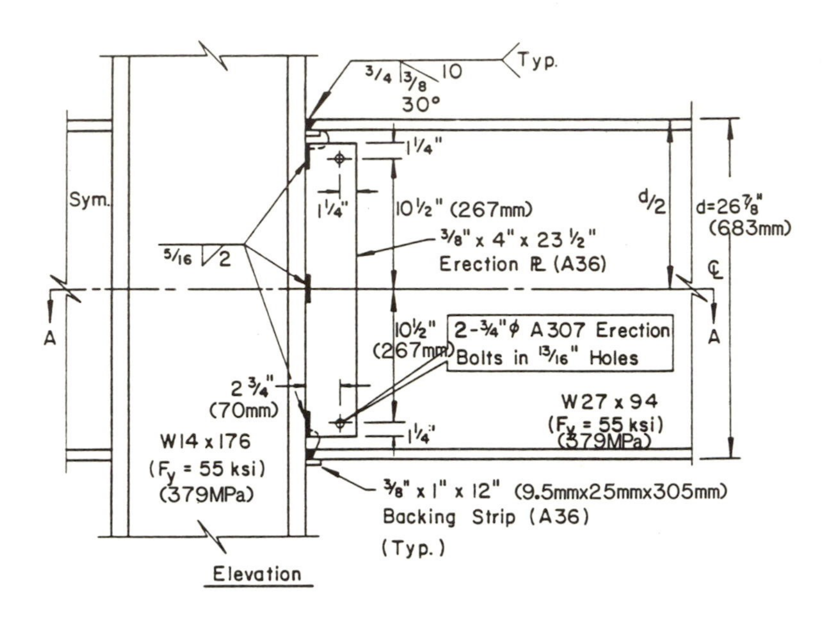

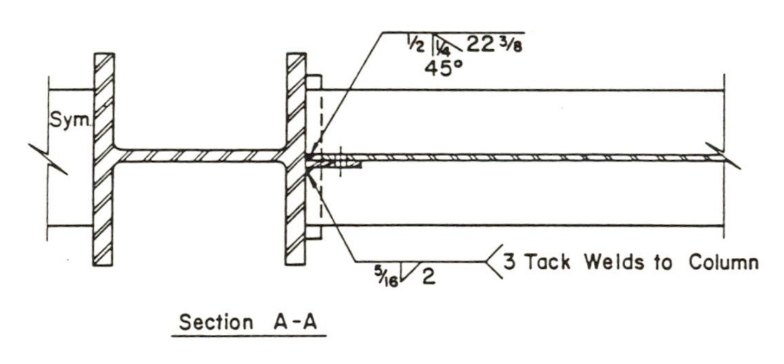

Scale: 0 127 254 mm
0 5 10 in

FIG. 6.11. Joint details of connection C12 (1 in. = 25·4 mm, 1 ksi = 6·895 MPa).

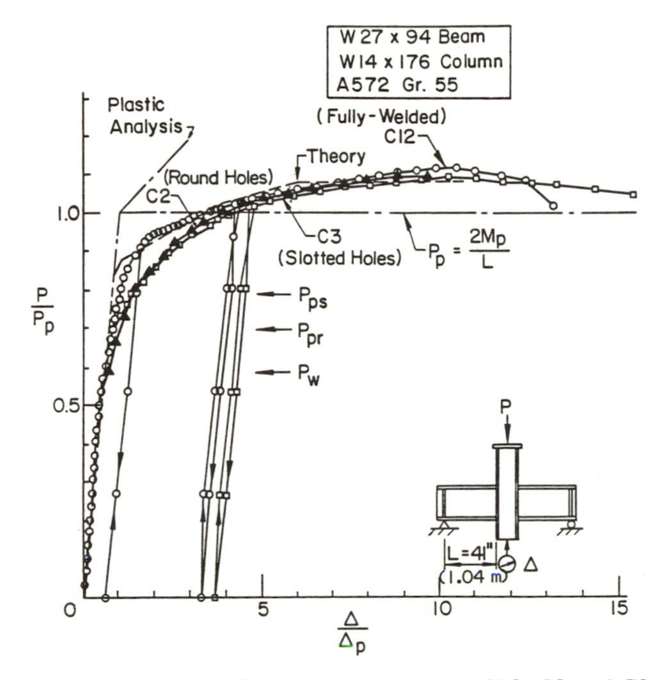

FIG. 6.12. Load–deflection curves of tests C12, C2 and C3.

Observations (1) and (2) provide the designer with the possibility of cost optimisation, because the use of bolting in place of field welding to connect the beam web to the column flange means an economy in fabrication, and the use of slotted holes instead of round holes means an ease of erection in the field. Both measures will tend to reduce the cost of the construction. For a more detailed discussion of these flange-welded web-bolted connections, readers are referred to the paper by Huang *et al.* (1973).

Another measure which a designer can consider in order to reduce the cost of construction is to use the beam seat to carry the shear. However, care must be exercised to prevent the buckling of the beam web. Beam web buckling can be prevented by the use of beam web stiffeners.

The use of flange-welded-only connections to carry both the shear and moment is not advisable due to the high possibility of excessive deformation and premature failure (failure before the attainment of the plastic limit load) of this type of connection.

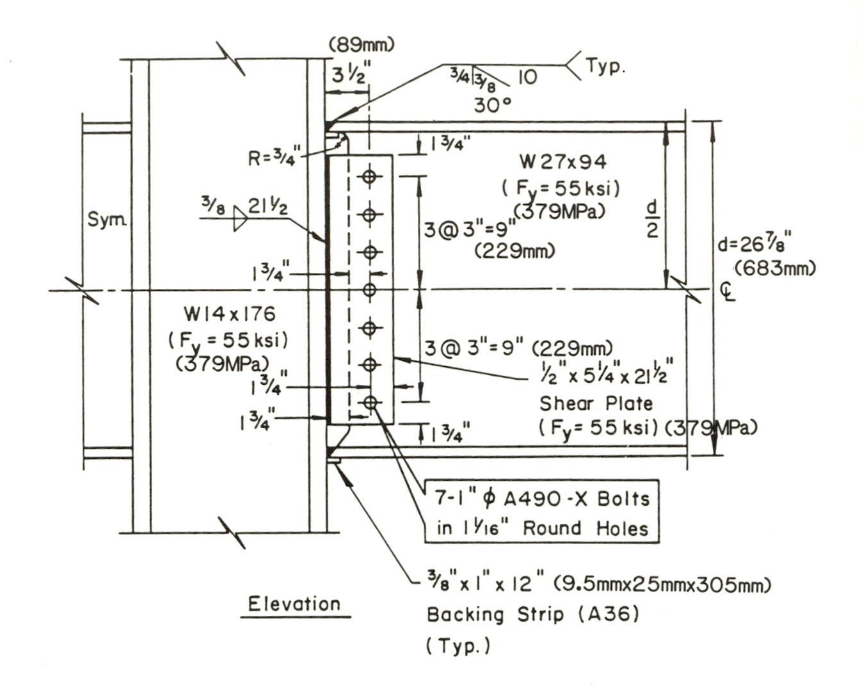

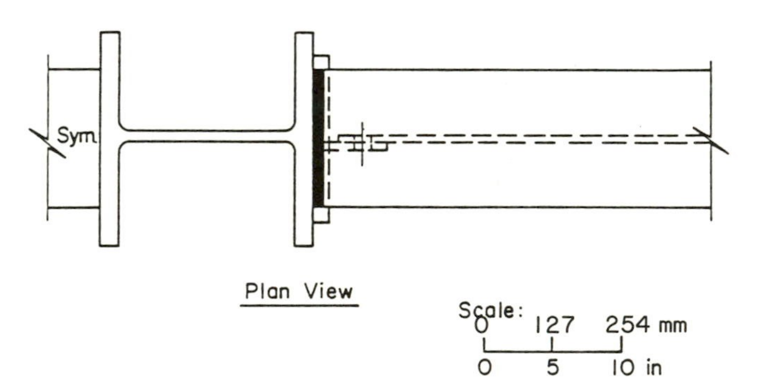

FIG. 6.13. Joint details of test C2 (1 in. = 25·4 mm, 1 ksi = 6·895 MPa).

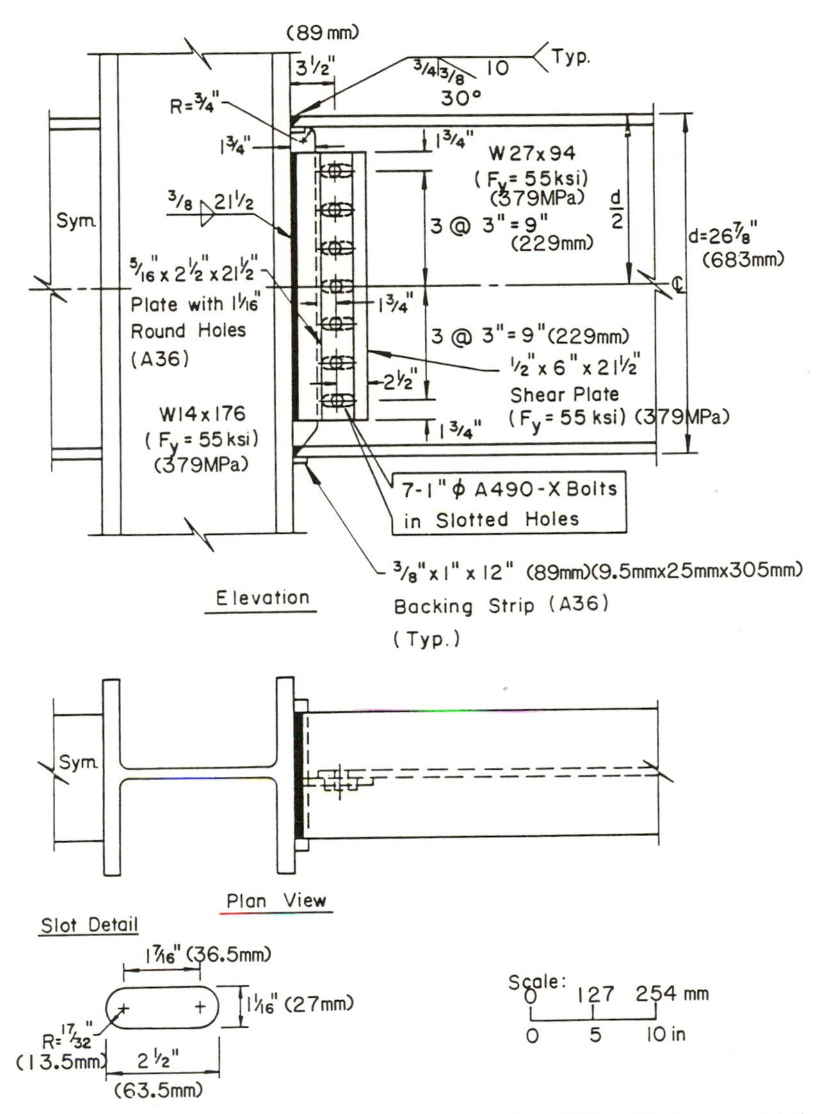

FIG. 6.14. Joint details of connection C3 (1 in. = 25·4 mm, 1 ksi = 6·895 MPa).

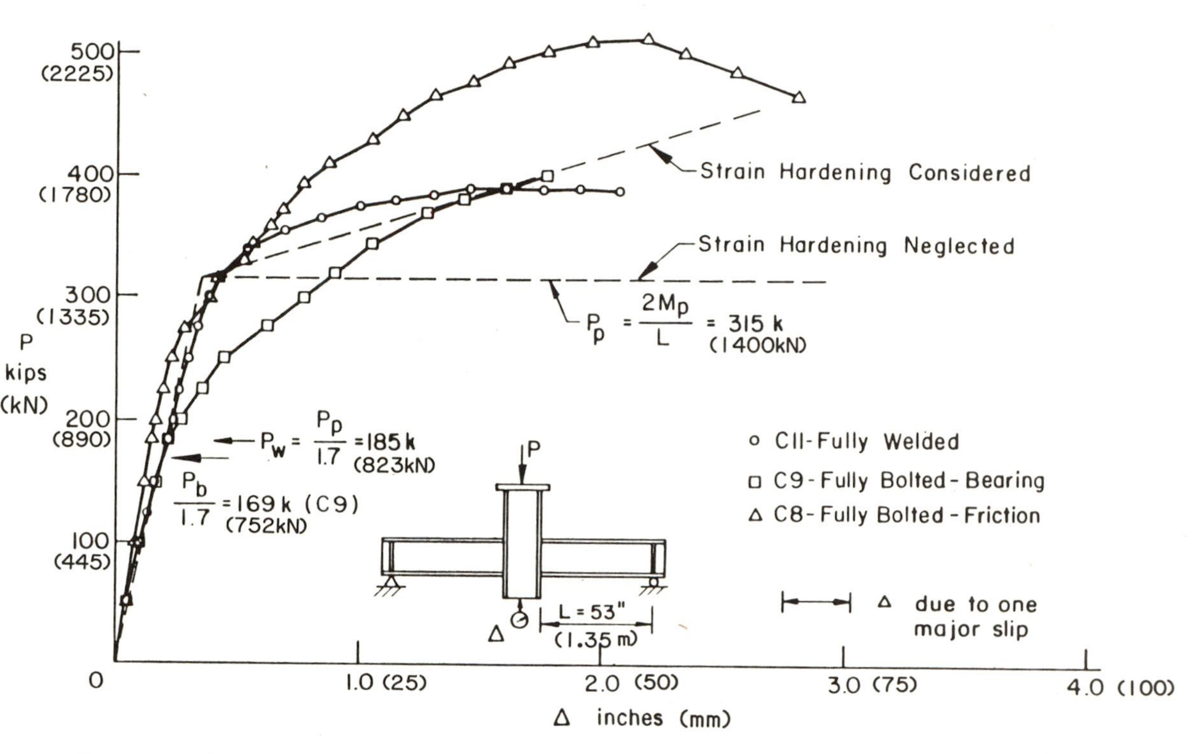

FIG. 6.15. Load–deflection curves of tests C11, C8 and C9 (1 kip = 4·448 kN, 1 in. = 25·4 mm).

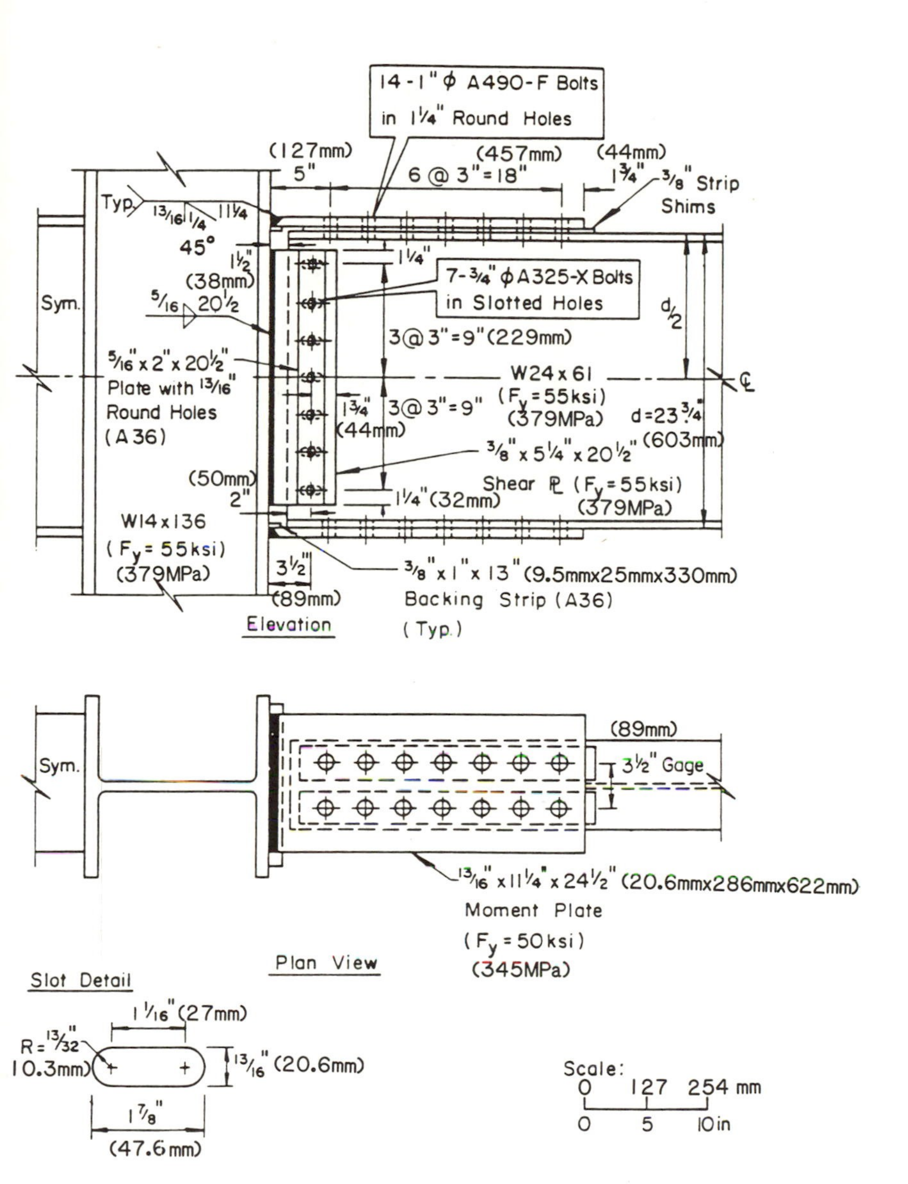

FIG. 6.16. Joint details of connection C8 (1 in. = 25·4 mm, 1 ksi = 6·895 MPa).

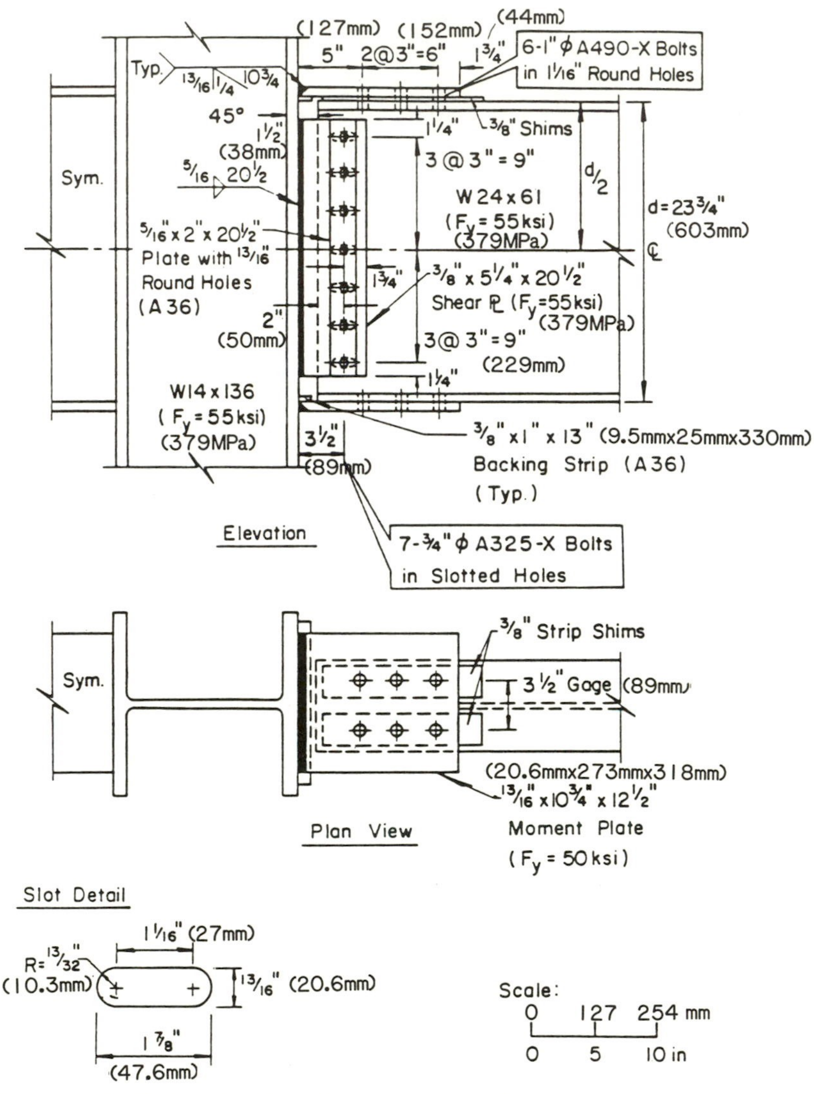

FIG. 6.17. Joint details of connection C9 (1 in. = 25·4 mm, 1 ksi = 6·895 MPa).

For some constructions, it is desirable to use fully-bolted connections. The behaviour of two fully-bolted connections C8 and C9, as well as their fully-welded control specimen C11, is shown in Fig. 6.15. Connection C8 was designed as a friction-type connection, having oversized holes in the moment plates and slotted holes in the shear plate (Fig. 6.16). Connection C9 is similar to Connection C8, with the only difference being that this connection was designed as a bearing-type connection with standard size round holes in the moment plate (Fig. 6.17).

The behaviour of a fully-bolted connection differs from that of its fully-welded counterpart in that a distinct shallower second slope is observed. This reduction in stiffness is due to the slippage of bolts into bearing. This phenomenon is more pronounced for a bearing-type connection (C9) and less significant for a friction-type connection (C8). The reduction of stiffness at loads less than the plastic limit load of the beam may be an important factor in affecting the stability of frames.

Insofar as the moment capacity and deformation capacity are concerned, fully-bolted connections are comparable to fully-welded connections. Some fully-bolted connections may show an increase in plastic moment capacity due to the strengthening effect of the moment plates needed to connect the beam flanges to the column flanges. Because of the added plate material in the lap area, plastic hinge may form at or beyond the end of the moment plates rather than at the face of the column. A more thorough discussion of these fully-bolted connections is given elsewhere (Standig *et al.*, 1976).

6.3.2 Moment Capacity of Flange Moment Connections

The moment capacity of a flange moment connection is said to have been reached when one or more of the following regions in the connection become critical.

(1) Yielding, buckling or crippling in the compression zone of the column web.
(2) Yielding of fracture in the tension zone of the column web and column flanges.
(3) Shear yielding due to the shearing effect in the panel zone.

6.3.2.1 Compression Zone

In analysing the compression region of the flange moment connection, the beam flange force T, which is calculated by dividing the beam

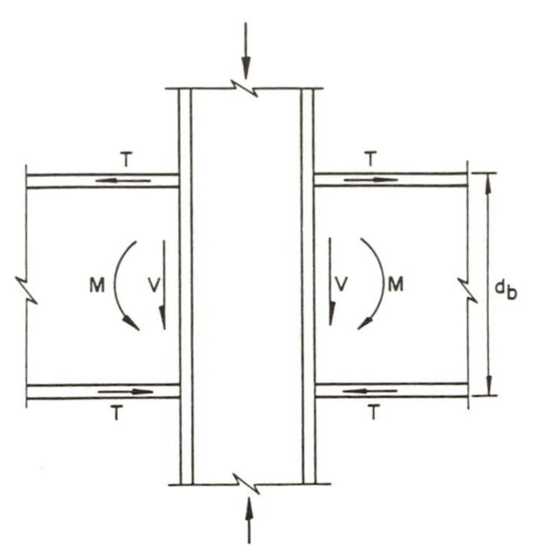

FIG. 6.18. Internal forces of an internal beam-to-column connection under symmetrical loads.

moment M at the column face by the beam depth d_b (Fig. 6.18), is assumed to be distributed on a 2·5 : 1 slope from the point of contact to the column 'k-line'. As a result, the beam flange force is resisted by a width of column web equal to $(t_b + 5k_c)$ at the column 'k-line', in which t_b = thickness of the beam flange and k_c = distance from outer face of flange to web-toe of fillet of rolled shape, or equivalent distance on welded section (Fig. 6.19). In order to prevent yielding, the resistance

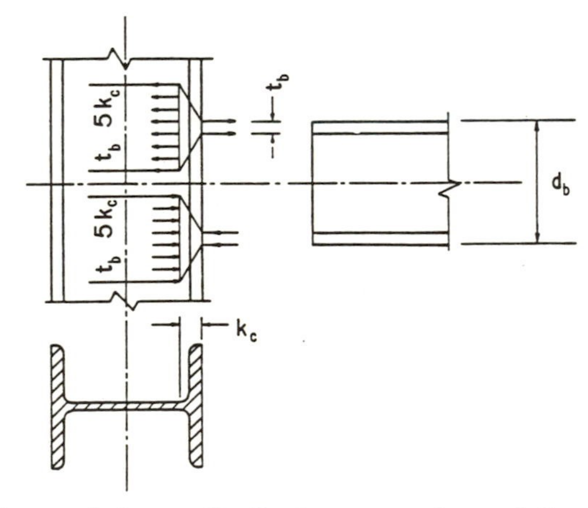

FIG. 6.19. Assumed force distributions on column k-line due to beam moment.

of the effective area of the column web must equal, or exceed, the applied concentrated force of the beam compression flange, i.e.

$$F_{yc}w_c(t_b+5k_c)\geqslant F_{yb}A_f \tag{6.3}$$

Rearranging

$$w_c\geqslant\frac{C_1A_f}{t_b+5k_c} \tag{6.4}$$

where w_c = thickness of the column web, C_1 = ratio of beam flange yield stress F_{yb} to column web yield stress F_{yc}, A_f = area of one flange of the beam.

To avoid column-web buckling, the following criterion (Chen and Newlin, 1973) must be satisfied

$$d_c'\leqslant\frac{10\,762w_c^3}{C_1A_f\sqrt{F_{yc}}} \tag{6.5}$$

where F_{yc} is in MPa, w_c in mm, A_f in mm^2 and d_c' is the depth of column web clear of fillets, in mm.

If either of eqn (6.4) or (6.5) is not satisfied, stiffeners should be provided on the column opposite to the compression flange of the beam.

An interaction equation integrating the strength and stability requirements to determine whether stiffeners are required was proposed by Chen and Newlin (1973) as

$$w_c\leqslant\frac{d_c'^2\sqrt{F_{yc}}+68C_1A_f}{77(F_{yc})^{1/4}d_c'} \tag{6.6}$$

where F_{yc} is in MPa, d_c' in mm and A_f in mm^2.

6.3.2.2 Tension Zone

To prevent yielding in the tension zone, the following criterion must be satisfied

$$w_c\geqslant\frac{C_1A_f}{t_b+5k_c} \tag{6.7}$$

where the nomenclature is defined as in eqn (6.4).

To prevent fracture between column web and column flange, the following criterion for minimum thickness of column flange needs to be satisfied

$$t_c\geqslant0{\cdot}4\sqrt{(C_1A_f)} \tag{6.8}$$

where t_c = thickness of the column flange in mm, A_f = area of one flange of the beam in mm^2.

Equation (6.8) was developed by Graham *et al.* (1959) using the yield line theory, under the assumption that the beam flange tensile force is carried by the column flange through (a) a direct resistance over the middle portion of the column flange, and (b) a bending resistance of the column flange plate.

If either eqn (6.7) or (6.8) is not satisfied, stiffeners should be provided on the column opposite to the tension flange of the beam. In most cases, eqn (6.8) will govern, and so it furnishes the prime criterion to determine whether or not stiffeners are needed.

6.3.2.3 Shear Stiffening

Figure 6.20 shows the possible system of forces acting on an interior beam-to-column flange moment-resisting connection. Under an anti-symmetrical loading, shear deformation of this connection will result. This deformation may lead to shear yielding. In order to avoid yielding in shear, the following criterion needs to be satisfied

$$w_c \geqslant \frac{\sqrt{3}}{F_{yc} d_c} \left(\frac{M_r + M_l}{d_b} - V_{av} \right) \tag{6.9}$$

where w_c = thickness of column web in panel zone, F_{yc} = yield stress of column web in panel zone, d_c = depth of column, d_b = depth of beam, M_r, M_l, $V_{av} = (V_a + V_b)/2$ are as defined in Fig. 6.20.

Equation (6.9) was developed by ignoring the effect of the column axial loads P_a and P_b on the shear yield stress of the web panel. If eqn (6.9) is not satisfied, then shear stiffening must be provided in the panel zone.

The inclusion of the effect of axial force on the shear capacity of the joint panel, to determine whether or not shear stiffening is required, was proposed by Fielding and Huang (1971). By using the Von Mises yield criterion for biaxial stress state in the panel zone and defining $P = (P_a + P_b)/2$, the criterion for providing shear stiffening is given as

$$w_c \geqslant \frac{\sqrt{3}\left(\dfrac{M_r + M_l}{d_b} - V_{av} \right)}{F_{yc} d_c \sqrt{\left(1 - \left(\dfrac{P}{P_y}\right)^2\right)}} \tag{6.10}$$

where P_y is the yield load of the column. The other symbols are

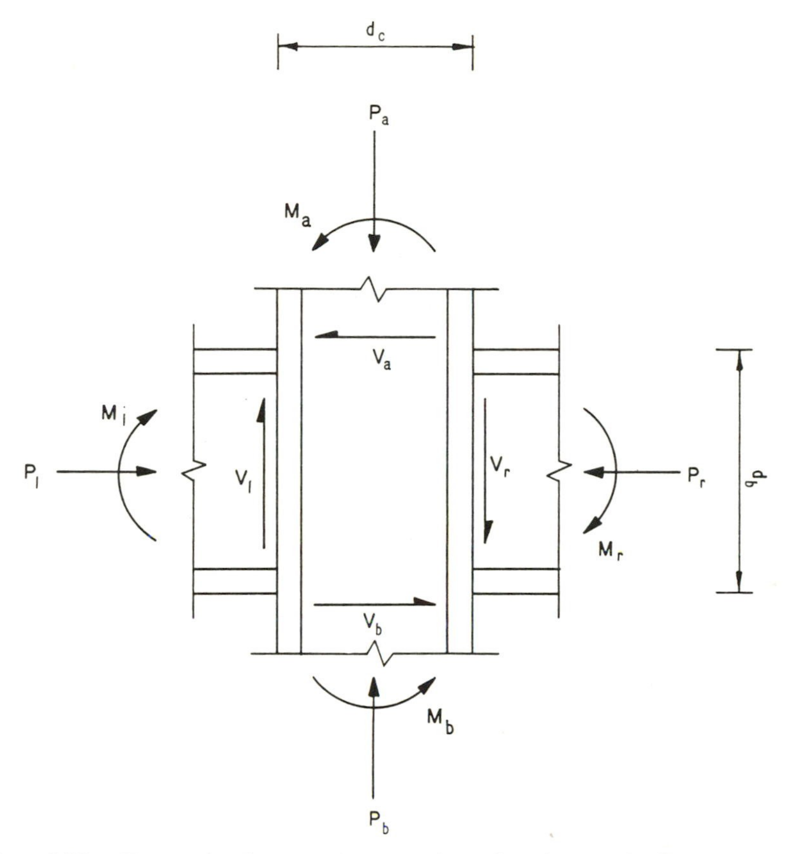

FIG. 6.20. Forces in the panel zone of an interior static flange moment connection under antisymmetric loads.

defined as before (eqn (6.9)). Equation (6.10) specifies the column web thickness w_c required to prevent general yielding under the action of antisymmetrical beam moments M_r and M_l and column load P.

6.3.3 Concluding Remarks

(1) A properly designed and detailed flange moment connection should have sufficient strength and ductility.

(2) Fully-welded connections are considered to be 'ideal' as they provide full continuity between the beams and the columns.

(3) The use of flange-welded web-bolted connections with round or slotted holes in the shear plate is justified, as they provide sufficient

strength and ductility that is comparable to fully-welded connections.

(4) Insofar as strength and ductility are concerned, fully-bolted connections designed as bearing or friction-type give satisfactory performance under static load. However, the reduction of stiffness at working load due to bolt slip may be undesirable.

(5) The load–deflection behaviour of fully-welded and flange-welded web-bolted connections can be predicted using simple plastic analysis if the shear force V is no more than $0{\cdot}6V_p$. For $V > 0{\cdot}6V_p$, a more sophisticated procedure, outlined in Section 6.3.1, needs to be used.

(6) The most critical region of a flange moment connection is the panel zone. Therefore, proper stiffening has to be provided to ensure satisfactory performance.

(7) The adequate performance of Connection C3 (flange-welded web-bolted connection with slotted holes in the shear plate) demonstrates the assumption that the flanges alone are able to develop the full plastic moment capacity of a wide-flange shape by strain-hardening (Fig. 6.8), provided that precautions are made to avoid beam-flange local buckling.

6.4 STATIC WEB MOMENT CONNECTIONS

A web connection is one in which the beam is attached to the column perpendicular to the plane of the column web. Upon application of the beam load, the column will bend about its weak axis. Because of the space restrictions imposed by the column flanges, the beam is usually connected to the column through flange connection plates. There are several ways by which the flange connection plates can be attached to the column. They are;

(a) the flange connection plates are welded to the column web as well as the inner face of the column flanges (Fig. 6.21(a)),
(b) the flange connection plates are welded to the inner face of the column flange only (Fig. 6.21(b)),
(c) the flange connection plates are welded to the column web only (Fig. 6.21(c)).

In order to investigate the behaviour of these different types of attachments, a programme designated as the pilot test programme was initiated at Lehigh University. In this test programme, eight simulated

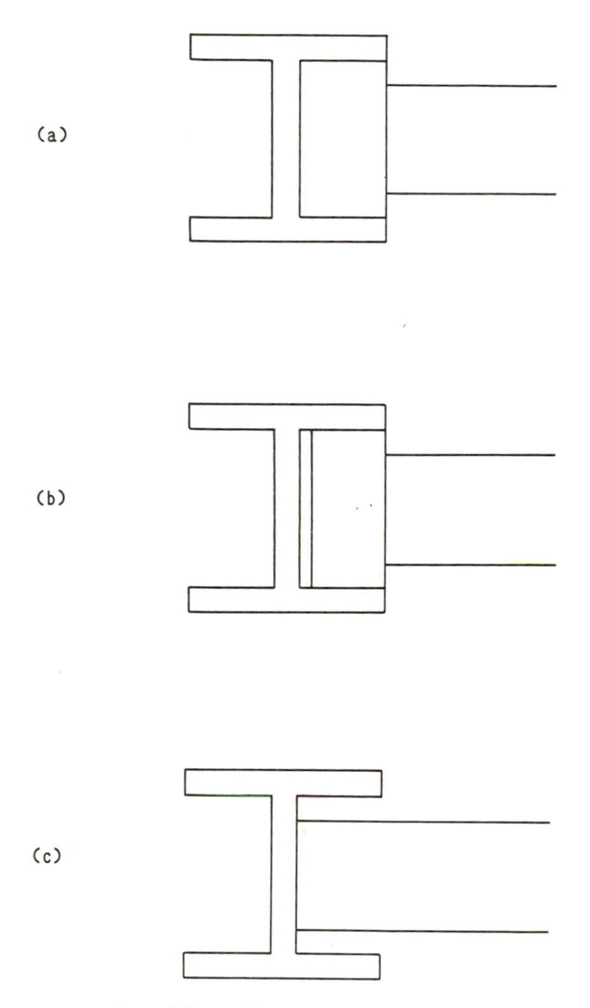

FIG. 6.21. Web connection types.

tests on web connection details were performed. The objective of this test programme is threefold.

(1) To investigate the behaviour and ultimate strength of the column web using different types of attachment details.

(2) To study the requirements for column web stiffening.

(3) To gain knowledge to attain the proper design of full-scaled specimens.

6.4.1 Pilot Test Programme

The setup for this test programme is shown in Fig. 6.22. Pairs of steel plates are welded to the column to simulate the tension and compression flanges of a beam. This tension–compression couple will simulate the bending moment of the beam. It can be seen in Fig. 6.22 that two tests were performed in one setup. Since the column (which was tested as a beam in the test setup) was not loaded axially, only the effect of bending delivered by the pair of tension–compression plates was considered. The effects of axial load and shear were not considered in these tests.

A total of eight tests were performed. Two different column sections were used (W12 × 106 and W14 × 184). Each was tested with four

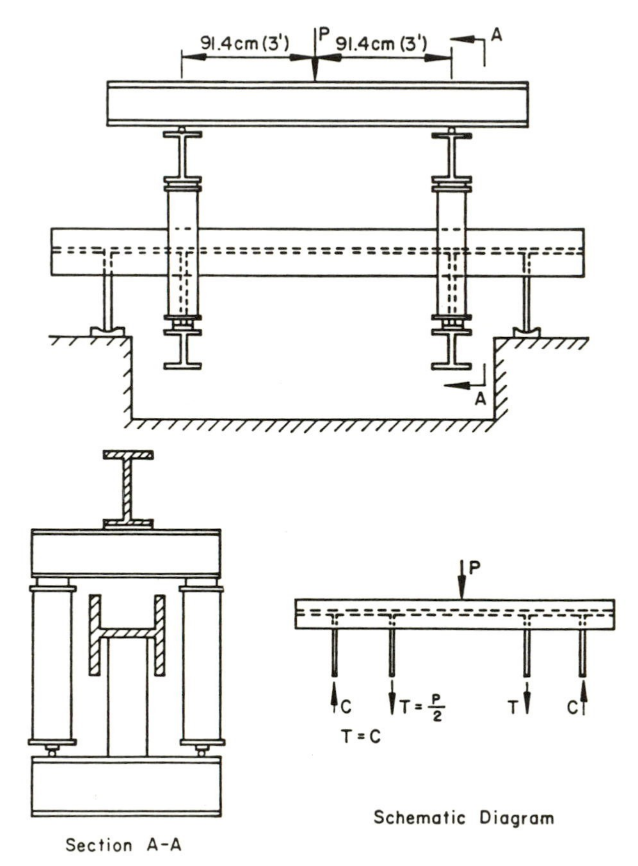

FIG. 6.22. Pilot test setup.

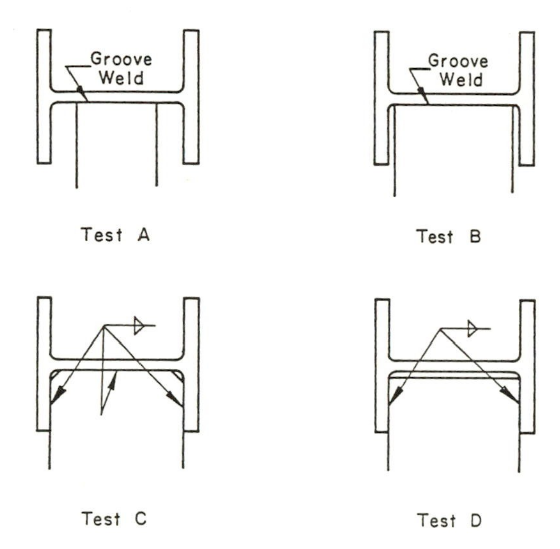

FIG. 6.23. Connection geometries of pilot tests.

different connection details, as shown in Fig. 6.23. These tests were designated as 12A, 12B, 12C, 12D and 14A, 14B, 14C, 14D. The numbers 12 and 14 refer to the column sections W12 × 106 and W14 × 184, respectively, and the letters A, B, C, D correspond to the connection details of Fig. 6.23.

For tests A and B (see Fig. 6.23), the plates were welded to the column web only by full penetration groove welds. The difference between them is that in test B, the width of the plates is equal to the clear distance between the column fillets, whereas in test A, the width of the plates is less than the distance between the column fillets. Thus, a yield line type of mechanism is expected of test A but not test B.

For test C, the plates were fillet welded to the column web as well as both the inner faces of the column flanges.

For test D, the plates were fillet welded to the inner faces of the column flanges only. Therefore, all the plate forces will be delivered to the column flanges.

The results of these tests (Rentschler *et al.*, 1982) indicated that both tests 12A and 14A failed to attain the predicted yield line load to initiate the yield line mechanism in the column web. Also, tests 12B and 14B failed to attain the predicted plastic limit load required to form a plastic hinge in the column. This is because all these connection details failed prematurely, by fracture of the column web near the edges of the tension plate. This fracture is caused by stress concentration near the edges of the tension plate. This stress concentration can be explained from the viewpoint of flexibility of the column web. Due to the constraint offered by the column flanges, the column web stiffness increases significantly near the column flanges in a phenomenon known as shear lag. This high stress concentration will cause yielding and finally lead to fracture of the material.

In addition to stress concentration, the out-of-plane deformation of the column web and flanges is quite substantial for tests 12A, 12B and 14A, 14B. This is attributed to the considerable flexibility to this type of web connection. This deformation, if coupled with a high axial force in the column, will easily cause local buckling of the column.

Insofar as stiffness and strength are concerned, connection details 12C, 12D and 14C, 14D are adequate. The load which corresponds to the formation of plastic hinge in the column was reached. No significant out-of-plane deformation of the column web and flanges was observed. The comparable performance of tests C and D is due to the fact that only a small portion of the plate forces is delivered to the column web, a large portion of the plate forces is carried by the column flanges. An elastic finite element analysis (Rentschler, 1979) has shown that the maximum force reaching the column web is approximately 10% of the total plate force being delivered to the column. Thus, the presence of a weld between the flange plate and the column web for test C does not attract an appreciable amount of the total flange plate force, and so the behaviour of test C is similar to test D, in which the flange plates were connected only to the column flanges.

Although the performance of tests C and D was adequate, the non-uniformity of stress distribution in the flange plate requires attention. The high stress concentrations at the edges of the flange plates that are connected to the column flanges may cause shear yielding of the flange plate adjacent to the flange weld. This shear yielding may cause problems in these types of connection details.

Based on the information obtained in these pilot tests, four full-sized specimens were designed and tested (Rentschler *et al.*, 1980). The results of these tests are discussed in the following section.

6.4.2 Behaviour of Web Moment Connections

The four specimens were designated as 14-1 to14-4; 14-1 and 14-2 are flange-welded web-bolted connections, 14-3 is a fully-bolted connec-

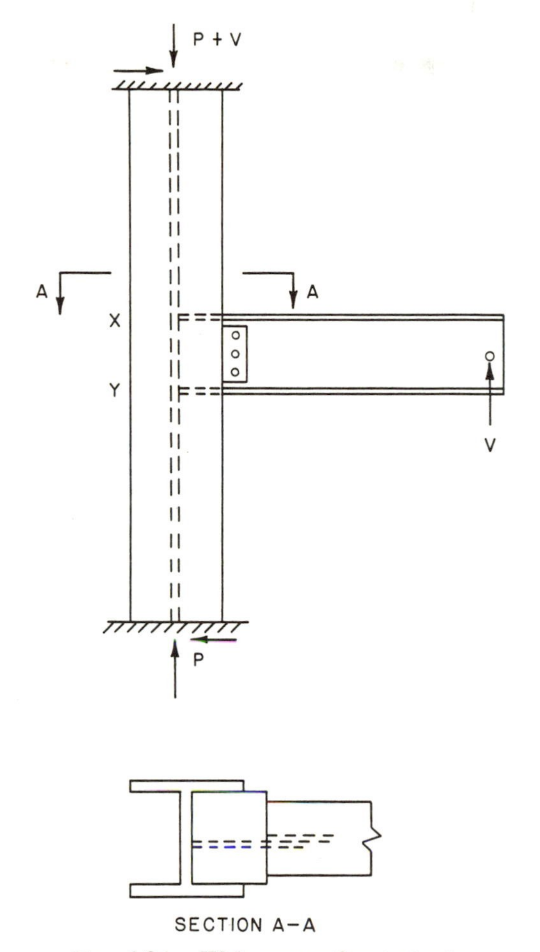

FIG. 6.24. Web connection test setup.

tion and 14-4 is a fully-welded connection. The test setup for these specimens is shown in Fig. 6.24. The connection assemblages were designed in such a way that, at the predicted plastic limit load, the beam would resist a beam plastic moment M_p and a beam shear V approximately equal to $0{\cdot}81\,V_p$, whilst the column would resist an axial load P approximately equal to $0{\cdot}50P_y$. The material used for these tests was A572 Grade 50 steel ($F_y = 55$ ksi (379 MPa)).

Figures 6.25 and 6.26 show the joint details of tests 14-1 and 14-2, respectively. In connection 14-1, the beam flanges are groove welded to the flange moment plates which in turn are fillet welded to the

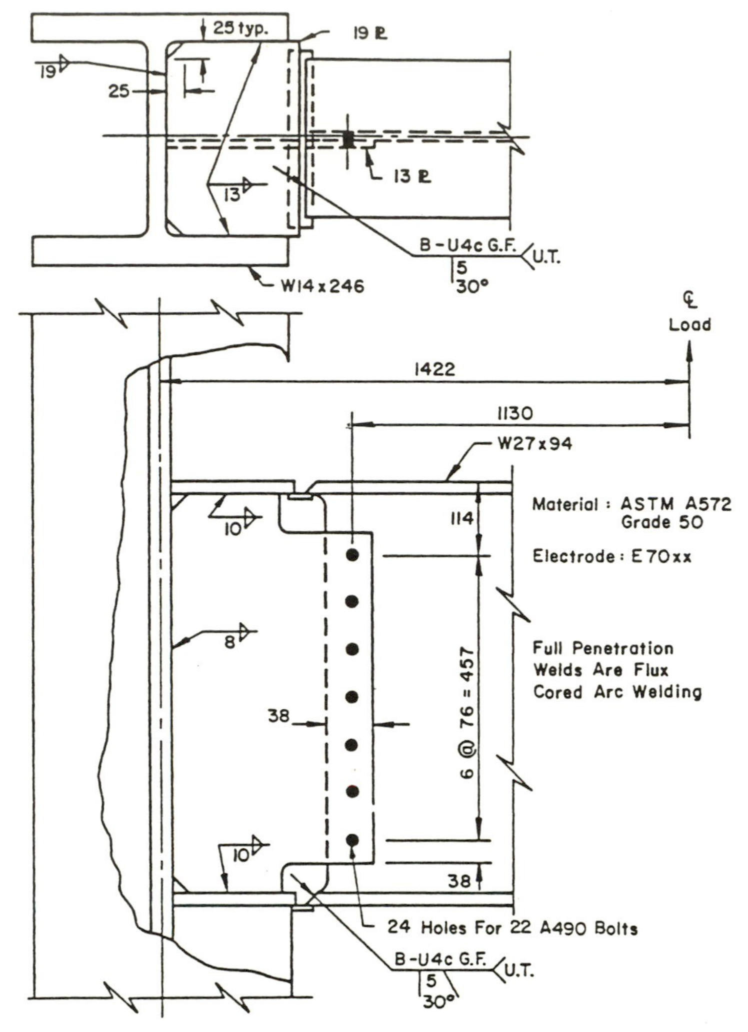

FIG. 6.25. Joint details of connection 14-1 (all dimensions in mm).

column flanges and web. The beam web is connected to a shear plate by seven 7/8-in (22 mm) diameter A490 high strength bolts. This shear plate is attached to the column web and flange moment plates by fillet welds.

For connection 14-2, the beam flanges are groove welded directly to the column web and the beam web is connected to the column web by

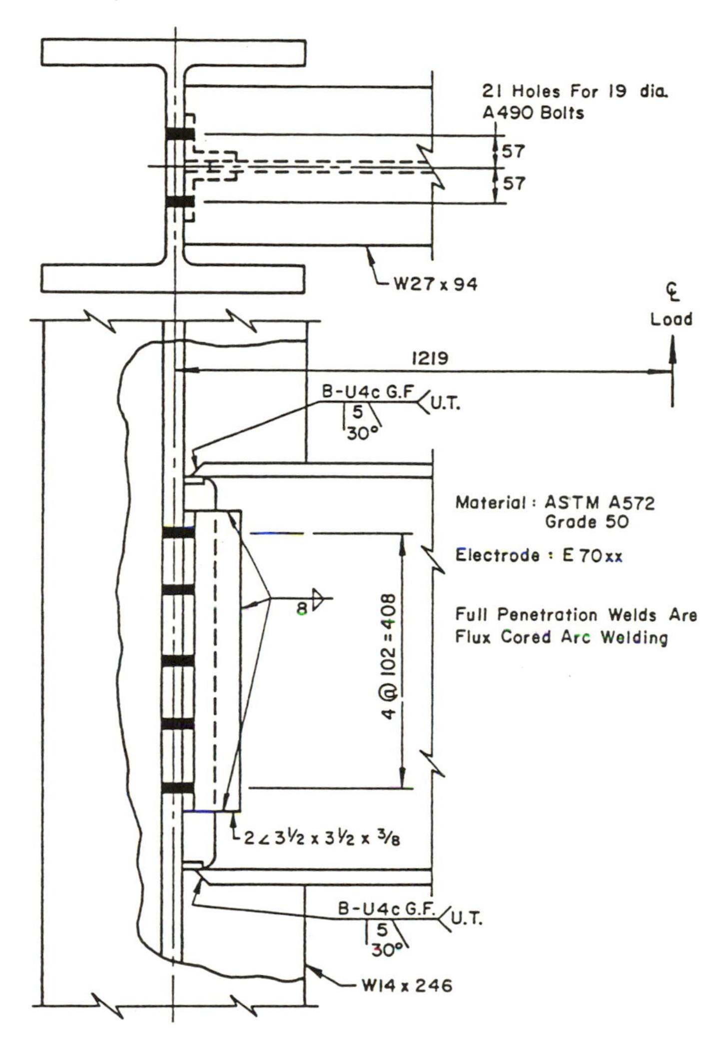

FIG. 6.26. Joint details of connection 14-2 (all dimensions in mm).

two structural angles of size $3\frac{1}{2}$ in × $3\frac{1}{2}$ in × $\frac{3}{8}$ in (89 mm × 89 mm × 9·5 mm). These angles are fillet welded to the beam web and bolted to the column web by eight 3/4-in (19 mm) diameter A490 bolts.

The load–deformation behaviour of these two connections is shown in Figs. 6.27 and 6.28, respectively. From Fig. 6.27, it can be seen that although connection 14-1 possesses the required strength and stiffness,

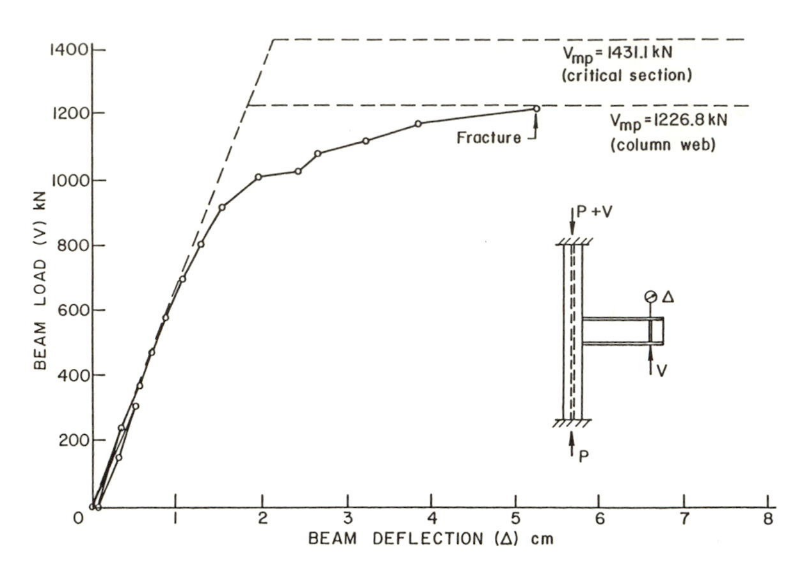

FIG. 6.27. Load–deflection curve of flange-welded web-bolted connection 14-1.

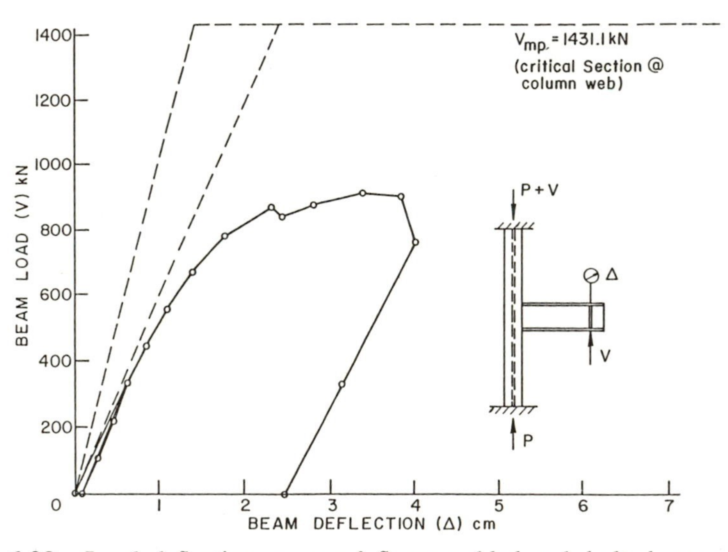

FIG. 6.28. Load–deflection curve of flange-welded web-bolted connection 14-2.

it lacks ductility due to fracture of the tension flange moment plate near the groove weld that joins with the tension beam flange. From Fig. 6.28, it is clear that connection 14-2 lacks both strength and ductility. This is attributed to the severe out-of-plane deformation of the column web and flanges. The stress concentration that builds up in the beam tension flange finally causes fracture of the column web. In order to reduce the out-of-plane deformation and to alleviate the stress concentration, back-up stiffeners and moment plates can be used. It has been shown (Rentschler, 1979) that the presence of back-up stiffeners on the other side of the column can reduce the column web deformation substantially and alleviate stress concentration as they attract forces from the beam. The use of moment plates alleviates the problem by delivering the beam forces to the column flanges as well as the web, so that both the column flanges and web can take part in resisting the beam forces.

In these figures, the symbol V_{mp} is the shear force required to produce the plastic moment M_p in the beam. V_{mp} at the critical section is not always equal to V_{mp} at the column web centreline, because the critical sections of these connections do not necessarily coincide with the column web centrelines. The critical section for connection 14-1 is at the juncture of the beam and the moment and shear connection plates. The critical section for connection 14-2 is at the column web. Therefore, V_{mp} (critical section) and V_{mp} (column web) have the same value for connection 14-2, but have different values for connection 14-1.

Connection 14-3, which is a fully-bolted connection, is shown in Fig. 6.29. In this connection, the beam is connected to the moment and shear connection plates by high strength bolts. The moment connection plate is connected to the column web and flanges by fillet welds, and the shear connection plate is connected to the column web and to the top and bottom moment plates also by fillet welds. The critical section of this connection is at the outer row of flange bolts.

The load–deformation behaviour of this connection is shown in Fig. 6.30. Note that two distinct elastic slopes are observed. The occurrence of the second shallower slope is due to bolt-slip into the bearing. After this second slope, the connection starts losing its stiffness due to local yielding of the assemblage elements. Failure of this connection is again due to fracture at the tip of the tension flange connection plate.

This fully-bolted connection is not proper for a plastically designed

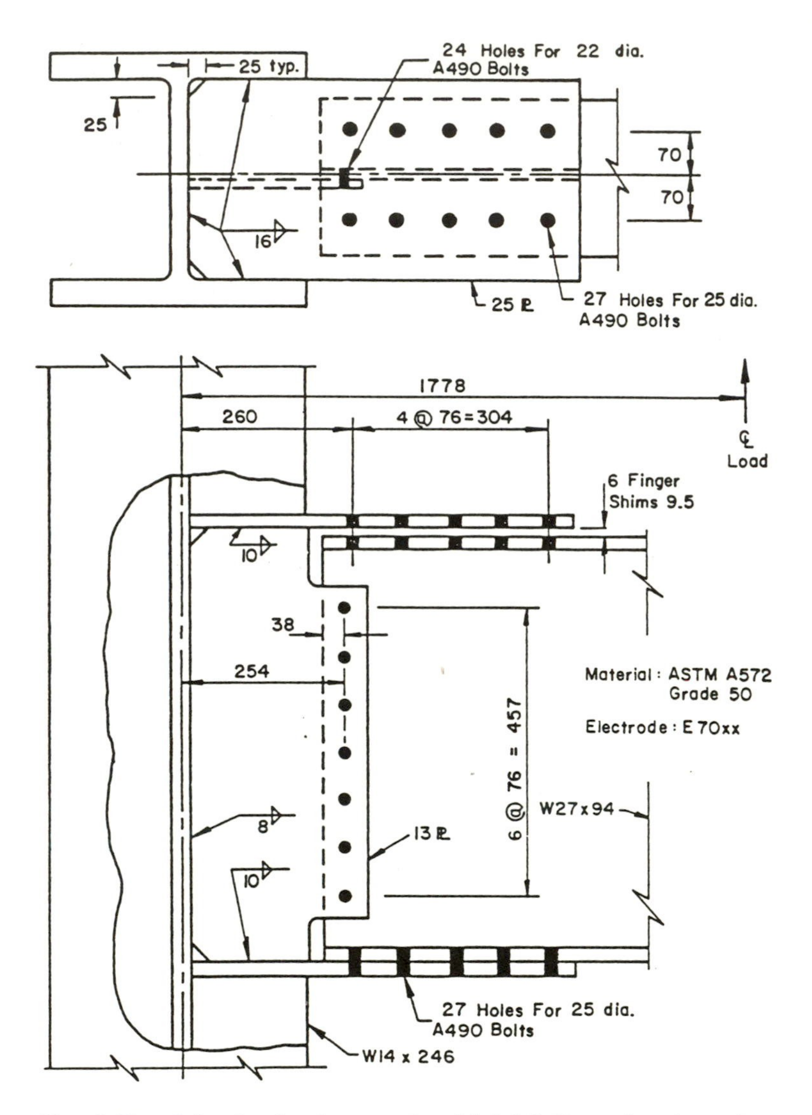

FIG. 6.29. Joint details of connection 14-3 (all dimensions in mm).

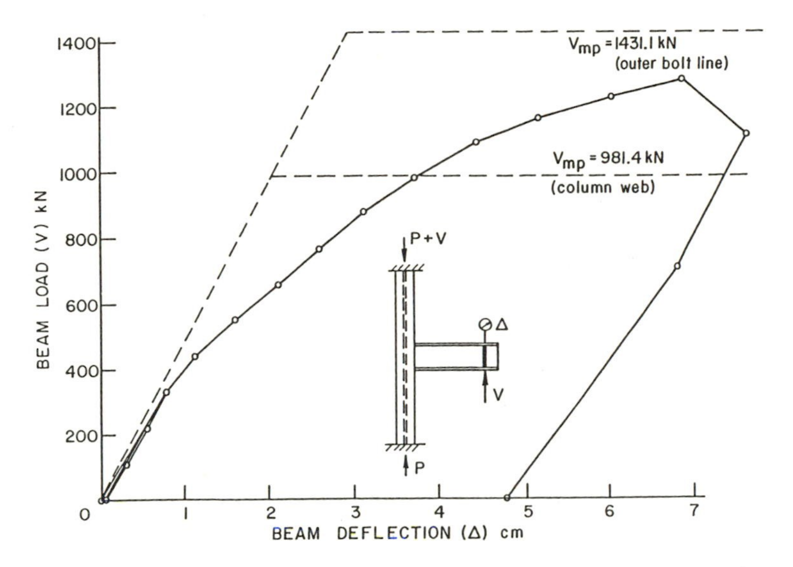

FIG. 6.30. Load–deflection curve of fully-bolted connection 14-3.

structure, because of the significant reduction in stiffness in the working load range due to bolt slip, and the inadequate ductility due to fracture in the tension flange connection plate. However, these shortcomings can be remedied by designing the joint as a friction-type connection and by using an extended moment flange connection plate. This bolt-slip phenomenon is less significant in a friction-type joint and the use of an extended moment plate will reduce the high stress concentration at the beam tension flange adjacent to the tips of the inner faces of the column flanges.

Figure 6.31 shows the joint details of the fully-welded connection 14-4. The beam flanges and web are groove welded to the moment and shear connection plates, which in turn are connected to the column by fillet welds. The critical section for this connection is at the column flange tips.

Figure 6.32 shows the load–deflection behaviour of this connection. It can be seen that this connection possesses the required stiffness and strength. Although local buckling of the beam compression flange and cracks in the area of the groove weld joining the tension flange of the beam to the flange moment plate were observed, the connection exhibited sufficient ductility and no fracture occurred.

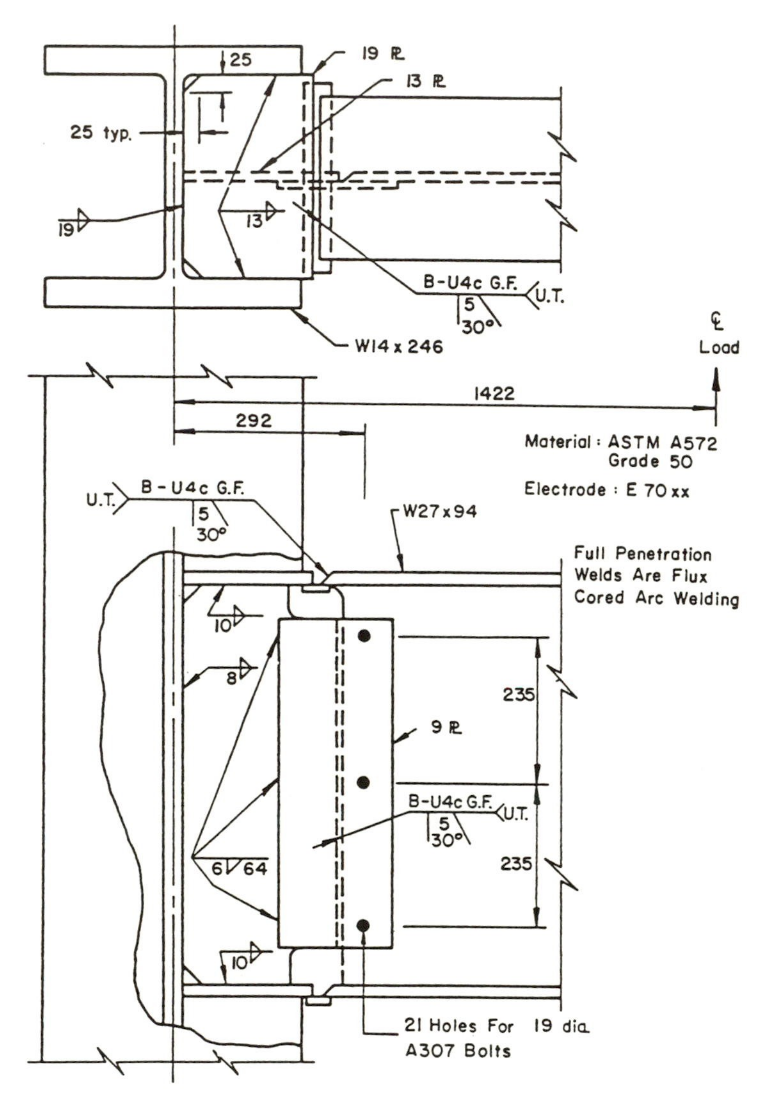

FIG. 6.31. Joint details of connection 14-4 (all dimensions in mm).

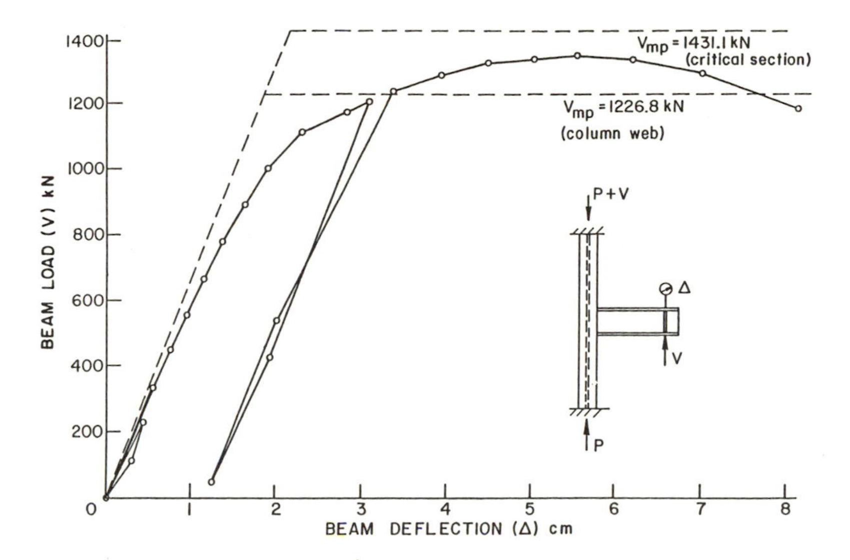

FIG. 6.32. Load–deflection curve of fully-welded connection 14-4.

6.4.3 Concluding Remarks

From the above discussions, it is clear that fracture at the junction of the tension beam flange and the moment plate near the tip of the column flange is the main problem which limits the ductility of a web moment connection. Fracture occurs as a result of a triaxial tensile stress state that develops there. Figure 6.33(a) shows that the longitudinal stress across the width of the beam flange at section A-A is highly non-uniform. This non-uniformity arises as a result of shear lag. The part of the moment plate that joins the inner face of the column flange is, relatively, stiffer than the part in the middle. Thus, stress tends to migrate to the edges of the moment plate. At section B-B (Fig. 6.33(b)), the stress is fairly uniform because the stiffness across the width of the beam flange is more or less constant. The shear stress along section C-C is also non-uniform (Fig. 6.33(c)). When a tensile force is applied to the moment plate, transverse and through-thickness strains will be induced due to the Poisson effect. However, because of the constraint of the column flanges, these strains can not be released along the welds of the moment plate and the column flanges. Consequently, a triaxial tensile stress state will develop there. This undesirable stress state, compounded by the high stress concentration at section

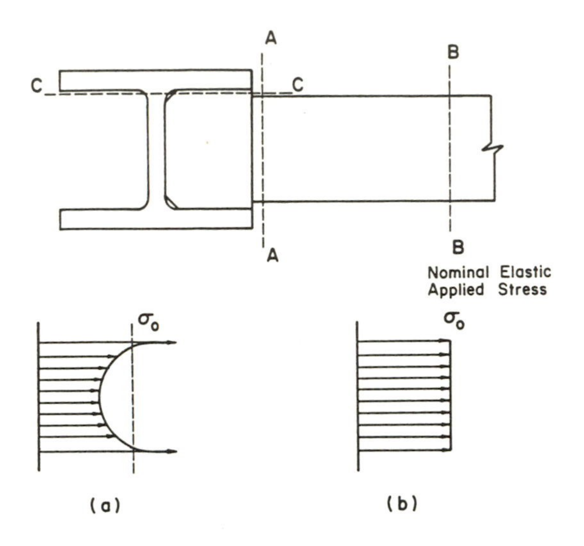

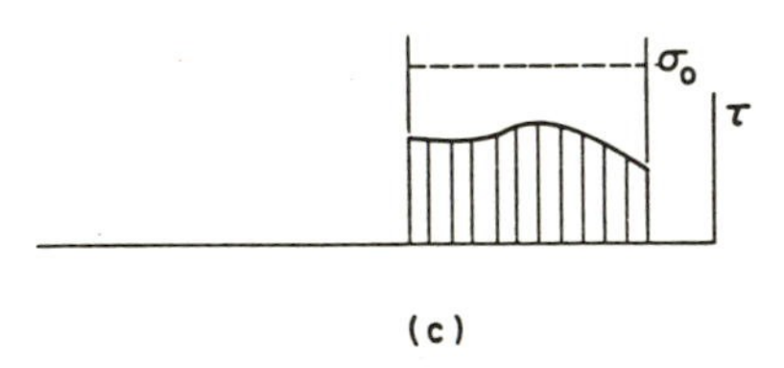

FIG. 6.33. Stress distributions (a) longitudinal stresses on section A–A; (b) longitudinal stresses on section B–B; (c) shear stresses on section C–C.

A-A and aggravated by the shear stress at section C-C, will ultimately cause fracture at the intersection of A-A and C-C.

To remedy this problem and to improve ductility for these types of web moment connections, the following suggestions were made (Driscoll and Beedle, 1982).

(1) Use oversized moment plates (Fig. 6.34(a)) to reduce the non-uniformity of tensile stresses across the width of the plate.

(2) Use a back-up stiffener (Fig. 6.34(b)) to reduce stress concentration at the column flange tip.

(3) Use an extended connection plate (Fig. 6.34(c), (d) and (e)) to avoid intersecting beam flange butt welds with the column flange fillet welds.

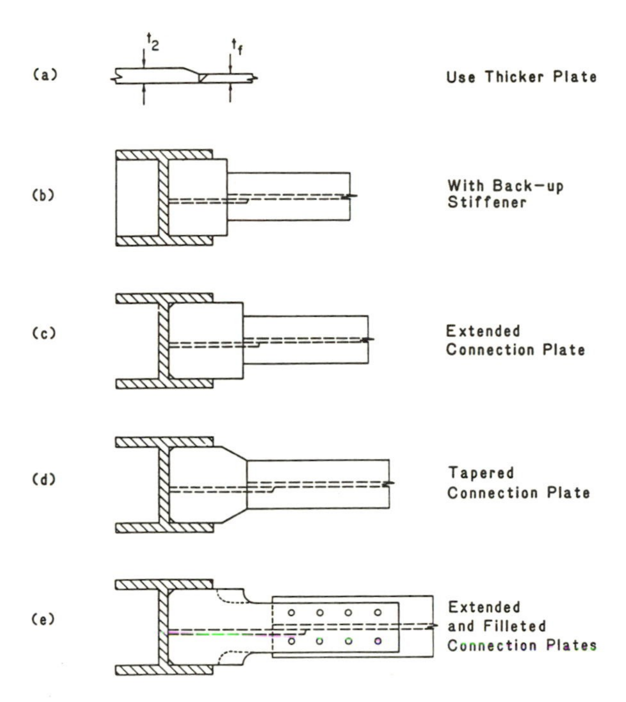

FIG. 6.34. Possible approaches, for use individually or in combination, for improving performance of tension flange connections to column web (Driscoll and Beedle, 1982).

(4) Use a tapered moment plate (Fig. 6.34(d)) or a moment plate with reduced width (Fig. 6.34(e)) to move the beam flange–moment plate juncture away from the critical section.

A test programme using the above suggested web connection details is now underway at Lehigh University. The results of these tests will give the designer more insight into the behaviour of web moment connections.

6.5 SEISMIC MOMENT CONNECTIONS

A seismic moment connection differs from a static moment connection in the following aspects:

(1) A seismic moment connection must be detailed to withstand forces which act in either direction.

(2) The problem of low cycle fatigue, which is associated with cyclic loadings at large plastic strain, becomes a factor to be considered in a seismic moment connection.
(3) Since seismic loadings are random in nature, a probabilistic rather than a deterministic approach of analysis is necessary to assess the behaviour of a seismic moment connection.

In order to warrant a probabilistic approach of analysis, a large amount of data is needed. Unfortunately, experimental data gathered up to the present time are not sufficient. Therefore, a quantitative evaluation of the reliability of seismic moment connections based on probabilistic concepts is not possible. Furthermore, as many of the experiments were carried out on small-scale samples under load or displacement control cyclic loadings, a direct correlation to full-size connections used in practice under random ground motions is questionable. In view of these difficulties, the discussion of seismic moment connections in the following sections will be confined to a qualitative rather than a quantitative description of connection behaviour. In particular, the behaviour of connections in a ductile moment-resisting frame will be discussed.

In a ductile moment-resisting frame, the members and joints are detailed to be capable of deforming well into the inelastic range without local failure or frame instability. In designing such frames, it is common practice to proportion the connections and the adjoining members in such a way as to confine plastic hinges to be developed in the beams. Although inelastic deformation in the panel zone of a connection is sometimes advisable, such deformation must be controlled so that deterioration, i.e. loss of strength and stiffness, of the connection is not significant.

6.5.1 Behaviour of Seismic Moment Connections

The behaviour of a seismic moment connection is represented by its hysteresis loops. An adequately designed moment connection should possess sufficient strength and ductility and should be strong enough to allow large rotation in the beams so that plastic hinges can be formed, providing ductility to the frame. Ductility is very important in seismic regions, to absorb and dissipate energy as well as to dampen the vibrations generated by ground motions. The area under the hysteresis loop is a measure of the ductility of the connection assemblage. Figure 6.35(a)–(d) shows four different stable hysteresis loops for a connection

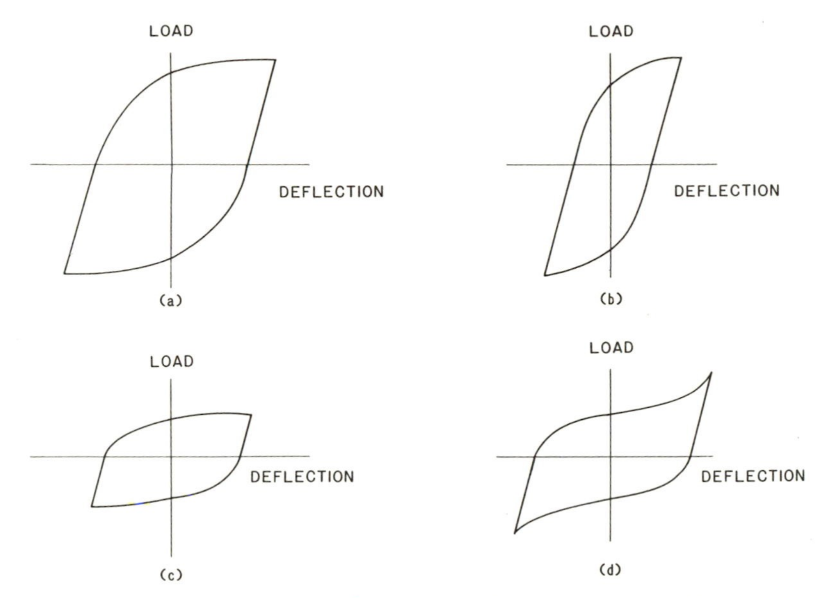

FIG. 6.35. Hysteresis loops of seismic moment connections.

assemblage. Connections which exhibit the behaviour of hysteresis loop A are desirable, whereas connections which exhibit the behaviour of hysteresis loops B, C and D are not desirable, due to lack of ductility, strength or stiffness.

The cyclic behaviour of moment-resisting connections and subassemblages was reported by Popov and Pinkney (1968, 1969), Popov and Stephen (1972), Bertero *et al.* (1972), Carpenter and Lu (1973), Krawinkler (1978) and Krawinkler and Popov (1982). Since most of the experiments were carried out on reduced-scale models, in which details of connections used in practice could not be simulated, the qualitative discussion of seismic moment connection behaviour will be drawn principally from the full-size specimens tested by Popov and Stephen (1972).

Figure 6.36 shows the joint detail and the hysteresis behaviour of a connection assemblage consisting of a W24×76 beam section. The connection is fully-welded. The beam flanges were connected to the column flange by full penetration welds and the beam web was connected to a shear plate, which was fillet welded to the column flange. A36 steel with a yield stress of 36 ksi (248 MPa) was used. It

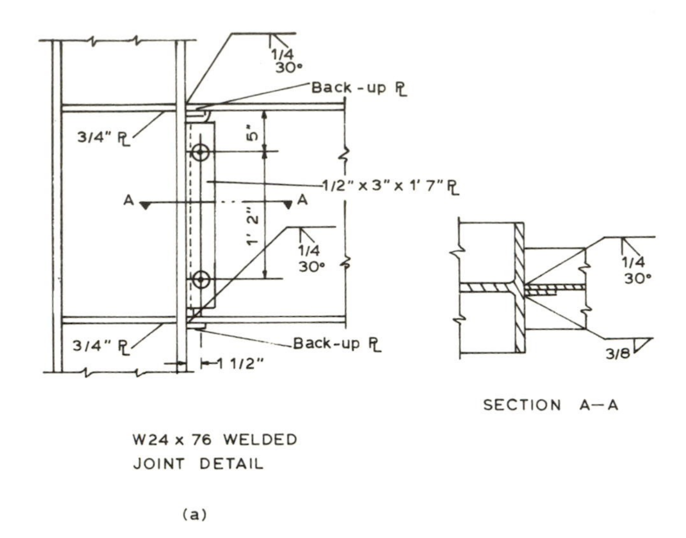

(a)

(b)

FIG. 6.36. (a) Joint details and (b) hysteresis loops for a fully-welded connection (1 in. = 25·4 mm, 1 kip = 4·448 kN).

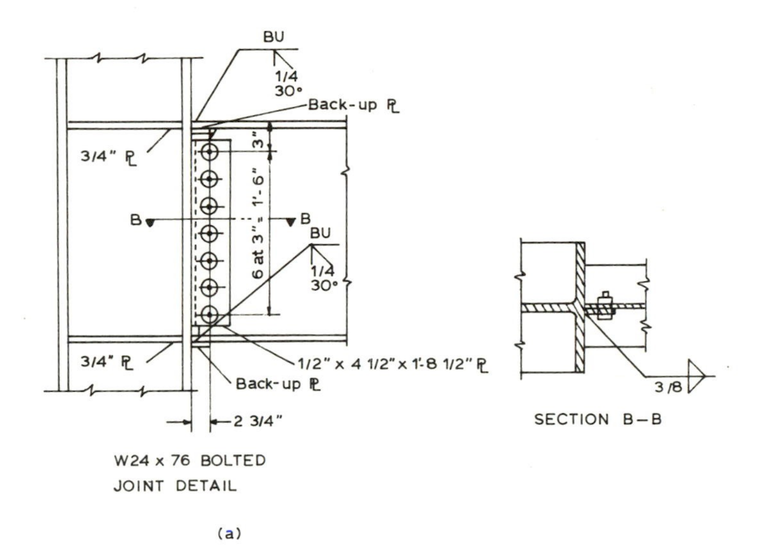

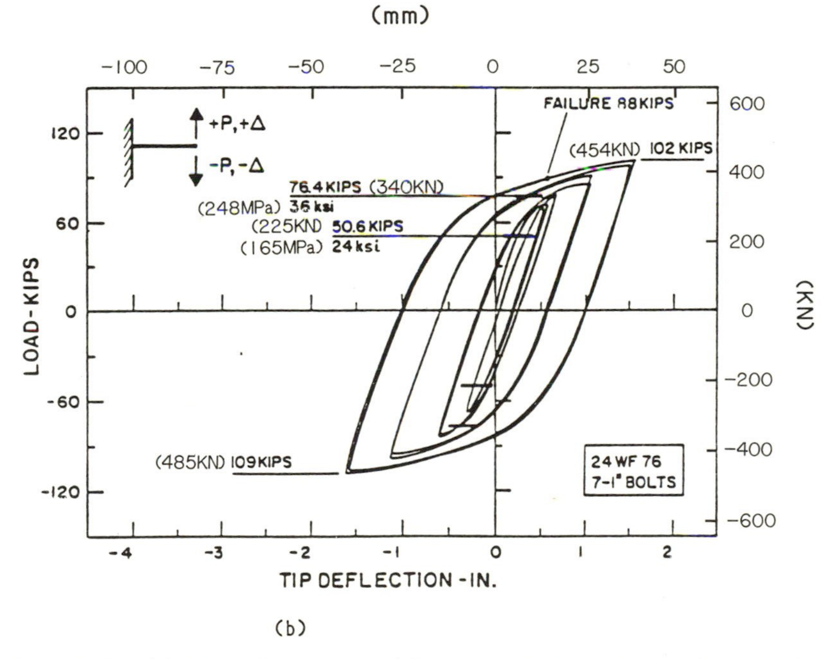

FIG. 6.37. (a) Joint details and (b) hysteresis loops for a flange-welded web-bolted connection (1 in. = 25·4 mm, 1 kip = 4·448 kN).

can be seen from the figure that adequate strength and good ductility are observed. The strength of the connection exceeded the plastic strength determined from simple plastic theory because of strain hardening of the steel.

Figure 6.37 shows the connection detail and the hysteresis behaviour of a flange-welded web-bolted connection assemblage. For this connection, the beam flanges were groove welded to the column flange. The beam web was attached to the shear plate with A325 high-strength bolts. Again, adequate strength and satisfactory ductility are observed. Although the ductility of this type of connection is somewhat inferior to that of a fully-welded connection, flange-welded web-bolted connections are favoured in practice for reasons of economy in fabrication and ease of erection.

Limited small-scale tests were performed on weak axis connections (Popov and Pinkney, 1967). In these connections, the beam flanges were butt-welded to horizontal column stiffeners and the beam web was fillet welded to vertical splice plates. The results of these tests showed a considerable variation in the deformation capacity between individual tests. The ductility of these web moment connections was significantly lower than that of flange moment connections due to stress concentration at critical sections, which ultimately led to fracture. In general, web moment connections are less reliable than flange moment connections under cyclic loadings.

6.5.2 Strength and Stiffness of Panel Zone Under Cyclic Loadings

Figure 6.38 shows the system of forces that acts on the panel zone of an interior beam-to-column seismic moment connection. Note that the forces act in either direction. Therefore, the column stiffeners must be capable of resisting both tensile and compressive forces. Furthermore, high shear force may develop in the panel zone which may cause excessive shear deformation. If that is the case, shear stiffening must be considered.

Shear stiffening of the panel zone is necessary for the following reasons.

(1) To strengthen the connection so that plastic hinges can be formed in the beams without premature failure in the connection.
(2) To limit the deformation of the panel zone so that drift and instability can be minimised.

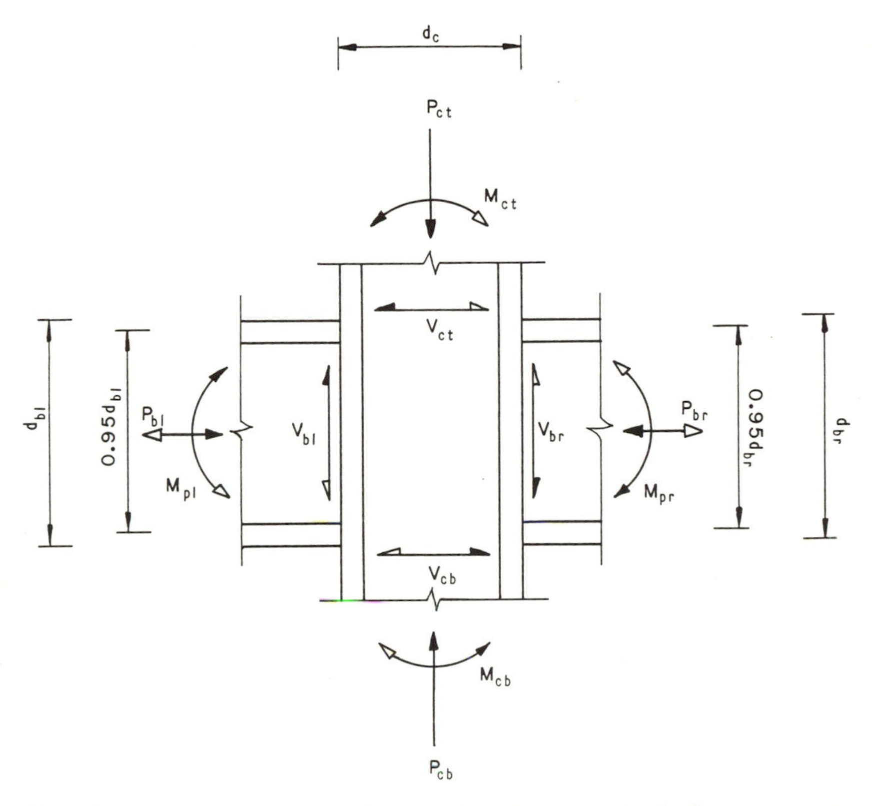

FIG. 6.38. Forces in the panel zone of an interior seismic flange moment connection.

In seismic moment connections, shear stiffening of the panel zone is usually provided by doubler plates. The suggested criterion for providing stiffeners was given by Krawinkler and Popov (1982).

If

$$\frac{M_{pl}}{0{\cdot}95d_{bl}}+\frac{M_{pr}}{0{\cdot}95d_{br}}-V_{col}>0{\cdot}55F_{yc}d_c w_c\left(1+\frac{3{\cdot}45b_c t_c^2}{d_b d_c w_c}\right) \qquad (6.11)$$

then stiffeners shall be provided.

In eqn (6.11), the symbols used are defined as follows: M_{pl} = plastic moment of the left beam, M_{pr} = plastic moment of the right beam,

d_{bl} = depth of the left beam, d_{br} = depth of the right beam, V_{col} = average column shear = $(V_{ct} + V_{cb})/2$, F_{yc} = yield stress of the column material, d_c = depth of the column, w_c = thickness of the column web, b_c = width of the column flange, t_c = thickness of the column flange, d_b = depth of the beam.

The design shear force (left hand side of eqn (6.11)) was calculated by assuming that the full plastic moment capacities of both the beams were developed under the most severe loading case, i.e. the moments on both sides of the connection are acting either both clockwise or both anti-clockwise. This assumption may be over conservative as there are cases where the development of two plastic hinges in the beams at the joints is not needed for the realisation of a mechanism motion, even in the severest earthquake. The shear capacity (right hand side of eqn (6.11)) was assessed by considering the contribution of the flexural resistance of the column flanges to the shear strength of the panel zone. The use of this shear capacity is valid if P/P_y is less than 0·5. It has been shown (Krawinkler, 1978) that this shear capacity is in good agreement with experimental results for joints with thin to medium thick column flanges.

In a series of experiments conducted by Bertero *et al.* (1973), it was found that in order to provide a more effective form of shear stiffening, the doubler plates should be placed as close to the column web as possible.

6.5.3 Fatigue Behaviour of Seismic Moment Connections

Deterioration of welded connections is usually a consequence of critical crack growth at welds under high stress concentrations. When a crack has grown to its critical size, fracture will occur. Therefore, an evaluation of the safety of these connections is a problem of low cycle fatigue (large plastic strain amplitudes) and elastic–plastic fracture mechanics. Because of the randomness of the applied loads and the strain history, a probabilistic approach should be used in assessing the safety of these connections. Unfortunately, due to the lack of a sizeable amount of data, a quantitative evaluation of the safety of connections is not possible. Nevertheless, the reader should bear in mind that fatigue crack growth will cause failure in seismic moment connections and so proper care must be exercised in designing and detailing these connections.

6.5.4 Concluding Remarks

The design of moment-resisting frames in seismic regions is usually based on two principles: serviceability under moderate earthquakes and safety under strong earthquakes. In order to satisfy these two criteria, the moment-resisting connections used in these frames should be detailed in such a way that they are strong enough to withstand deformation without loss in stiffness, and ductile enough to undergo inelastic deformation without premature failure. Based on experiments, it was found that fully-welded flange connections and flange-welded web-bolted flange connections were adequate as far as strength and ductility are concerned. Fully-bolted flange connections and web connections are not very reliable and so extra care should be exercised in their use.

In many cases, high shear force may develop in the panel zone of a seismic flange moment connection. As a result, shear stiffening must be considered.

An evaluation of the safety of these seismic moment connections should be based on probabilistic models because of the randomness in nature of the applied loads and strain history. However, a probabilistic approach is not justified at the present time because of the lack of experimental data. Further research, therefore, is needed in this respect.

6.6 SUMMARY AND CONCLUSIONS

A qualitative examination of the behaviour of moment-resisting beam-to-column connections under static and seismic loadings has been presented in this chapter. Based on full-sized connection tests, it was found that flange-welded web-bolted flange moment connections would give comparable performance to fully-welded connections. For static loadings, the use of slotted holes in the shear plate is satisfactory. For fully-bolted connections, bolt-slip at or under working load will decrease the stiffness of the connection and will affect the overall stability of the frame.

Web moment connections under static or seismic loadings are not very reliable and so further research is needed.

The evaluation of the safety of seismic moment connections should be based on probabilistic models, since the applied loads and strain histories are random in nature. In order to accomplish this, more experimental data are needed. Consequently, more experiments should be conducted before a quantitative analysis of the behaviour of these connections can be made.

REFERENCES

BERTERO, V. V., POPOV, E. P. and KRAWINKLER, H. (1972) Beam–column subassemblages under repeat loading. *Journal of the Structural Division, ASCE*, **98,** ST5, 113–59.

BERTERO, V. V., KRAWINKLER, H. and POPOV, E. P. (1973) Further studies on seismic behavior of steel beam–column subassemblages. EERC Report 73–27, Earthquake Engineering Research Center, University of California, USA.

CARPENTER L. D. and LU, L.-W. (1973) Reversed and repeated load tests of full-scale steel frames. Bulletin No. 24, American Iron and Steel Institute, New York.

CHEN, W. F. and NEWLIN, D. E. (1973) Column web strength in beam-to-column moment connections. *Journal of the Structural Division, ASCE*, **99,** ST9, 1978–84.

DRISCOLL, G. C. and BEEDLE, L. S. (1982) Suggestions for avoiding beam-to-column web connection failure. *AISC Engineering Journal*, **19**(1), 16–19.

FIELDING, D. J. and HUANG, J. S. (1971) Shear in steel beam-to-column connections. *The Welding Journal*, **50,** 3135–265.

GRAHAM, J. D., SHERBOURNE, A. N., KHABBAZ, R. N. and JENSEN, C. D. (1959) Welded interior beam-to-column connections. AISC Publication, A.I.A. File No. 13-C.

HUANG, J. S. and CHEN, W. F. (1973) Steel beam-to-column moment connections. ASCE National Structural Engineering Meeting, April 9–13, San Francisco, Pre-print No. 1920.

HUANG, J. S., CHEN, W. F. and REGEC, J. E. (1971) Test program of steel beam-to-column connections. Fritz Engineering Laboratory, Report No. 333.15, Lehigh University, Bethlehem, USA.

HUANG, J. S., CHEN, W. F. and BEEDLE, L. S. (1973) Behaviour and design of steel beam-to-column moment connections. WRC Bulletin No. 188.

KRAWINKLER, H. (1978) Shear in beam–column joints in seismic design of steel frames. *AISC Engineering Journal*, **15**(3), 82–91.

KRAWINKLER, H. and POPOV, E. P. (1982) Seismic behaviour of moment connections and joints. *Journal of the Structural Division, ASCE*, **108,** ST2, 373–91.

POPOV, E. P. and PINKNEY, R. B. (1967) Behaviour of steel building connections subjected to repeated inelastic strain reversal. University of California, Structural Engineering Laboratory, Report No. SESM 67–30.

POPOV, E. P. and PINKNEY, R. B. (1968) Behaviour of steel building connections subjected to inelastic strain reversals—experimental data. Bulletin No. 14, American Iron and Steel Institute.

POPOV, E. P. and PINKNEY, R. B. (1969) Cyclic yield reversal in steel building joints. *Journal of the Structural Division, ASCE*, **95,** ST3, 327–53.

POPOV, E. P. and STEPHEN, R. M. (1972) Cyclic loading of full-size steel connections. Bulletin No. 21, American Iron and Steel Institute.

RENTSCHLER, G. P. (1979) Analysis and design of steel beam-to-column web connections. PhD Dissertation, Department of Civil Engineering, Lehigh University, Bethlehem, USA.

RENTSCHLER, G. P., CHEN, W. F. and DRISCOLL, G. C. (1980) Tests of beam-to-column web moment connections. *Journal of the Structural Division, ASCE*, **106,** ST5, 1005–22.

RENTSCHLER, G. P., CHEN, W. F. and DRISCOLL, G. C. (1982) Beam-to-column web connection details. *Journal of the Structural Division, ASCE*, **108,** ST2, 393–409.

STANDIG, K. F., Rentschler, G. P. and CHEN, W. F. (1976) Tests of bolted beam-to-column moment connections. WRC Bulletin No. 218.

WITTEVEEN, J. W., STARK, J. W. B., BIJLAARD, F. S. K. and ZOETEMEIJER, P. (1982) Welded and bolted beam-to-column connections. *Journal of the Structural Division, ASCE*, **108,** ST2, 433–55.

Chapter 7

FLEXIBLY CONNECTED STEEL FRAMES

KURT H. GERSTLE

Department of Civil Engineering, University of Colorado, Boulder, USA

SUMMARY

After reviewing the historical practice, this chapter discusses the effects of connection flexibility on frame behaviour; in general it seems advantageous to assume linear elastic behaviour under working loads for computational purposes.

The extent to which the moments on the beams and columns are affected by connection flexibility is explored by studying a number of examples; it is shown that the neglect of connection rotation leads to serious underestimation of frame deflections.

A nonlinear analytical procedure is suggested, but appears to be unsuitable for routine design office practice; however, computerised design of flexibly connected building frames is possible and appears useful in the design office.

NOTATION

A	Cross-sectional area of beam flange
C_i	Connection coefficients (Frye and Morris, 1975)
d	Beam depth
EI	Flexural beam stiffness
F_y	Yield strength
K	Structure stiffness; Moment scaling factor (Frye and Morris, 1975)
k	Rotational connection stiffness

k_i	Characteristic connection stiffnesses
k^i	Stiffness of ith element
L	Beam length
l	Fictitious connection element length
M	Moment applied to connection
M_{el}, M_{pl}, M_u, M_y	Characteristic connection moments
p	Dimensionless load parameter
Q	Dimensionless stiffness coefficient
S	Member force
U	Ultimate load
w	Uniform beam load
X	Nodal force
X^F	Fixed-end force
Δ	Nodal displacement
θ	Connection rotation
ψ	Dimensionless connection stiffness parameter

7.1 INTRODUCTION

7.1.1 Effects of Connection Flexibility on Frame Behaviour

Beam-to-column connections are flexible, ranging in stiffness from those almost rigid to some types close to pinned. This connection flexibility can be characterised by the moment–rotation curves shown in Fig. 7.1 (Frye and Morris, 1975) in which the slope of the curve represents the rotational stiffness k of the connection. Conversely, $1/k$ indicates its rotational flexibility.

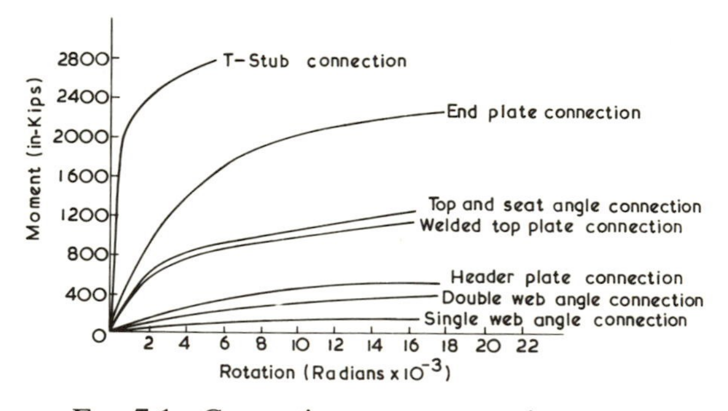

FIG. 7.1 Connection moment–rotation curves.

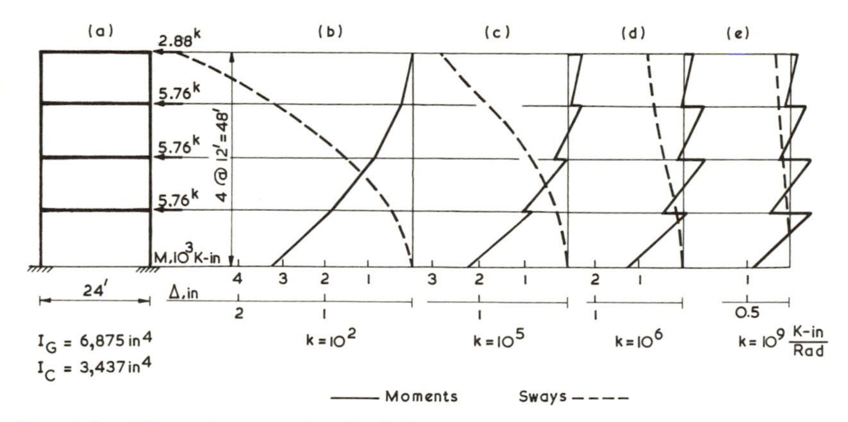

FIG. 7.2 Effect of connection flexibility on frame sway and column moments.

To illustrate the effects of connection flexibility on the behaviour of unbraced multi-storey frames, we consider a four-storey, single-bay frame under lateral load (Kahl, 1976), as shown in Fig. 7.2(a). Figures 7.2(b)–(e) show column moments and deflected shapes of the frame, for four different girder–column connection stiffnesses ranging from near-pinned to near-rigid. As the connections increase in stiffness, the frame response changes from flexural cantilever behaviour to the shear-type response generally associated with rigid frames. It is apparent that the assumption of perfectly rigid joints may lead to serious underestimation of both column moments and frame sways.

The effects of this rotational connection flexibility on unbraced frames are thus twofold:

(1) The joint rotation contributes to the overall frame deformations, in particular the frame sway under lateral loads. This reduction in frame stiffness will also affect the natural period of vibration and, therefore, the dynamic response to earthquake motions.

(2) The joint rotation will affect the distribution of internal forces and moments in girders and columns. An analysis which neglects connection deformation may thus be unable to arrive at realistic predictions of stresses and deflections.

7.1.2 Historical Review and Current Practice

Early analyses of frames with flexible connections were carried out by Batho and Rowan (1931), Rathbun (1936), Lothers (1960) and others. Modern matrix analysis was applied by Goble (1963), Romstad and

Subramanian (1970) and many others. Although relatively straightforward, flexibly-connected frame analysis by matrix methods has rarely been used in practice, possibly because of lack of appropriate computer programs and insufficient information about actual connection stiffnesses.

Current US practice is summarised in the *Specifications of the American Institute of Steel Construction* (AISC, 1980). Section 1.2 of these specifications lists three 'Types of Construction' (which should, more appropriately, be called 'Types of Analysis').

Type 1, or 'rigid frame' analysis; this has traditionally been considered the 'exact' method in structural practice.

Type 2, or 'simple framing', which accounts for the effects of connection flexibility in an approximate sense by assuming the girders simply supported under gravity loads, while assuming rigid girder–column connections for lateral loads.

Type 3, or 'semi-rigid framing', which calls for rational analysis including the effects of connection rotation.

Type 1 and Type 2 procedures have been widely used and have resulted in serviceable building frames, but of unknown stiffness, strength, and economy. Type 3 analysis, extending over the elastic and inelastic ranges, is needed to answer questions regarding these factors.

7.1.3 Outline of Approach

In this chapter, we will present analysis and design methods capable of realistic predictions of unbraced frame response with flexible connections, and attempt to delineate conditions under which connection flexibility should be included in analysis. Modern methods of analysis allow consideration of connection flexibility without placing an undue burden on the designer.

Accordingly, the sequence will be to present in the next section some information about connection properties, followed by a simple linear matrix method for computer analysis to predict frame response under working loads, following traditional working stress approaches.

Determination of frame strength requires more refined analyses which consider both material nonlinearity (plastification) and geometric nonlinearity (frame or member instability). Such analyses are probably beyond office routine; we will discuss them only shortly and present some results, along with experimental verification. Lastly, the possibilities of automated design procedures for optimised, flexibly-connected unbraced building frames will be discussed. The entire presentation is intended for the professional designer with the aim of allowing more rational predictions, which, when verified experimentally, may lead to more reliable design procedures and more economic frames.

Significant Conclusions
Connection flexibility has significant effects on frame behaviour. For realistic predictions, it should be included in analysis and design.

7.2 CONNECTION BEHAVIOUR

7.2.1 Available Data

It is interesting to note that in spite of various attempts (Frye and Morris, 1975; Krishnamurthy, 1979) no reliable method for prediction of connection response has been accepted by the profession. It provides food for thought that in spite of all analytical progress of recent years, such a basic problem still escapes our understanding.

In the absence of analytical solutions, reliance must be placed on test results; connection testing has been carried out only sporadically since the 1930s (Batho and Rowan, 1931; Hechtman and Johnson, 1947; Chesson and Munse, 1958). Complete, systematic test programmes of specific connection types covering a full range of sizes and conditions are rare (Krishnamurthy, 1979). In particular, experimental data on the behaviour of modern high-strength bolted connections are sadly lacking. Some of the data which form the basis of Figs. 7.1 and 7.3 are of riveted joints! New connection research is needed to establish reliable stiffness data for use by analysts.

7.2.2 Nonlinear Moment–Rotation Behaviour

Figure 7.1 shows the typical nonlinear moment–rotation behaviour of several connection types. To be used in analysis, these curves must be represented analytically.

Frye and Morris (1975) used experimental results from all available investigations to represent all of the curves in Fig. 7.1 in the form of a power series relating the rotation θ to the scaled moment KM

$$\theta = C_1(KM) + C_2(KM)^3 + C_3(KM)^5$$

The coefficients C_i are specified by Frye and Morris (1975) for the different connection types considered. The scaling factors K are also given as a function of connection and fastener size, proportion and material thickness.

While a thorough documentation of the accuracy of this representation is still outstanding, it appears at this time to provide the best tool for prediction of the response of a wide variety of connection types to monotonic loadings.

7.2.3 Linearisation of Connection Behaviour

It has been observed that after loading of connections along the nonlinear path shown in Fig. 7.1, unloading and moderate moment reversal will take place along a linear path, of slope similar to the initial stiffness of the loading curve, and that thereafter the connection response to load variations at the working level will proceed elastically; that is, the connection will 'shake down' to the elastic state. Thus, the choice of the initial elastic stiffness for linear representation of the connection response seems appropriate.

An alternative choice for the linear connection stiffness (as, for instance, taken in Fig. 7.11(b)) is the secant stiffness from the origin to the point on the curve representing the working, or allowable, connection moment. Because of variability of actual connection behaviour, and the inevitable scatter due to fabrication practices, extreme care in the choice of linear stiffness appears unjustified; a fair approximation is probably sufficient.

The validity of such linearisation will be documented later in this chapter.

7.2.4 Strength–Stiffness Relations

While the Frye and Morris representation of the connection response enables determination of its stiffness once it has been selected, the structural designer should be able to anticipate this stiffness during the preliminary design phase. For this purpose, it seems desirable to relate connection stiffness to the required connection strength.

To do this, we consider an idealised moment-resistant connection with effective cross-sectional area A concentrated at each girder flange; for a girder depth d, the ultimate moment capacity M_u is approximately

$$M_u = \frac{A}{2} \,.\, F_y \,.\, d$$

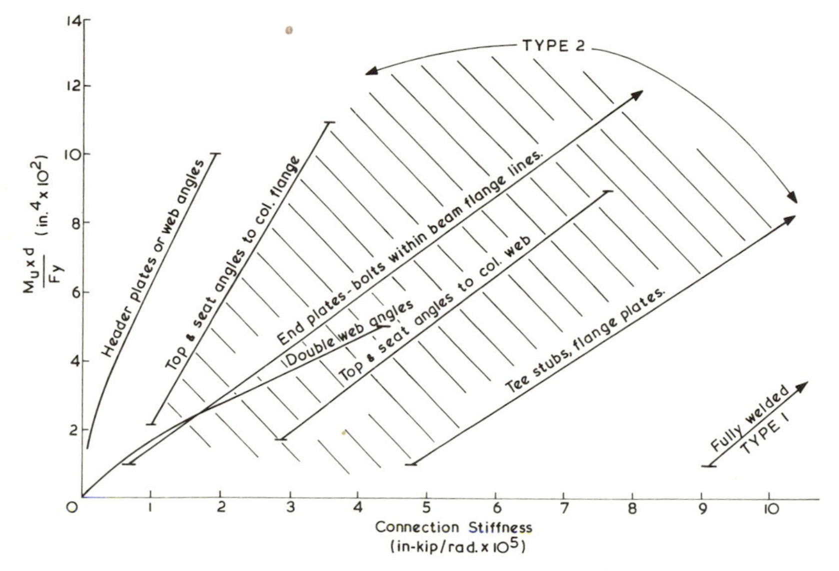

FIG. 7.3. Relation between connection strength and stiffness.

and the linearised bending stiffness is proportional to the square of the depth

$$k \approx A\,.\,(d/2)^2$$

Eliminating A, we obtain the proportionality

$$k \approx \frac{M_u\,.\,d}{F_y}$$

This relationship between strength and stiffness, as inferred from available experimental results (Ackroyd and Gerstle, 1977) for several connection types, is shown in Fig. 7.3, which allows estimation of connection stiffness once the required moment and girder depth is known. This enables the designer to include the connection stiffness in refined analysis as soon as an estimate of the joint moments is available during preliminary design.

7.2.5 Moment–Rotation Behaviour under Load Cycles

Systematic connection tests under load cycles are even scanter than those under monotonic loading, but some information is available (Popov and Pinkney, 1969; Marley, 1982) on the basis of which the

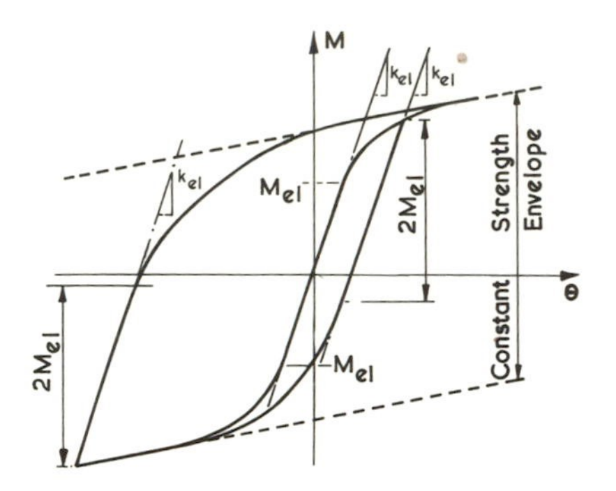

FIG. 7.4. Connection behaviour under load cycles.

characteristics shown in Fig. 7.4 may be assumed (Moncarz and Gerstle, 1981).

The following information can be obtained from connection tests under monotonically increasing moment: (1) initial modulus k_{el}, (2) proportional limit M_{el}, (3) shape of nonlinear portion of curve, and (4) asymptotic linear strain-hardening envelope.

For prediction of the response under load histories, these additional assumptions can be made, based on experimental curves (Marley, 1982): (5) elastic unloading with modulus k_{el}, (6) constant elastic range of extent $2M_{el}$, (7) equal positive and negative strength envelopes. A moment–rotation curve following these assumptions is shown in Fig. 7.4.

7.2.6 Experimental Verification

A test program was carried out to confirm the validity of the above formulation (Marley, 1982) for the particular case of top-and-seat angle connections under load cycles. Figure 7.5 shows measured and analytical moment–rotation curves for a pair of $4\,\text{in} \times 4\,\text{in} \times \frac{1}{2}\,\text{in}$ ($102 \times 102 \times 13$ mm) angles, 5 in (127 mm) long, connected by $\frac{3}{4}$ in (19 mm) diameter A-325 bolts to W5 × 16 members (i.e., 127 mm deep weighing 23·7 kg/m). This plot shows excellent coincidence between test results and the piecewise-linear formulation to be presented later, which follows the above hypotheses. It is also observed that linearly-elastic behaviour can be assumed with fair accuracy up to an applied moment equal to about one-half of the ultimate strength; that is, over most of the service range of the connection.

These results cannot necessarily be extrapolated to other connection

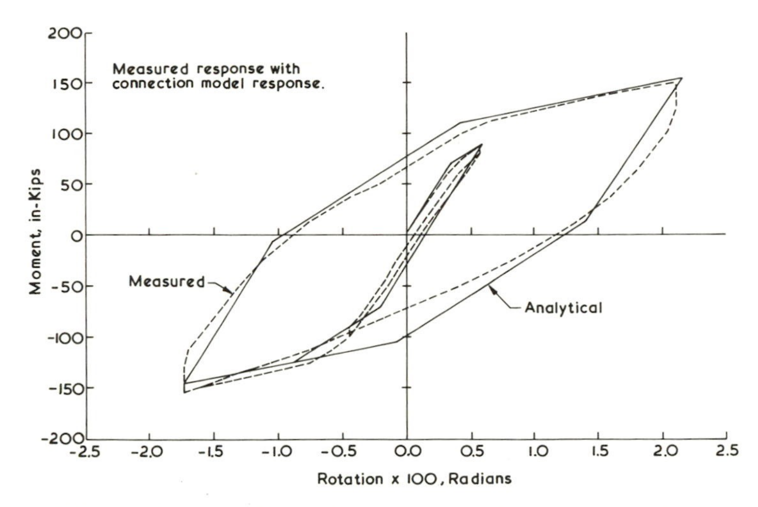

FIG. 7.5. Analytical and experimental connection response to load cycles.

types, or to situations in which member distortion contributes significantly to the rotation. For such cases, the proposed formulation can only be considered to be a hypothesis in the absence of further test results.

7.2.7 Effects of Panel Shear (Kato *et al.*, 1984)

In addition to connection rotation, which is of importance in lightly welding or field-bolted joints, distortion of the column web due to shear yielding, as shown in Fig. 7.6(a), should be considered. Figure 7.6(b) (Kato *et al.*, 1984) indicates that such distortions are significant only after the shear stress in the panel zone has attained its yield value. Therefore, if this panel zone has been properly designed by providing adequate stiffeners and web thickness (possibly by doubler plates), panel distortion is not important under working loads. Little is known about its effects on frame strength, and we will not consider it any further.

Significant Conclusions

(1) *Insufficient connection information is available at this time.*

(2) *Linearly-elastic connection behaviour may be assumed for prediction of frame behaviour under working loads.*

(3) *Joint panel distortion can be neglected in frame design.*

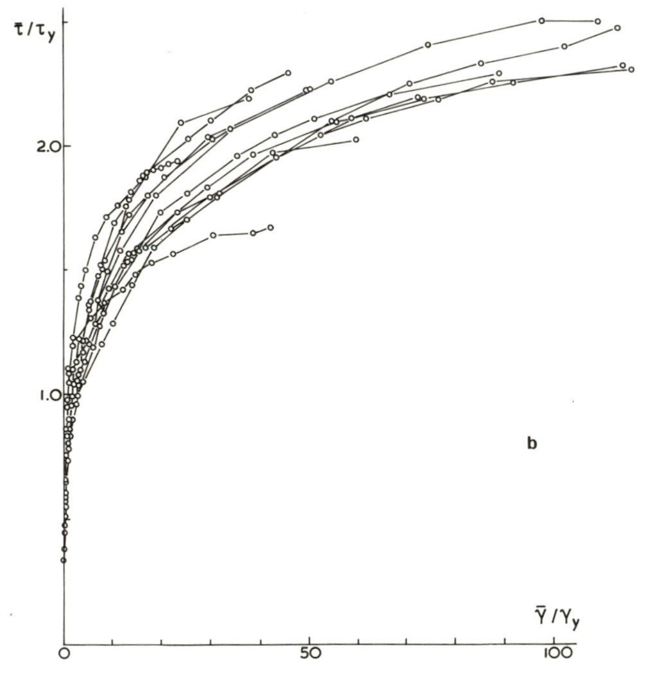

Fig. 7.6. Effects of panel shear: (a) joint distortion; (b) panel shear stress–strain curves.

7.3 FLEXIBLY-CONNECTED FRAME ANALYSIS—LINEAR APPROACH

7.3.1 Introduction

Since connection flexibility in unbraced multi-storey building frames can be a primary determinant of both deformations and internal force distribution, it appears important for professional designers to have the capability of analysing flexibly-connected frames. The linearly-elastic method outlined in this Section appears well suited for current office practice, since it requires only minor modifications of commonly used procedures, and can be understood by anyone familiar with modern matrix methods of structural analysis.

In common with other linearly-elastic analyses, this approach can only predict the response of the structure working level with confidence. In order to predict strength and conditions of failure, nonlinear analyses are required which consider the effects of nonlinear connection behaviour, member plastification and second-order deformations leading to buckling. Some of these nonlinear effects, whose rational analysis is probably beyond routine office capability, will be considered in a further section.

7.3.2 The Displacement Method

The displacement method states that the forces $\{X\}$ applied to numbered nodes of the structure are balanced by those associated with the nodal displacements $\{\Delta\}$, $[K]\{\Delta\}$, and those due to loads acting on fixed-ended members $\{X^{f}\}$:

$$\{X\}=[K]\{\Delta\}+\{X^{f}\} \tag{7.1}$$

The structure stiffness matrix $[K]$, a square matrix equal in order to the number of degrees of freedom of the structure, is assembled from the stiffness matrices $[k^{i}]$ of the individual elements.

Equation 7.1 is solved for the unknown displacements $\{\Delta\}$, which are then used to determine the forces $\{S\}$ on each individual element; for the ith element, for instance, the member end forces $\{S^{i}\}$ are given by

$$\{S^{i}\}=[k^{i}]\{\Delta\}-\{X^{fi}\} \tag{7.2}$$

thus completing the analysis.

This method is described in many texts (Gerstle, 1974; Weaver and Gere, 1980) and displacement method software constitutes the mainstay of structural analysis in modern office practice. Most of these programs are intended for rigid-frame analysis and do not permit the inclusion of connection flexibility without modifications. As outlined in the following, such modifications are reasonably straightforward and would enhance the analytical capabilities of offices engaged in steel frame analysis considerably.

7.3.3 Element Stiffness and Fixed-End Moments

The effects of connection flexibility are modelled by attaching rotational springs of moduli k_1, k_2 to the girder ends, as shown in Fig. 7.7(a). By classical methods, the following girder stiffnesses are obtained (Moncarz and Gerstle, 1981)

$$k_{33}=\frac{-\left(\frac{L}{EI}+\frac{1}{k_1}+\frac{1}{k_2}\right)}{L^2\left[\left(\frac{L}{2EI}+\frac{1}{k_1}\right)^2-\left(\frac{L}{3EI}+\frac{1}{k_1}\right)\left(\frac{L}{EI}+\frac{1}{k_1}+\frac{1}{k_2}\right)\right]}$$

$$k_{43}=k_{34}=\frac{\left(\frac{L}{2EI}+\frac{1}{k_1}\right)}{L\left[\left(\frac{L}{2EI}+\frac{1}{k_1}\right)^2-\left(\frac{L}{3EI}+\frac{1}{k_1}\right)\left(\frac{L}{EI}+\frac{1}{k_1}+\frac{1}{k_2}\right)\right]} \tag{7.3}$$

$$k_{44}=\frac{-\left(\frac{L}{3EI}+\frac{1}{k_1}\right)}{\left(\frac{L}{2EI}+\frac{1}{k_1}\right)^2-\left(\frac{L}{3EI}+\frac{1}{k_1}\right)\left(\frac{L}{EI}+\frac{1}{k_1}+\frac{1}{k_2}\right)}$$

The remaining stiffnesses can be found by statics and by symmetry of the stiffness matrix.

The fixed-end forces on the member ends due to a uniform downward load w are

$$x_3^{\mathrm{f}}=\frac{wL}{2}\times\frac{\left(\frac{L}{2EI}+\frac{1}{k_1}\right)\left(\frac{L}{3EI}+\frac{1}{k_1}\right)-\left(\frac{L}{4EI}+\frac{1}{k_1}\right)\left(\frac{L}{EI}+\frac{1}{k_1}+\frac{1}{k_2}\right)}{\left(\frac{L}{2EI}+\frac{1}{k_1}\right)^2-\left(\frac{L}{3EI}+\frac{1}{k_1}\right)\left(\frac{L}{EI}+\frac{1}{k_1}+\frac{1}{k_2}\right)}$$

$$x_4^{\mathrm{f}}=-\frac{wL^2}{2}\times\frac{\left(\frac{L}{3EI}+\frac{1}{k_1}\right)^2-\left(\frac{L}{4EI}+\frac{1}{k_1}\right)\left(\frac{L}{2EI}+\frac{1}{k_1}\right)}{\left(\frac{L}{2EI}+\frac{1}{k_1}\right)^2-\left(\frac{L}{3EI}+\frac{1}{k_1}\right)\left(\frac{L}{EI}+\frac{1}{k_1}+\frac{1}{k_2}\right)} \tag{7.4}$$

The remaining fixed-end forces can be found by statics.

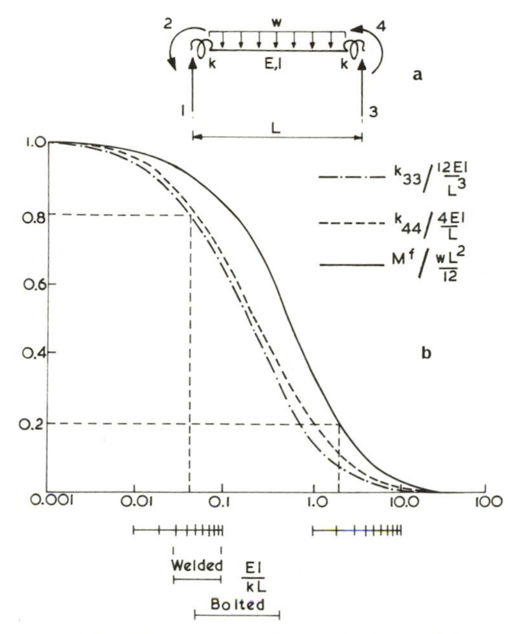

FIG. 7.7. Connection flexibility as girder end effect: (a) beam element with flexible connections; (b) effective range of connection flexibility.

Equations 7.3 and 7.4 are for the general case of unequal connection stiffnesses k_1 and k_2. Usual steel building frames will have identical connections at both girder ends (although exterior and interior connections may act differently), and the analysis will then deal with equal stiffnesses $k_1 = k_2 = k$, in which case the element stiffnesses and fixed-end forces under uniform load simplify to

$$k_{33} = \frac{12EI}{L^3} \cdot \frac{1}{Q}\left[1 + 2\frac{EI}{kL}\right] \tag{7.5a}$$

$$k_{43} = \frac{6EI}{L^2} \cdot \frac{1}{Q}\left[1 + 2\frac{EI}{kL}\right] \tag{7.5b}$$

$$k_{44} = \frac{4EI}{L} \cdot \frac{1}{Q}\left[1 + 3\frac{EI}{kL}\right] \tag{7.5c}$$

$$x_3^f = \frac{wL}{2} \tag{7.6a}$$

$$x_4^f = \frac{wL^2}{12} \cdot \frac{1}{1 + 2\dfrac{EI}{kL}} \tag{7.6b}$$

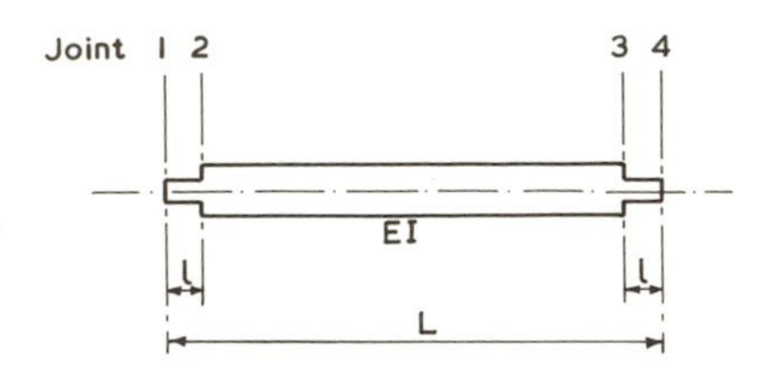

FIG. 7.8. Flexible connections as beam haunches.

in which

$$Q = 12\left(\frac{EI}{kL}\right)^2 + 8\left(\frac{EI}{kL}\right) + 1 \tag{7.7}$$

Thus, rigid-frame analysis programs can be enhanced by replacing the standard stiffnesses and fixed-end forces by these modified values. The only additional input information in this case is the appropriate connection moduli k.

The connection flexibility can also be included in computer programs for rigid frames with prismatic members, by simulating the flexible connection as a short, fictitious, soft elastic beam element of length and stiffness EI_{con}, as shown in Fig. 7.8. The total rotational stiffness of the soft element is to be equal to the rotational connection stiffness k

$$M/\theta = \frac{EI_{con}}{l} = k \tag{7.8}$$

The element should be short, and its shear stiffness should be set very large. One additional node is then needed for each flexibly-connected girder end.

Significant Conclusions

(1) *Inclusion of connection flexibility in linearly-elastic frame analysis is conceptually and computationally simple.*

(2) *At some computational expense, rigid-frame computer programmes can be used for the analyses of flexibly-connected frames.*

7.4 BEHAVIOUR OF FLEXIBLY-CONNECTED FRAMES UNDER WORKING LOADS

7.4.1 Introduction

The validity of the linearly-elastic approach will be documented in the next section by comparison with nonlinear analysis results. Pending this validation, we will use linear analysis including connection flexibility to explore how these connection rotations affect the frame response under working loads.

7.4.2 Range of Effective Connection Flexibility (Ackroyd and Gerstle, 1982)

Equations 7.5–7.7 show that element stiffnesses and fixed-end forces reflect connection flexibility through only one non-dimensional parameter, EI/kL, which may be interpreted as the ratio of rotational member stiffness to the rotational connection stiffness. Accordingly, this ratio can serve as an index to determine the importance of connection flexibility in structural analysis.

Figure 7.7(b) plots the ratios of the stiffnesses k_{33} and k_{44}, and that of the fixed-end moment X_4^f of the flexibly-connected member to those of the corresponding rigidly-connected member, as a function of the parameter EI/kL. These ratios vary from zero for very soft connections to unity for rigid connections. For values of $EI/kL < 0{\cdot}05$, the values of these quantities will be within 20 per cent of those for rigid joints, and connection flexibility can reasonably be neglected. For values of $EI/kL > 2{\cdot}0$, these quantities will be within 20 per cent of those for the ideal pin-end, so that this condition can well be assumed.

It follows that the effects of connection flexibility should be considered for cases in which the ratio of girder to connection stiffness falls within the range

$$0{\cdot}05 < EI/kL < 2{\cdot}0 \tag{7.9}$$

A review of typical building frame designs (Ackroyd and Gerstle, 1977) indicated that the range of girder stiffnesses EI/L is from $0{\cdot}5 \times 10^5$ to $1{\cdot}0 \times 10^5$ in-kips/radian. The range of connection stiffnesses typically found in field-bolted frames is from $2{\cdot}0 \times 10^5$ to $10{\cdot}0 \times 10^5$ in-kips/radian, while for welded frames the range is from $10{\cdot}0 \times 10^5$ to $50{\cdot}0 \times 10^5$ in-kips/radian. These ranges translate into stiffness ratios of

$$0{\cdot}05 < EI/kL < 0{\cdot}5 \qquad \text{for bolted frames}$$

$$0{\cdot}02 < EI/kL < 0{\cdot}1 \qquad \text{for welded frames}$$

It appears that field bolted, or lightly welded, frames should be analysed as flexibly-connected, but frames with reasonably heavy welded connections might be assumed rigid with fair accuracy.

7.4.3 Internal Moments in Flexibly-Connected Frames

To explore the extent to which column and girder moments are affected by connection flexibility, a series of frames was designed according to common US office practice (Ackroyd and Gerstle, 1977). A preliminary Type 2, or 'simple framing' analysis (according to Section 1.2 of the AISC Specifications) was carried out for an extensive series of building frames of various proportions subject to specified floor- and wind-loads, and members were sized in accordance with AISC specifications. Connection stiffnesses for the specified type of joint were selected from Fig. 7.3.

With member and connection stiffnesses known, linear analyses including the effects of connection flexibility were performed and critical moments in girders and columns were determined (this analysis will be called 'rigorous' or Type 3). The assumption of various connection types, ranging from floppy to rigid, permitted assessment of both the validity of preliminary analysis methods as well as the effects of connection flexibility.

Of the numerous results, only those for a three-bay, five-storey, unbraced frame of storey height 12 ft, bay width 20 ft, spaced at 30 ft centres, will be presented; these data are quite representative of the totality. Three feasible connection types were considered; very soft top-and-seat-angle-connections, fairly rigid flange plate connections, and, finally, perfectly rigid joints representing fully-welded connections, leading to results as would be obtained from classical rigid-frame analysis.

Figure 7.9 shows results in non-dimensional form. The moments

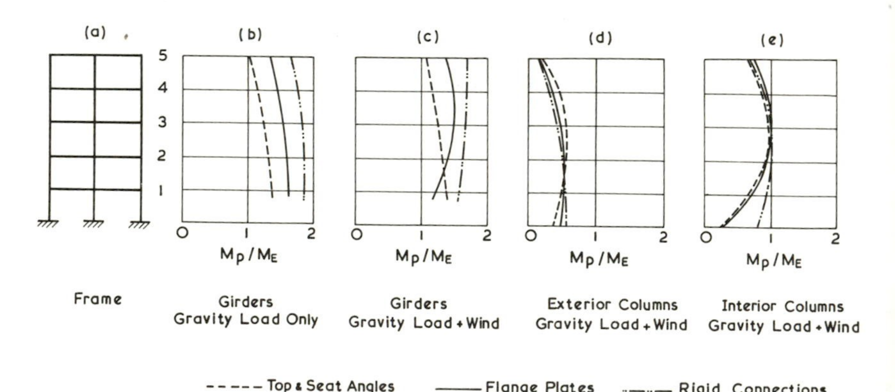

FIG. 7.9. Ratio of preliminary to exact design moments for five-storey frame.

from preliminary Type 2 analysis, which were used to size the members, were divided by the critical moments obtained from rigorous analysis, and entered along the abscissas of Fig. 7.9; storey level was plotted along the ordinate. Positive girder moment ratios (due to gravity load), and moment ratios for exterior and interior columns (due to gravity and lateral loads) are plotted separately in this figure. A moment ratio $M_{prel}/M_{rigorous}$ of 1 would indicate perfect agreement between preliminary and rigorous analysis; values less than 1 indicate unconservative, and greater than 1 overconservative, member sizing.

The results indicate that girder design moments are highly sensitive to connection flexibility; column moments are much less dependent on connection behaviour. Type 2 analysis consistently overestimates girder moments and underestimates column moments. A 'Type 2 frame', therefore, is characterised by girders which are too large, and columns that are too small, when checked by Type 3 analysis methods.

7.4.4 Sway of Flexibly-Connected Frames

The top-storey sways of the family of frames, which ranged in height from five to twenty-five 12-ft storeys, were among the primary results of the linear analyses which included connection flexibility (Ackroyd and Gerstle, 1982). They are shown in non-dimensional form in Fig. 7.10. The ordinates represent the ratio of top-storey drift to building height, while the abscissae represent the aspect ratios or slenderness of the building. Three different connection types are considered, the stiffnesses of which were taken from Fig. 7.3 for the corresponding design moments; floppy top-and-seat angle-connections, fairly stiff flange plates, and perfectly rigid joints representing fully welded joints.

The curves of Fig. 7.10 indicate the importance of the effect of connection flexibility on frame sway; when moderately flexible flange plates are used, the sway will increase about 40 per cent over that predicted by rigid-frame analysis; when softer top-and-seat angles are used, the sway can exceed that of a rigid frame by from 100 to 200 per cent; that is to say, connection rotations account for one third–two thirds of the total sway; elastic member distortions may be responsible for only a minor amount of the total deflections.

By drawing a horizontal line in Fig. 7.10 opposite the specified maximum sway ratio, the permissible slenderness can be obtained for unbraced frames with various types of joints. However, it may well be that the widely-used sway ratio of 1/400 was adopted in full realisation that actual sway might exceed that predicted by rigid-frame analysis; it

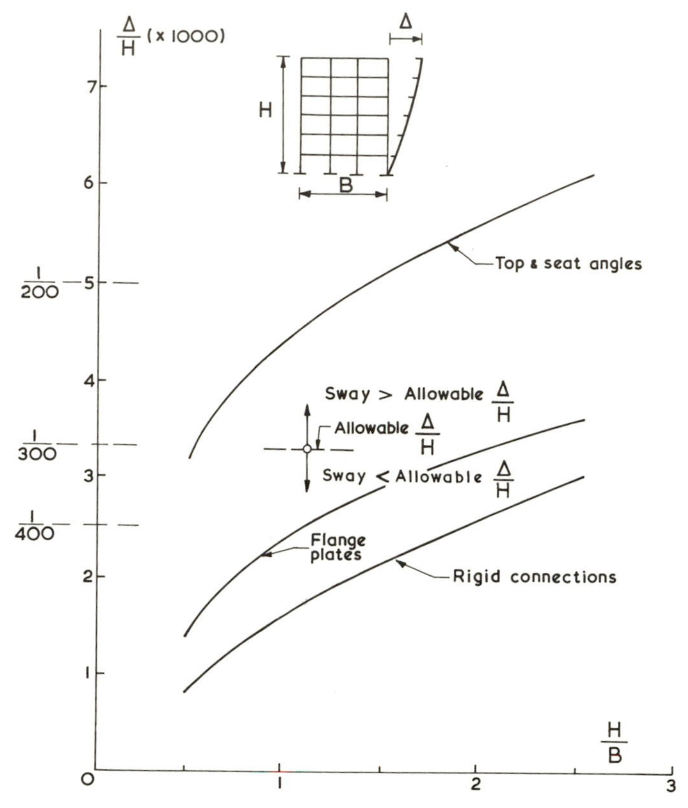

FIG. 7.10. Frame drift versus frame slenderness for different connection flexibilities.

is quite likely that with more realistic analysis, these maximum allowable sway ratios may be increased, say to 1/300 or more.

Significant Conclusions

(1) *The ratio of rotational connection to girder stiffness offers a useful criterion for assessing the importance of connection flexibility.*

(2) *Internal member forces may be strongly affected by connection rotations.*

(3) *Neglect of connection rotation may lead to serious underestimation of frame deflections.*

7.5 FLEXIBLY-CONNECTED FRAME ANALYSIS—NONLINEAR APPROACH

7.5.1 Introduction

The nonlinear approaches discussed in this section are probably well beyond routine office practice. The results of such analyses, however, can be useful in assessing the range of validity of the simpler analysis methods, to understand the importance of various secondary effects and to determine strength and failure modes of structures.

Here, we will shortly outline an analytical approach, and then discuss its application to three different nonlinear effects; those of connection nonlinearity and response to load cycles, those of member plastification, and those of geometric nonlinearity leading to member and frame instability.

Because it is unrealistic to expect such analyses to be performed routinely, it is important that available results be cast in such form that general conclusions may be drawn for design. We will attempt to summarise some of the available results in such a fashion in subsequent sections.

7.5.2 Analytical Approach (Ackroyd and Gerstle, 1983; Cook, 1983)

A powerful way of analysing nonlinear structural behaviour is by means of the step-by-step, or piecewise-linear, approach. The load is applied in suitably small increments; at the beginning of each load increment, the current stiffness of each element of the structure and the current structure displacements, which may include second order or time-dependent contributions, and, in the case of dynamic analyses, accelerations, are known from the preceding load step; these are used to carry out a linear analysis for the additional forces and displacements occurring during the current load increment, to be added to those carried forward from previous load steps. Iterations may be carried out to improve solutions and, in most cases, any chosen accuracy can be attained by suitable choice of step size and iteration scheme.

Such calculations, consisting of a large number of sequential analyses, are demanding of computer storage and time. Furthermore, such highly specialised analyses require custom-made programs unavailable in professional practice, and the interpretation of results may be by no means straightforward. At this time, no nonlinear analysis can be considered a routine operation in the sense of linearly-elastic analysis. In fact, all nonlinear results presented here have been obtained in the course of research done in a university setting.

7.5.3 Effects of Connection Nonlinearity (Moncarz and Gerstle, 1981)

As discussed earlier, the relation between applied moment and resulting rotation of connections is, in general, curvilinear, as shown by the smooth curves labelled 1 and 3 in Fig. 7.11(b), which are those obtained by Hechtman and Johnson (1947) for their top-and-seat angle specimens 25 and 23. These were used for the lower, and upper, beam-to-column connections of the single-bay, two-storey, flexibly-connected frame shown in Fig. 7.11(a).

As a simple example of a piecewise-linear analysis, these M–θ curves were tri-linearised, as shown by the solid straight-line segments of Fig. 7.11(b), labelled 1 and 3. A small-deformation analysis of the frame response to loads shown in Fig. 7.11(a), increasing monotonically, was carried out in the way described in the preceding section, and, among other quantities, the maximum girder moment was determined under increasing lateral load w; it is plotted versus lateral load in Fig. 7.12(a) under the label 'Real connections', up to $w = 40$ lb/ft^2. Similar results are plotted in Fig. 7.12(b) for the column design moments.

The same frame was also analysed using small-deformation theory according to three other assumptions, the results of which are also plotted in Figs. 7.12(a) and (b):

(1) As a flexibly-connected frame with linearly-elastic connections of the stiffnesses shown by the dashed and dash-dotted lines labelled 2 and 4 in Fig. 7.11(b).

(2) As a rigid-jointed frame.

(3) According to AISC Type 2 analysis, which assumes pinned connections under gravity loads, but moment-resistant connections against lateral loads.

A comparison of these results sheds light on the validity of some of the engineering simplifications used in practice (girder moments appear nonlinear only because of shifting critical location):

(1) Rigid frame analysis underestimates the girder moments, but overestimates the column moments.

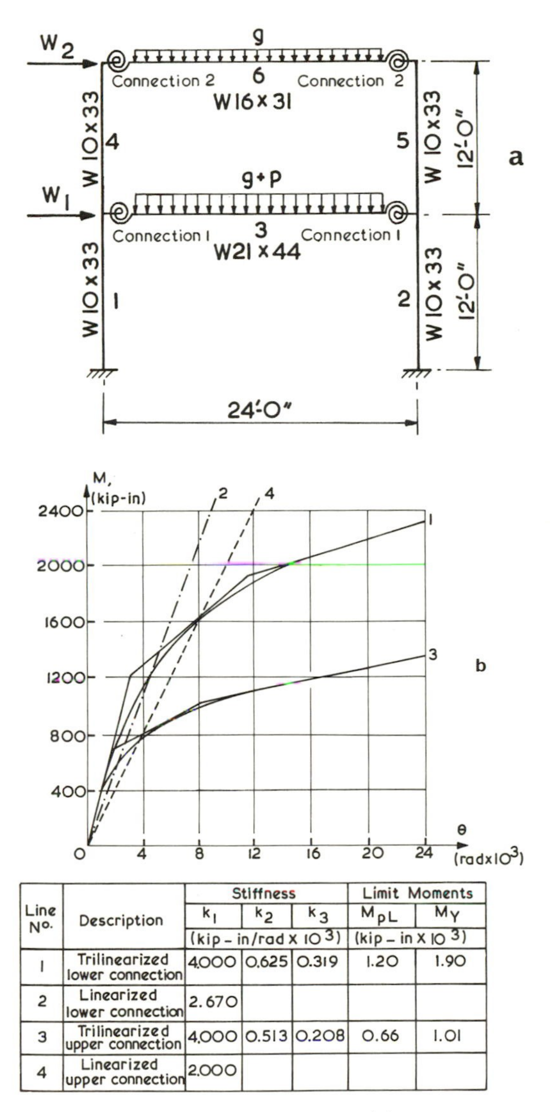

Line No.	Description	Stiffness			Limit Moments	
		k_1	k_2	k_3	M_{pL}	M_Y
		(kip – in/rad × 10³)			(kip – in × 10³)	
1	Trilinearized lower connection	4.000	0.625	0.319	1.20	1.90
2	Linearized lower connection	2.670				
3	Trilinearized upper connection	4.000	0.513	0.208	0.66	1.01
4	Linearized upper connection	2.000				

FIG. 7.11. (a) Flexibly-connected frame; (b) connection flexibilities.

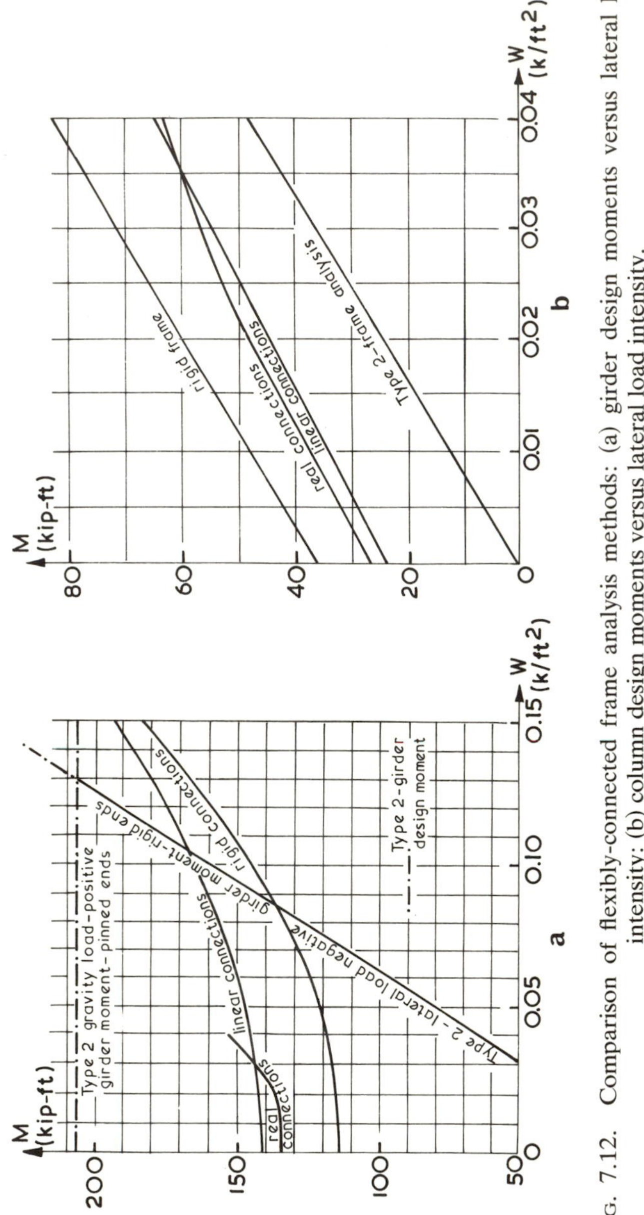

FIG. 7.12. Comparison of flexibly-connected frame analysis methods: (a) girder design moments versus lateral load intensity: (b) column design moments versus lateral load intensity.

(2) Type 2 analysis overestimates the girder moments, but underestimates the column moments.

(3) The assumption of linearly-elastic connection behaviour leads to design moments which are very close to the more exact values based on nonlinear connection behaviour.

(4) Point 3 confirms the validity of the results obtained earlier by linearly-elastic analysis for the working load behaviour of flexibly-connected steel frames.

The piecewise-linear analysis was extended to explore the frame response under cyclic load applications, and in particular the stabilisation of connection behaviour to the elastic state which had been hypothesised in justification of the validity of Type 2 analysis (Disque, 1975). To this end, the cyclic connection moment–rotation curves shown in Fig. 7.4 were piecewise-linearised as shown in Fig. 7.13, and applied to a flexibly-connected subassemblage shown in Fig. 7.14 that had been discussed earlier by Disque (1975). The assumed behaviour of the connections in this frame is shown in Fig. 7.14(b). The calculated sway under successive cycles of applied working loads is shown in Fig. 7.15(a), and the behaviour of one connection in Fig. 7.15(b). The following conclusions can be drawn from these Figures for this particular structure:

(1) The storey sways stabilise; deflection stability (shakedown) occurs.

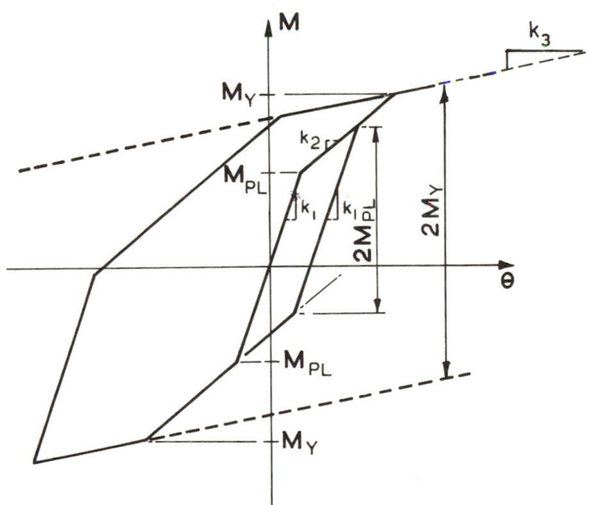

FIG. 7.13. Tri-linearised moment–rotation curve under load cycles.

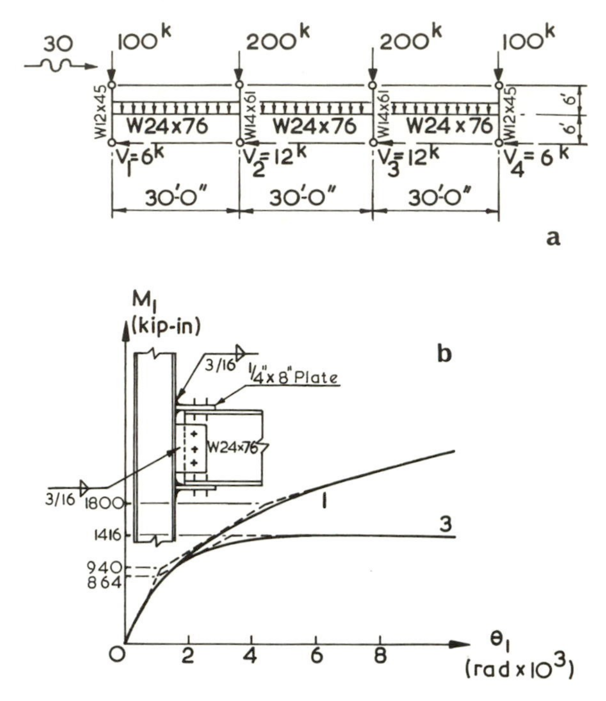

FIG. 7.14. (a) Disque's subassemblage; (b) assumed connection behaviour.

(2) While the connection behaviour consists of stable loops, alternating plasticity may take place in the connections.

These conclusions are probably conservative because it will be shown in the next Section that repeated cycles at design load levels are very unlikely to occur during the lifetime of the structure. Strain-hardening may also play a beneficial role in the connection.

7.5.4 Likelihood of Occurrence of Nonlinear Connection Cycling

(Cook and Gerstle, 1981)

The connection and frame behaviour under nonlinear load cycles, as shown in Figs. 7.4 and 7.13, appears complex. However, a probabilistic study based on statistical data on live load distribution and wind loads occurring in the US and their combinations, has led to the conclusion that, for frames designed according to current US practice, the likelihood of reversed connection plasticity during the postulated 100-year lifespan of a structure is extremely slight. Figure 7.16 (Cook and

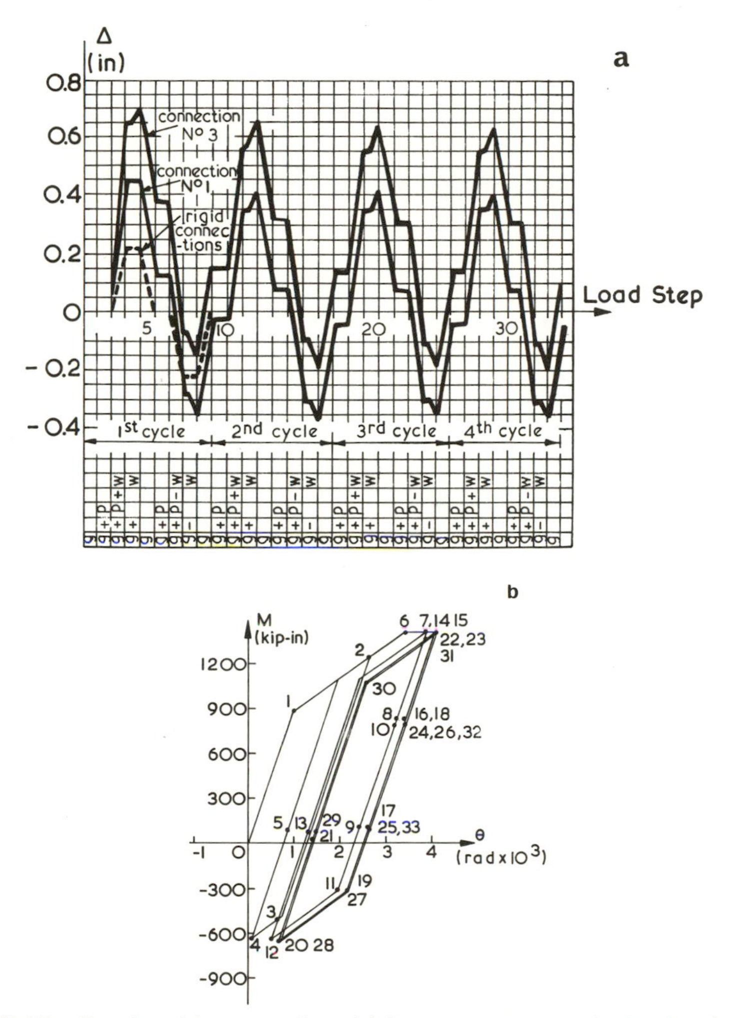

FIG. 7.15. Results of frame cycling: (a) first-storey sway under load cycles; (b) moment–rotation history for windward connection.

Gerstle, 1981) shows the results of this study; one load reversal may be expected at levels about equal to the working stress design load. Two or more load reversals are likely to occur only at even lower loads, well below any values resulting in reversed plasticity in the connections.

Excluding seismic events, we can therefore conclude that cyclic connection plasticity is not of importance in assessing either frame

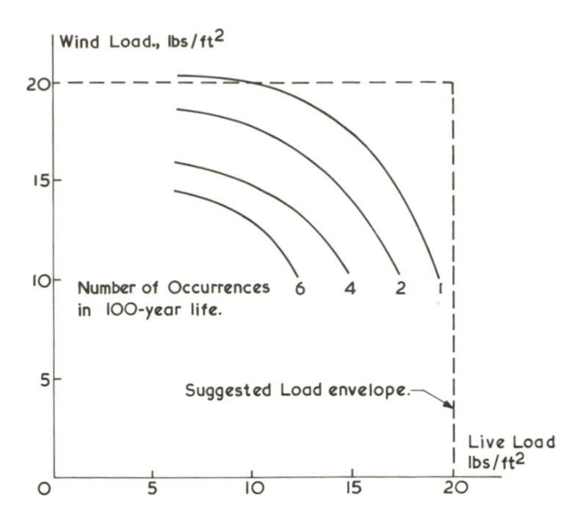

FIG. 7.16. Probability of load cycles at various intensities.

response to working loads, nor the frame strength under extreme loads. Consideration of the connection and frame behaviour shown in Figs. 7.4 and 7.13 is thus not necessary in the design of steel frames, unless earthquake resistance is required.

7.5.5 Strength of Flexibly-Connected Frames (Ackroyd and Gerstle, 1983; Cook, 1983)

Accurate determination of the strength of structures requires nonlinear analysis, because collapse usually occurs only after members and connections have plastified and after second-order deformations have led to buckling instabilities of members or frames.

In particular, the connection flexibility in steel frames affects both the moment transfer into the column, as well as the column restraint or effective length. The axial forces and moments in the column, in conjunction with the initial stress, may lead to partial member plastification, which will in turn diminish its buckling strength. Similarly, connection flexibility, which according to the nonlinear M–θ curves shown earlier may diminish radically under increasing moment, may rob the column of its rotational support, thus increasing its effective length. Lastly, stability of unbraced frames is strongly affected by connection stiffness and must be investigated.

Several recent studies have included these factors in a nonlinear, computer-based analysis, in order to predict the strength of a family of multi-storey, flexibly-connected steel frames designed according to commonly used Type 1 and Type 2 analyses. The following assumptions and formulations were made:

(1) Initial stresses of realistic magnitude in the columns.
(2) Elastic behaviour in girders.
(3) Nonlinear connection behaviour, according to Frye and Morris.
(4) Plastification of columns under combined axial load and bending (material nonlinearity).
(5) Column instability effects are included by basing equilibrium on the deformed shape of the discretised column (geometric nonlinearity).

The element stiffnesses, which would vary from load step to load step, were calculated, assembled for each load step, and solved for incremental forces and displacements by the direct stiffness method. Instability was indicated when the determinant of the structure stiffness matrix approached a value of zero.

A series of unbraced steel building frames, ranging in height from three to four 12-foot storeys and in width from two to four 20-foot bays, was designed according to Type 2 methods and checked for strength using the nonlinear analysis outlined above (taller frames were eliminated as too flexible in sway). Joints were designed with top-and-seat angle, or header plate, connections of appropriate size to resist the girder end moments, using standard practice. Both strong- and weak-axis orientation of the columns was considered.

To represent the great variety of possible gravity and lateral load combinations, three load paths were selected for analysis, shown in Fig. 7.17. All load paths begin with proportional gravity and wind

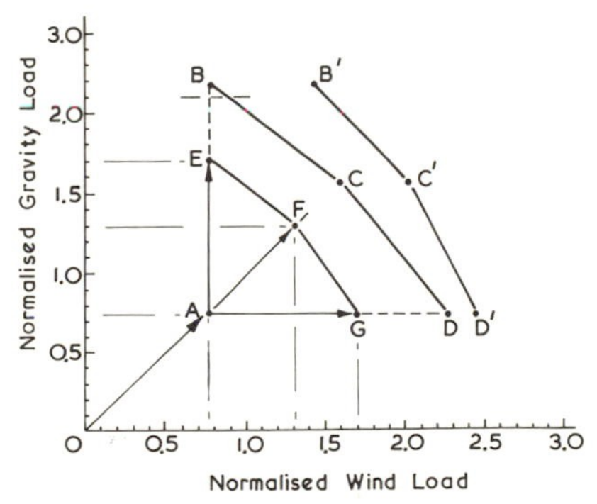

FIG. 7.17. Strength envelope of three-storey, two-bay frame with flexible connections.

loading to 75 per cent of their AISC working load values (point A). Thereafter, path 1 consists of increasing gravity load to failure under constant wind load; for path 2, gravity and wind loads increase in proportion to failure; in path 3, only the wind load increases to failure. The end points of these three paths denote failure loads which serve to define an actual strength envelope for the frame under analysis, to be compared to Part 2 (plastic design) of the AISC Specifications.

Figure 7.17 indicates the results for a three-storey, two-bay frame in strong-axis column bending, which are representative of all results. Envelope E–F–G represents the strength required by AISC, Part 2

$$U = 1{\cdot}7 \times \text{gravity load}$$
$$U = 1{\cdot}3 \times (\text{gravity} + \text{lateral load})$$

(Points E and G are a conservative extrapolation of these criteria.) Any frame whose predicted strength exceeds Envelope E–F–G is considered safe.

The calculated strength of this frame, considering both material and geometric nonlinearities as outlined above, is given by Envelope B–C–D. The P–Δ effect due to frame sway is included among the geometric nonlinearities. If this effect were neglected, the apparent strength of the frame would be as shown by Envelope B′–C′–D′, a considerable overestimate. Because of the considerable sway of the flexibly-connected frame, the P–Δ effect at failure becomes significant even for these relatively low frames.

In any case, it appears from these and other comparisons that the predicted frame strength is greater than required by these specifications by a factor of from 10 to 40 per cent. It follows, assuming sway can be held to a permitted value, that considerable savings can be effected by rational analyses which include actual connection behaviour.

7.5.6 Effect of Connection Flexibility on Frame Strength (Ackroyd and Gerstle, 1983)

In the light of the complex interactions affecting frame strength which were discussed in the preceding section, questions arise regarding the effect of connection flexibility on frame strength—specifically, can overstiff connections lead to a decrease of frame strength?

To answer this question, a number of subassemblages representing critical portions of typical unbraced multi-storey steel frames, each

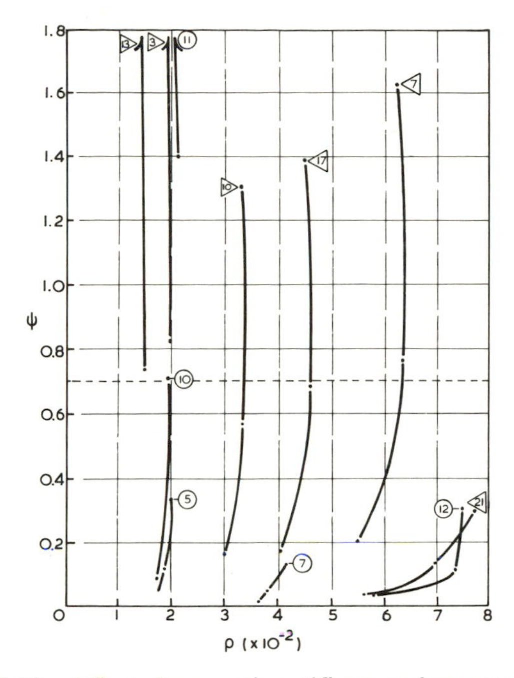

FIG. 7.18. Effect of connection stiffness on frame strength.

with three different flexible connections, were analysed. In particular, the effect of different connection stiffnesses on the frame strength was studied. Results are, in part, summarised in Fig. 7.18, which documents the effect of variation of connection stiffness on frame strength. A non-dimensional strength parameter $p=$(total load/total Euler buckling load) is plotted along the abscissa, as a function of a non-dimensional frame-loading-connection parameter

$$\psi = \frac{I_G}{I_C}\frac{l}{h}\frac{P_2 l}{M_w + Qh}\frac{k_0 l}{EI_G}\left(\frac{r}{h}\right)^2$$

plotted along the ordinate. This parameter depends not only on the connection stiffness k_0, but also on the relative stiffness of the frame components (I_G/I_C), on the frame aspect ratio (l/h), column slenderness (r/h), and relative magnitude of gravity to lateral loads ($[PL/M_w]+Qh$). The total Euler buckling load is the sum of the buckling loads of all columns of the floor under consideration.

It is observed from Fig. 7.18 that, in general, the frame strength increases with increasing connection stiffness k_0, other quantities remaining constant. Only for conditions of high values of ψ, as might arise in long-span frames, only a few storeys high, frame strength might be reduced by providing overstiff connections. However, this effect appears to be so slight that it can probably be neglected in practice.

7.5.7 Experimental Verification

An extensive series of two-bay, single-storey, and single-bay, two-storey frames, flexibly-connected with top-and-seat angle connections, was tested to verify the validity of the proposed frame analysis (Stelmack, 1983). No column instability, out-of-plane action, or large deflections were included in either analysis or tests, so that the results give information only about the effects of connection nonlinearity.

Figure 7.19 presents analytical predictions and measured sway for one of the two-storey frames with 1/2 in (13 mm)-thick top-and-seat angle connections. The proposed formulation and analysis appears to capture the basics of the frame behaviour satisfactorily for cyclic loads up to failure. No testing seems to have been carried out on flexibly-

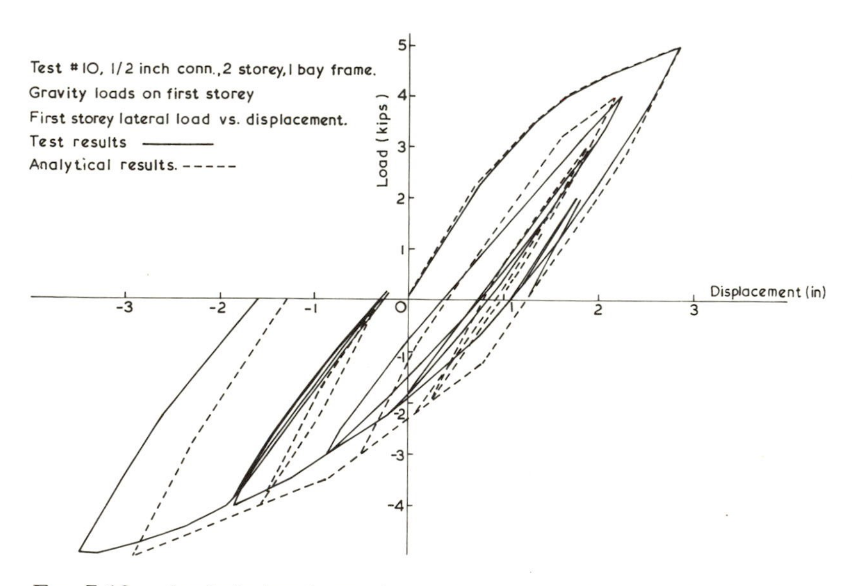

FIG. 7.19. Analytical and experimental response of frame to load cycles.

connected frames, in which failure is initiated by member or frame buckling.

Significant Conclusions

(1) *Nonlinear frame analysis is a complex procedure not suitable for routine design practice.*

(2) *Linear analysis of flexibly-connected frames is suitable for working-stress analysis and design.*

(3) *Type 2 design of flexibly-connected frames will lead to strengths in excess to those required by US specifications.*

(4) *In general, frame strength increases with increasing connection stiffness.*

(5) *Load histories leading to inelastic stress reversals are unlikely to occur during the lifetime of steel building frames, barring earthquakes.*

7.6 AUTOMATIC DESIGN OF FLEXIBLY-CONNECTED STEEL FRAMES (Ackroyd, 1977)

In the preceding Sections, *analysis* of flexibly-connected steel frames has been discussed. A much more difficult, and potentially more rewarding, problem is a procedure which will permit efficient, rational *design* of such frames in a professional environment. In the following, a computer-based, iterative design program for unbraced frames is outlined to this end.

This program follows essentially the sequence which would be followed in conventional longhand design:

(1) A preliminary, 'quick and dirty' analysis is carried out, requiring nothing more than frame geometry and loads for input.

(2) Members and connections are sized to resist the internal forces resulting from this preliminary analysis, and their stiffnesses recorded.

(3) The member and connection stiffnesses of step 2 are used for a more exact, linearly-elastic analysis.

(4) Members and connections are resized, based on the internal forces of step 3, and the process continued to convergence.

Such a program requires a number of key subroutines; preliminary and exact analyses, column and girder sizing according to current specifications, determination of connection stiffnesses, and a list of available member sizes and properties, all interconnected by a logic permitting analysis–design iterations.

A program of this type is available (Ackroyd, 1977). The required

TABLE 7.1
RESULTS OF ITERATIVE DESIGNS[a]

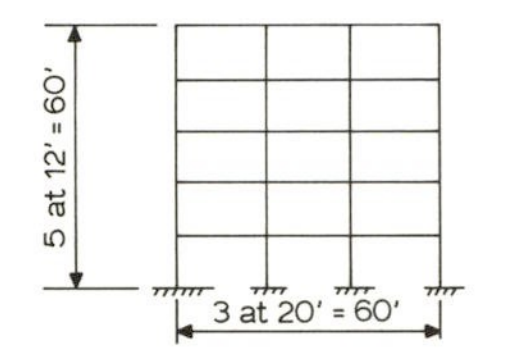

Storey	Member	Type 1 construction			Type 2 construction				
			Fully welded			Flange plates		T + S angles	
		Preliminary	Iteration 1	Iteration 2	Preliminary	Iteration 1	Iteration 2	Iteration 1	Iteration 2
	Ext. column	W14×68	W14×68	W14×74	W14×61	W14×68	W14×74	W14×74	W14×74
1	Int. column	W14×78	W14×78	W14×84	W14×78	W14×84	W14×84	W14×87	W14×87
	Girder	W21×44	W18×40	W18×40	W21×49	W18×40	W18×40	W21×44	W18×40
	Ext. column	W14×61	W14×61	W14×61	W14×53	W14×53	W14×61	W14×61	W14×61
2	Int. column	W14×68	W14×68	W14×68	W14×68	W14×68	W14×68	W14×68	W14×68
	Girder	W18×45	W14×40	W18×40	W21×49	W18×40	W18×40	W21×44	W18×40
	Ext. column	W14×53	W14×48	W14×48	W14×30	W14×38	W14×43	W14×38	W14×43
3	Int. column	W14×61	W14×48	W14×48	W14×61	W14×53	W14×53	W14×53	W14×53
	Girder	W18×40	W18×35	W18×35	W21×49	W18×40	W18×35	W21×44	W21×44
	Ext. column	W14×43	W14×38	W14×34	W14×30	W14×34	W14×34	W14×34	W14×34
4	Int. column	W14×30	W14×34	W14×34	W14×30	W14×34	W14×34	W14×34	W14×34
	Girder	W16×36	W18×35	W18×35	W21×49	W18×40	W18×35	W21×44	W21×44
	Ext. column	W14×30	W14×34	W14×34	W14×30	W14×34	W14×34	W14×34	W14×34
5	Int. column	W14×30	W14×34	W14×34	W14×30	W14×34	W14×34	W14×34	W14×34
	Girder	W16×36	W18×35	W18×35	W21×49	W18×40	W18×40	W21×49	W21×44
Total weight (lb)		24 588	23 364	23 556	26 004	24 000	23 856	25 908	25 248

[a] In all cases, the members are specified by using two numbers, the first denoting the depth and the second the weight; e.g. W16×36 denotes a member 16 in deep weighing 36 lb per foot. (Note: 1 in = 25·4 mm, 1 lb = 0·454 kg.)

input information is identical to that needed to carry out any conventional design, with the additional specification of the connection type to be used.

After a moderate number of iterations, usually no more than three, a design is obtained in which all members and connections are stressed to the same factor of safety within the scope of Part 1 (working-stress design) of the AISC Specifications.

The technique provides the option of a Type 1 (rigid-frame) or Type 2 (simple-framing) preliminary analysis. In the former case, the resulting design will converge on a big column–small girder combination; in the latter case the result will be the converse. Floor, wall and wind loads can be accommodated, and the AISC reduction for gravity–wind load combinations is incorporated. Automatic frame mesh generation is also provided for regular frames. Candidate member groups can be specified selectively.

An example will demonstrate the details, nature of output and potential of the method. Table 7.1 shows the three-bay, five-storey unbraced, flexibly-connected frame to be designed. It is subjected to specified floor, wall and wind loads. Girders are to be selected from among the W14–27 series (i.e. 356–675 mm deep), columns from among the W14 series (356 mm deep). Three different concepts were explored; fully-welded connections, with type 1 preliminary analysis, and flange plate connections and top-and-seat angle connections, both with type 2 preliminary analysis.

Input for the design according to one of these alternatives consisted of 33 punch cards. The output is summarised in Table 7.1, showing member sizes for each iteration for each alternative, thus permitting easy comparison. Of particular interest are the total weight figures for one frame; fully welded construction would save about 7 per cent steel over the top-and-seat angle alternative. For the flange plate alternative, about 8 per cent of steel is saved by carrying out a linearly-elastic 'exact' analysis over the preliminary Type 2 design; this is pure savings since no extra connection expense is involved.

The potential of such a design tool seems impressive.

Significant Conclusions

(1) *Computerised design of flexibly-connected building frames is possible and appears useful for design practice.*

(2) *Such computerised design can lead to increased economy in steel building design.*

REFERENCES

ACKROYD, M. H. (1977) Automatic design of steel frames with flexible connections. Report to AISI, University of Colorado, USA.

ACKROYD, M. H. and GERSTLE, K. H. (1977) Strength and stiffness of Type 2 steel frames. Report to AISI, University of Colorado, USA.

ACKROYD, M. H. and GERSTLE, K. H. (1982) Strength and stiffness of Type 2 steel frames. *Trans. ASCE, J. Struct. Div.*, **108,** ST7, 1541–56.

ACKROYD, M. H. and GERSTLE, K. H. (1983) Strength of flexibly-connected steel frames. *Engineering Structures*, **5,** 31–8.

AMERICAN INSTITUTE OF STEEL CONSTRUCTION (1980) *Specifications, Manual of Steel Construction*, Section 1.2, AISC, Chicago.

BATHO, C. and ROWAN, H. C. (1931) Investigation on beam and stanchion connections. Report of the Steel Structures Committee of the Division of Scientific and Industrial Research, **1** and **2,** 1931–1934.

CHESSON, E., JR. and MUNSE, W. H. (1958) Behavior of riveted truss-type connections. *Trans. ASCE*, **123,** 1087–1128.

COOK, N. E., JR. (1983) Strength of flexibly-connected steel frames under load histories. PhD Thesis, University of Colorado, USA.

COOK, N. E., JR. and GERSTLE, K. H. (1981) Appropriate repeated wind and live loads for application to buildings. Unpublished Report, CEAE Dept, University of Colorado, USA.

DISQUE, R. O. (1975) Directional moment connections; a proposed design method for unbraced steel frames. *AISC Engineering Journal*, First Quarter, 14–18.

FRYE, M. J. and MORRIS, G. A. (1975) Analysis of flexibly-connected steel frames. *Canadian Journal of Civil Engineering*, **2,** 280–91.

GERSTLE, K. H. (1974) *Basic Structural Analysis*, Prentice-Hall, Inc., New Jersey, USA.

GOBLE, G. G. (1963) A study of the behavior of building frames with semi-rigid joints. Report submitted to the AISC and The Ohio Steel Fabricators Association, Case Institute of Technology.

HECHTMAN, R. A. and JOHNSON, B. G. (1947). Riveted semi-rigid beam-to-column building connections. Progress Report Number One, AISC Publication, Appendix B.

KAHL, T. L. (1976) Flexibly-connected steel frames. Report to AISI, University of Colorado, USA.

KATO, B., CHEN, W. F. and NAKAO, M. (1984) Effect of joint-panel shear deformations. *Monograph on Beam–Column Connections*, Chapter 10, ASCE.

KRISHNAMURTHY, N., HUANG, H.-T., JEFFREY, P. K. and AVERY, L. K. (1979) Analytical M–θ curves for end-plate connections. *Trans. ASCE, J. Struct. Div.*, **105,** ST1, 133–45.

LOTHERS, J. E. (1960) *Advanced Design in Structural Steel*, Prentice-Hall, Inc., New Jersey, pp. 367–405.

MARLEY, M. J. (1982) Analysis and tests of flexibly-connected steel frames. Report to AISI, University of Colorado, USA.

MONCARZ, P. D. and GERSTLE, K. H. (1981) Steel frames with nonlinear connections. *Trans. ASCE, J. Struct. Div.*, **107,** ST8, 1427–41.

Popov, E. P. and Pinkney, R. B. (1969) Cyclic yield reversal in steel building connections. *Trans. ASCE, J. Struct. Div.*, **95,** ST3, 327–53.

Rathbun, J. C. (1936) Elastic properties of riveted connections. *Trans. ASCE,* **101,** 524–63.

Romstad, K. M. and Subramanian, C. V. (1970) Analysis of frames with partial rigidity. *Trans. ASCE, J. Struct. Div.*, **96,** ST11, 2283–300.

Stelmack, T. W. (1983) Analytical and experimental response of flexibly-connected steel frames. Report to AISI, University of Colorado, USA.

Weaver, W. Jr. and Gere, J. M. (1980) *Matrix Analysis of Framed Structures, 2nd Edn*, D. Van Nostrand Co., New York.

Chapter 8

PORTAL FRAMES COMPOSED OF COLD-FORMED CHANNEL- AND Z-SECTIONS

G. J. HANCOCK

School of Civil and Mining Engineering, University of Sydney, New South Wales, Australia

SUMMARY

The mode of failure of portal frames composed of cold-formed members generally involves inelastic local buckling. A method of analysis based on the finite strip method is presented for calculating the inelastic local buckling loads of thin-walled sections.

A method of matrix displacement analysis of thin-walled structures is presented which accounts for non-uniform torsion, monosymmetric and asymmetric thin-walled sections, eccentric restraints and joints peculiar to thin-walled structures. The analysis is applied to a study of pitched roof portal frames composed of channel- and Z-section members and subjected to vertical load. The stresses and deformations of structures of this type are described.

Simple design methods for Z- and channel-section portals based on plane frame linear elastic analyses are described.

NOTATION

A	Area of cross-section
a_x, a_y	Coordinates of shear centre relative to origin
B_z	Bimoment calculated with respect to shear centre axis
$B_{Z'}$	Bimoment calculated with respect to origin

b_x, b_y	Coordinates of centroid relative to origin
C	Section constant in cross-section transformation matrix
E	Young's modulus
E_t	Tangent value of E in yielded zones
$[ER]$	Eccentric restraint matrix
e_X, e_Y, e_Z	Distances in X, Y and Z directions, respectively, of point of eccentric restraint relative to origin of joint
F_R	Force acting at an eccentric restraint
F_x, F_y, F_z	Forces in x, y and z directions, respectively
f	Longitudinal (normal) stress
G	Shear modulus
G_y	Value of G in yielded zones
$[G]$	Cross-section transformation matrix
I_w	Warping section constant
I_x, I_y	Second moment of area of cross-section
J	Torsion section constant
$[K]$	Frame stiffness matrix
$[K_{mX'}]$	Member stiffness matrix for one member in member origin coordinates
$[K_{mx}]$	Member stiffness matrix for one member in principal axis coordinates
L	Length of an element or member
M_x, M_y, M_z	Moments about x, y and z axes, respectively
s	Distance along centreline of a cross-section
u_s, v_s	Displacement of shear centre in x-, y-directions, respectively
W	Total vertical load on one rafter
$\{W\}$	Vector of joint loads
$\{W_{mX'}\}$	Vector of loads acting on the end of a member in member origin coordinates
$\{W_r\}$	Vector of joint loads at an eccentric restraint
w	Displacement of centroid in z-direction
X, Y, Z	Frame or global axes
X', Y', Z'	Member origin axes
x, y	Member principal axes (located in cross-section)
z	Centroidal axis along member
α	Orientation of member principal axes with respect to member origin axes (see Fig. 8.4(a))
α_w	Sectorial coordinate in the web of a Z-section

α_z	Sectorial coordinate defined with respect to the shear centre axis
α_{z0}	Value of α_z at $s=0$
$\alpha_{Z'}$	Sectorial coordinate defined with respect to the member origin axis Z'
$\{\delta\}$	Vector of joint deformations in frame coordinates
$\{\delta_{mX'}\}$	Vector of joint deformations of the end of a member in member origin coordinates
$\{\delta_R\}$	Vector of joint displacements of an eccentric restraint
$\delta_{XR}, \delta_{YR}, \delta_{ZR}$	Displacements in X, Y and Z directions at an eccentric restraint
$\delta_x, \delta_y, \delta_z$	Displacements in x, y and z directions, respectively
$\theta_{XR}, \theta_{YR}, \theta_{ZR}$	Rotations about X, Y and Z axes, respectively, at an eccentric restraint
$\theta_x, \theta_y, \theta_z$	Rotations about x, y and z axes, respectively
θ'_z	Twist rotation per unit length about z axis
$\rho_{Z'}$	Perpendicular distance from Z' axis to tangent to centreline of a cross-section
ρ_z	Perpendicular distance from shear centre axis to tangent to centreline of a cross-section

All moments and rotations are right hand rule positive.

Subscripts
A and B refer to the ends of an element or member.

8.1 INTRODUCTION

A recent development in metal structures has been the manufacture of rigid-jointed portal frames composed of cold-formed members. Structures of this type are inexpensive to fabricate, transport and erect, thus producing considerable economies in labour. Many designers have chosen to use cold-formed channels attached back-to-back. However, despite the more severe analytical difficulties, structurally efficient portal frames can be produced using single channel- or Z-section members. The designer of such a structure requires a knowledge of both the stiffness and strength of structural systems of this type. This chapter presents methods for analysing and designing structural systems composed of thin-walled channel- and Z-section members.

A method of structural analysis which includes the effects of cross-section monosymmetry or asymmetry, non-uniform torsion, eccentric restraints as well as joint types peculiar to thin-walled members is required. The most important contributions to the theory of bending, torsion and buckling of thin-walled elastic members of open cross-section have been made initially by Timoshenko (1945), who studied non-uniform torsion of an I-beam, and by Vlasov (1959) who extended the concept, resulting in a consistent general theory for the behaviour of thin-walled members. Renton (1962), Krahula (1967) and Krajcinovic (1969) have produced versions of the matrix displacement method which incorporate non-uniform torsion. In all three cases, the stiffness matrices used in the method were based on the solution of the differential equations rather than using the approximate Rayleigh–Ritz method. This latter approach has been used by Barsoum and Gallagher (1970), and Bazant and El Nimeiri (1973). It has the disadvantage that it requires subdivision of thin-walled members into elements to achieve an accurate solution for non-uniform torsion.

Cross-section monosymmetry and asymmetry have been incorporated in the matrix displacement analysis by Bazant and El Nimeiri (1973), Rajasekaran (1977), and Baigent and Hancock (1980, 1982*a*). Lateral restraint positioning, eccentric from the shear centre, and the specific nature of joint connections have also been included by Baigent and Hancock (1980, 1982*a*) and confirmed by testing of channel-section portal frames. A matrix displacement analysis of frames composed of thin-walled members, which incorporates all of the phenomena described above, is presented in Section 8.2 of this chapter. The method is then applied to determine the stress-resultants, stresses and deflections of pitched roof portal frames composed of Z- and channel-section members in Section 8.3 of this chapter.

The results of the elastic analyses just cited produce estimates of stress and deflection in thin-walled structures. However, they provide no real estimate of the ultimate load or collapse behaviour of a structure. The most significant contribution towards the prediction of the inelastic collapse loads of steel frameworks composed of hot-rolled members was made by Baker and his associates at Cambridge and has been reported in *The Steel Skeleton, Vol. II* by Baker, Horne and Heyman (1956). Structures composed of cold-formed members are not likely to develop complete collapse mechanisms of the type described for structures composed of hot-rolled members, as a result of the instability of the thin-plate elements forming the sections.

However, after first yield, they may carry a substantial increase in load before failure if the plate elements are not too slender. In this case, inelastic local buckles develop at critical cross-sections in the frame. Baigent and Hancock (1982*b*, 1982*c*) have described methods of predicting this phenomenon. However, at this time, design codes do not permit utilisation of this additional capacity, unless the plate slenderness is sufficiently low to satisfy those for a compact section, defined in Section 3.9 of the most recent edition of the AISI specification (1980). In this case, some plasticity is permitted. A method for determining the inelastic local buckling capacity of thin-walled sections, using the finite strip method of analysis developed by Cheung (1976), is described in Section 8.4.

Designers do not generally have access to computer programs which can perform structural analyses of portal frames composed of cold-formed members. Hence, a simplified procedure based on a plane frame linear elastic analysis is required. A method has been described by Hancock (1983) for Z-section portal frames and is summarised in Section 8.5. It produces conservative estimates of first yield in Z-section portal frames. A simplified procedure for channel section portals is also described and discussed in Section 8.5.

8.2 MATRIX DISPLACEMENT ANALYSIS OF FRAMES COMPOSED OF THIN-WALLED MEMBERS

8.2.1 Member Stiffness Matrix

In the matrix displacement method, a set of equations is derived to relate the displacements of the joints of a structure to the forces applied at the joints. The equations are based on the assembly of the stiffness equations for the individual members or elements of the structure, using the equations of equilibrium and assuming compatibility at the joints. The resulting stiffness equations can be solved to determine the displacements of the joints. The displacements are then substituted into the member stiffness equations to derive the stress resultants for the members. The analysis of prismatic structures using the matrix displacement method is extended to include structures composed of thin-walled elements, as follows.

There are seven actions with corresponding displacements at each end of a thin-walled element. They include the three forces (F_x, F_y, F_z) and the three moments (M_x, M_y, M_z) shown in Fig. 8.1(i) and normally

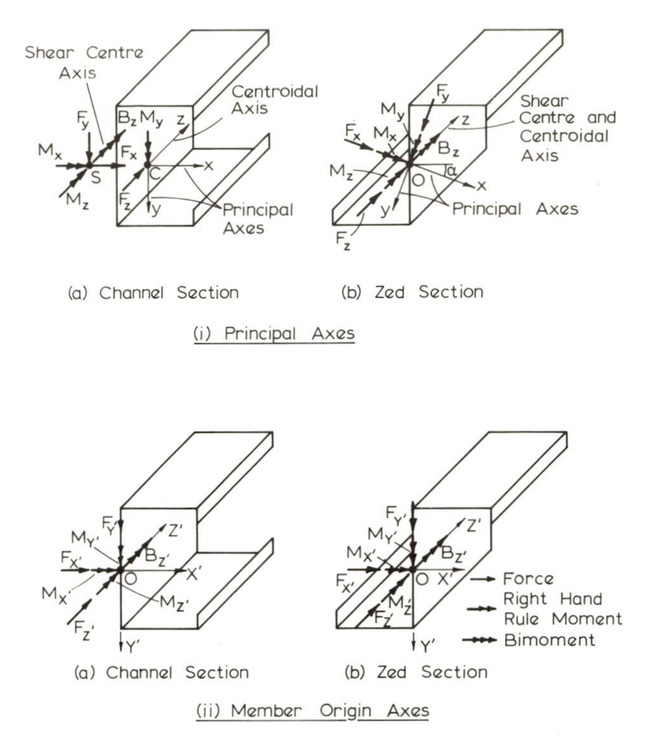

FIG. 8.1. Actions aligned with principal and member origin axes.

used in the analysis of prismatic member structures. In the case of thin-walled elements, they do not necessarily act through a common point. The shear forces (F_x, F_y) act through the shear centre (S) in directions parallel with the principal axes, as shown in Fig. 8.1(i). However, the axial force (F_z) acts along the centroidal axis (z). The moments (M_x, M_y) act about the principal axes (x, y) but in planes containing the shear centre. The double headed arrows representing the moments M_x, M_y in Fig. 8.1(i) are located such that the moments lie in planes perpendicular to the directions of the arrows and located at their tips. The torque (M_z) acts about the shear centre axis.

For a thin-walled member, a bimoment (represented by the triple headed arrow on the shear centre axis in Fig. 8.1(i)) can also be applied at each end of a thin-walled element. Vlasov (1959) defined the bimoment (B_z), calculated with respect to the shear centre axis, as the product of the longitudinal stress (f) with the sectorial coordinate (α_z) integrated over the area of the cross-section.

$$B_z = \int_A f \,.\, \alpha_z \,.\, \mathrm{d}A \tag{8.1}$$

The sectorial coordinate (α_z) is the longitudinal warping displacement resulting from unit negative twist ($\theta_z' = -1$) about the shear centre axis. It is calculated from

$$\alpha_z = \alpha_{z0} + \int_A \rho_z \,.\, \mathrm{d}s \tag{8.2a}$$

and

$$\int_A \alpha_z \,.\, \mathrm{d}A = 0 \tag{8.2b}$$

where ρ_z is the perpendicular distance from the shear centre axis to the tangent to the centreline of the cross-section at an element ds located an arc distance s along the centreline of the cross-section. The sectorial coordinates (α_z) for a channel- and a Z-section are shown in Fig. 8.2.

The longitudinal stress distribution resulting from a bimoment (B_z)

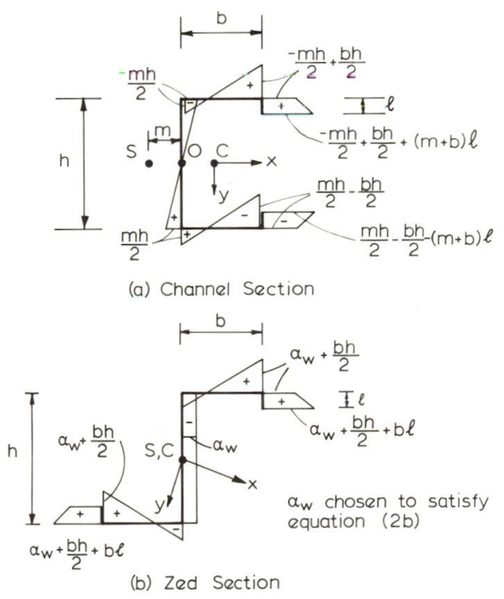

FIG. 8.2. Sectorial coordinate (α_Z).

acting at a section in a thin-walled member can be calculated from the sectorial coordinate using

$$f = \frac{B_z \, . \, \alpha_z}{I_w} \tag{8.3a}$$

and

$$I_w = \int_A (\alpha_z)^2 \, . \, dA \tag{8.3b}$$

where I_w is the section warping constant. From eqn (8.3a), it can be seen that the bimomental stress distribution has the same pattern in the cross-section as the sectorial coordinate. Equation (8.2b) ensures that the stress distribution equivalent to a pure bimoment produces no net axial force when integrated over the section area.

The stiffness matrix for the thin-walled elements shown in Fig. 8.1(i) can be developed by considering the stiffness relationships between the actions shown in Fig. 8.3, which are aligned with the principal axes

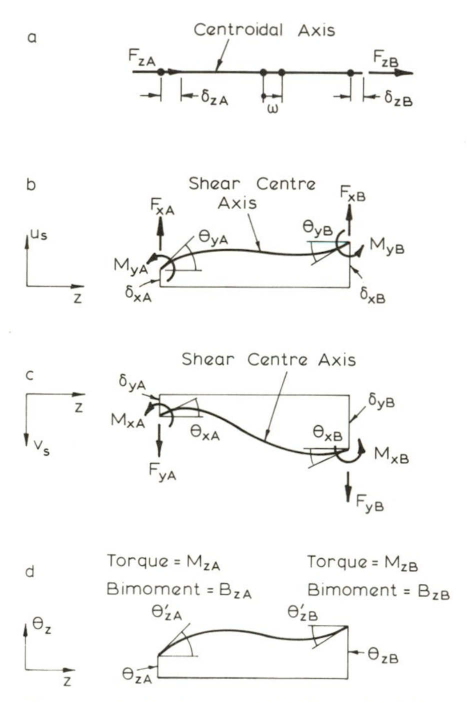

FIG. 8.3. End actions and displacements aligned with centroidal and shear centre axes.

(x, y), the centroidal axis (z) and the shear centre axis, and the corresponding displacements of these axes as shown in Fig. 8.3. Since the differential equations describing the linear axial, flexural and torsional behaviour are uncoupled when referred to the axes shown in Fig. 8.3, then the stiffness relationships will be uncoupled.

The axial forces at ends A and B (F_{zA}, F_{zB}, respectively) can be related to the axial displacements of the centroidal axis at A and B (δ_{zA}, δ_{zB}, respectively) by linear stiffness relationships. These relationships are simply derived by integrating the axial differential equation (eqn (8.4)) accounting for end conditions.

$$EA\frac{dw}{dz} = F_z \tag{8.4}$$

The stiffness relationships are given in Appendix 1(a).

The forces and moments (F_{xA}, M_{yA}, F_{xB}, M_{yB}), associated with flexure of the shear centre axis in a plane parallel with the x–z plane, can be related to the corresponding displacements and rotations (δ_{xA}, θ_{yA}, δ_{xB}, θ_{yB}) of the shear centre axis at ends A and B by the stiffness relationships developed by Livesley (1956). These relationships are simply derived by integrating the flexural differential equation (eqn (8.5)) accounting for end conditions.

$$-EI_y\frac{d^2u_s}{dz^2} = M_y \tag{8.5}$$

Similarly, the actions (F_{yA}, M_{xA}, F_{yB}, M_{xB}), associated with flexure of the shear centre axis in a plane parallel with the y–z plane, can be related to the corresponding displacements and rotations (δ_{yA}, θ_{xA}, δ_{yB}, θ_{xB}) of the shear centre axis at ends A and B by the stiffness relationships derived by integrating eqn (8.6) accounting for end conditions.

$$-EI_x\frac{d^2v_s}{dz^2} = M_x \tag{8.6}$$

The stiffness relationships are given in Appendix 1(b).

The torques and bimoments (M_{zA}, B_{zA}, M_{zB}, B_{zB}) associated with torsion about the shear centre axis can be related to the corresponding torsional rotations and negative rate of change of angle of twist about the shear centre axis (θ_{zA}, $-\theta'_{zA}$, θ_{zB}, $-\theta'_{zB}$) by the stiffness relationships developed by Krahula (1967). These relationships are simply derived by integrating the torsion differential equation (eqn (8.7)) accounting for end conditions.

$$GJ\frac{d\theta_z}{dz}+\frac{dB_z}{dz}=M_z \tag{8.7}$$

The stiffness relationships are given in Appendix 1(c). In eqns (8.4)–(8.7), EA, EI_y, EI_x and GJ are the axial, y-flexural, x-flexural and torsional rigidities, respectively. In solving eqn (8.7), the equation (eqn (8.8)) derived by Vlasov for relating the bimoment (B_z) to the twist angle (θ_z) is required.

$$B_z=-EI_w\frac{d^2\theta_z}{dz^2} \tag{8.8}$$

where EI_w is the warping rigidity.

The stiffness relationships in Appendix 1 are represented in matrix notation by

$$\{W_{mx}\}=[K_{mx}]\cdot\{\delta_{mx}\} \tag{8.9}$$

where $[K_{mx}]$ is the thin-walled member stiffness matrix. The vectors $\{W_{mx}\}$ and $\{\delta_{mx}\}$ represent the end actions and end displacements, respectively, of the element.

8.2.2 Cross-Section Transformations

The forces and displacements shown in Fig. 8.1(i) can be transformed to a common member axis system (X', Y', Z') called the *member origin axes*. These axes can be chosen arbitrarily, but are generally located on that flat element of the cross-section which is used for connection at a joint. In the case of the channel- and Z-sections shown in Fig. 8.1, the member origin axes are located on the centre of the web and are aligned parallel with, and perpendicular to, the web, as shown in Fig. 8.1(ii). The stiffness matrix for a thin-walled member relating the loads $\{W_{mX'}\}$ aligned with the member origin axes to the corresponding displacements $\{\delta_{mX'}\}$ is given by

$$\{W_{mX'}\}=[K_{mX'}]\cdot\{\delta_{mX'}\} \tag{8.10}$$

A cross-section transformation matrix $[G]$ transforms the end actions aligned with the principal axes, centroidal axis and shear centre axis to those actions aligned with the member origin axis system within the cross-section (Fig. 8.1). The cross-section transformation matrix is derived for the general cross-section shown in Fig. 8.4, as follows.

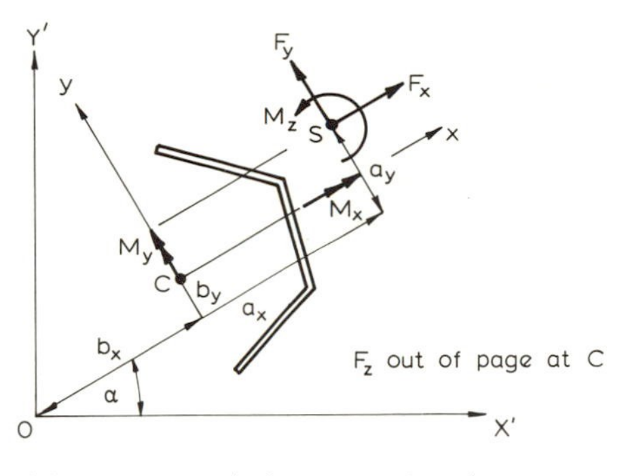

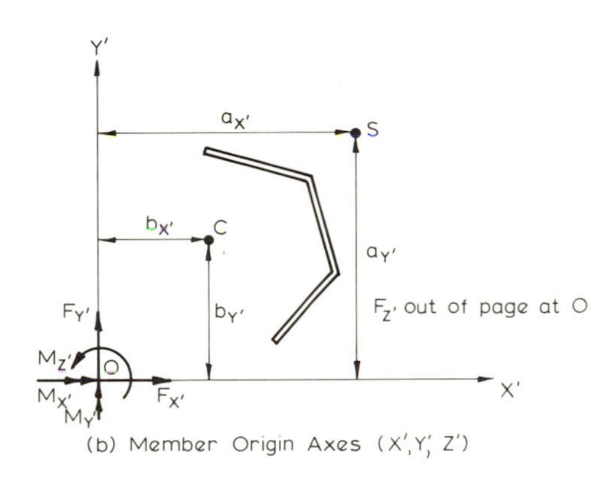

FIG. 8.4. Cross-section forces and moments.

Using force equilibrium, the three forces $F_{X'}$, $F_{Y'}$ and $F_{Z'}$ aligned with the member origin axes are related to the forces F_x, F_y and F_z aligned with the centroid and shear centre axes. The relationships are given by:

$$F_{X'} = F_x \cos\alpha - F_y \sin\alpha \tag{8.11}$$

$$F_{Y'} = F_x \sin\alpha + F_y \cos\alpha \tag{8.12}$$

$$F_{Z'} = F_z \tag{8.13}$$

The moments $M_{X'}$, $M_{Y'}$ and $M_{Z'}$ are related to the forces F_x, F_y, F_z, the moments M_x, M_y, M_z and the distances from the member axis system origin to the shear centre and centroid (a_x, a_y) and ($b_{X'}$, $b_{Y'}$), respectively, using moment equilibrium. The expressions are given by:

$$M_{X'} = M_x \cos\alpha - M_y \sin\alpha + F_z \,.\, b_{Y'} \tag{8.14}$$

$$M_{Y'} = M_x \sin\alpha + M_y \cos\alpha - F_z \,.\, b_{X'} \tag{8.15}$$

$$M_{Z'} = M_z - a_y \,.\, F_x + a_x \,.\, F_y \tag{8.16}$$

In addition, the bimoment $B_{Z'}$ defined with respect to the member origin axis Z' can be expressed in terms of the bimoment B_z defined with respect to the shear centre axis, the moments about the major principal axes of the cross-section (M_x, M_y) and the axial force (F_z), by

$$B_{Z'} = B_z + a_x \,.\, M_x + a_y \,.\, M_y + F_z \,.\, C \tag{8.17}$$

where

$$B_{Z'} = \int_A f \,.\, \alpha_{Z'} \,.\, \mathrm{d}A \tag{8.18}$$

Equation (8.17) was derived by Vlasov to transform a bimoment from one axis system to another. Equation (8.17) can be derived by substituting the equation for longitudinal stress

$$f = \frac{F_z}{A} + \frac{M_x \,.\, y}{I_x} - \frac{M_y \,.\, x}{I_y} + \frac{B_z \,.\, \alpha_z}{I_w} \tag{8.19}$$

and the equation derived by Vlasov for transforming the sectorial coordinate

$$\alpha_{Z'} = \alpha_z + a_x \,.\, y - a_y \,.\, x + C \tag{8.20}$$

where

$$\alpha_{Z'} = \alpha_{Z'0} + \int_0^s \rho_Z \,.\, \mathrm{d}s \tag{8.21}$$

into eqn (8.18). In eqn (8.21), $\rho_{Z'}$ is the perpendicular distance from the member origin axis (Z') to the tangent to the centreline of the cross-section at an element ds. The constant C is so chosen that the value of $\alpha_{Z'}$ is zero at the member origin O. For the channel-section α_z (Fig. 8.2(a)) $a_x \,.\, y$ and $a_y \,.\, x$ are all zero at O, hence C is zero. However for the Z-section, although $a_x \,.\, y$ and $a_y \,.\, x$ are zero at O, α_z is equal to $-\alpha_w$ (Fig. 8.2(b)) and hence C equals α_w.

Equations (8.11)–(8.18) are represented in matrix notation by

$$\{W_{mX'}\} = [G]\{W_{mx}\} \tag{8.22}$$

The matrix $[G]$ is given in Appendix 2.

A corresponding set of displacement transformations can be derived to transform the end displacements associated with the principal, centroidal and shear centre axes (δ_x, δ_y, δ_z, θ_x, θ_y, θ_z, $-\theta'_z$) to those associated with the member origin axes ($\delta_{X'}$, $\delta_{Y'}$, $\delta_{Z'}$, $\theta_{X'}$, $\theta_{Y'}$, $\theta_{Z'}$, $-\theta'_{Z'}$) as

$$\{\delta_{mx}\}=[G]^{\mathrm{T}}\{\delta_{mX'}\} \tag{8.23}$$

These relationships are simply the transpose of those in eqn (8.22) as a consequence of Maxwell's reciprocal theorem of structural analysis. Elimination of $\{W_{mx}\}$ and $\{\delta_{mx}\}$ from eqns (8.9), (8.22) and (8.23) produces

$$[K_{mX'}]=[G][K_{mx}][G]^{\mathrm{T}} \tag{8.24}$$

For the channel-section in Fig. 8.1, α, C, a_y and $b_{Y'}$ are all zero, and only a_x and $b_{X'}$ are non-zero. For the Z-section shown in Fig. 8.1, a_x, a_y, $b_{X'}$, $b_{Y'}$ are all zero, and only α and C are non-zero.

8.2.3 Frame Stiffness Matrix

The 14 end actions defined by $\{W_{mX'}\}$ and the 14 end displacements defined by $\{\delta_{mX'}\}$ are aligned with the member origin axes (X', Y', Z') of the thin-walled element. Before these actions can be included in the analysis of equilibrium at a joint (node) within a frame, they must be transformed to be parallel with the frame (global) axis system (X, Y, Z). The transformation of the forces and moments from the member origin axes to the frame axes simply involves a rotation of the actions from one axis system to another. The transformation of the forces and moments at one end of a member has been described in detail by Harrison (1980) using an Eulerian transformation.

Bimoment equilibrium and warping compatibility at a joint depend on the geometry of the joint. A detailed discussion follows in Section 8.2.4. In general, however, the bimoment is not transformed but is regarded as being continuous at a node within a thin-walled member unless constrained.

The stiffness matrix for the complete frame can be derived by summing the actions from all the members or elements at each joint or node after transforming to the global axis system. The resulting set of linear equations is expressed by

$$\{W\}=[K]\{\delta\} \tag{8.25}$$

where $[K]$ is the frame stiffness matrix and $\{W\}$ and $\{\delta\}$ are the vectors of joint forces and displacements for the complete system.

8.2.4 Joints

Joint details for thin-walled structures fall into two main categories. The first includes the case where all component plate elements of one member are attached to the other. An example of this type of joint is shown in Fig. 8.5(a). The second type of joint detail involves only a partial attachment of the member and this detail is shown in Fig. 8.5(b) for the case of a joint in which channel-sections are connected through the webs by a stiffened plate.

The most common type of joints in rigid-jointed structures belongs to the first category. Several authors have discussed the effect of warping continuity at these joints. Vacharajittiphan and Trahair (1974) carried out a thorough study to relate the warping at joints to the distortion at joints. They included the effect of joint angle and different types of stiffeners at joints. However, their work was generally confined to the effect of the stiffener configuration on the warping of the member and did not consider the transfer of member forces across the joint. Morrell (1980) carried out an experimental investigation into the influence of joint detail on the torsional behaviour of axially discontinuous structures. He concluded that warping effects are transmitted around an axially discontinuous structure, regardless of joint angle. He also recognised that the detail of the joint influenced both the magnitude and the sign of the transmitted bimoment.

The second category of joint construction was not common until the recent introduction of framed structures composed of cold-formed members. As shown in Fig. 8.5(b), the members are attached by the web to a stiffened joint plate. For this type of joint, the flanges are free to warp with respect to the web of the member and the bimoment is discontinuous across the joint. For the analysis of structures in which these joints are used, it is possible to assume that the joint plates are prismatic members. Hence, the joint plate cannot transmit a bimoment

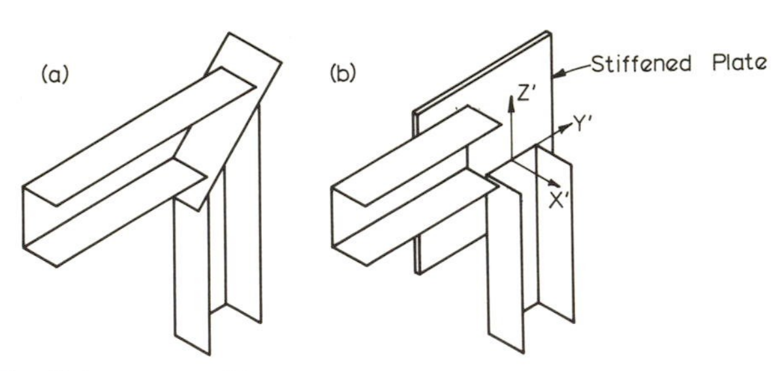

FIG. 8.5. Joint types: (a) complete attachment; (b) partial attachment.

and all of the torque is transmitted by pure torsion. Consequently, the bimoment on the end of the channel calculated with respect to the X', Y' and Z' axes is zero. Using eqn (8.17), combined with the fact that $B_{Z'}$ is zero, leads to the expression

$$B_z = -a_x \, . \, M_x \tag{8.26}$$

Therefore a bimoment B_z calculated with respect to the shear centre of the cross-section is equivalent to the major-axis moment multiplied by the distance from the shear centre to the plane of the moment.

Similarly, the bimoment is zero on the end of a Z-section bolted by its web to a stiffened plate and calculated with respect to the X', Y', Z' axes located on the web. Using eqn (8.17), combined with the fact that $B_{Z'}$ is zero, leads to the expression

$$B_z = -F_z \, . \, C = -\alpha_w \, . \, F_z \tag{8.27}$$

Therefore, a bimoment B_z calculated with respect to the shear centre of the cross-section is equivalent to the axial force multiplied by the sectorial coordinate at the centre of the web.

8.2.5 Restraints

For a general matrix displacement analysis of thin-walled structures, a total of seven restraining actions may be applied at any particular joint. The restraining actions take the form of displacement restraints in three directions, rotation restraints about the three axes and a warping restraint.

In most structural analysis problems, the restraint is usually located at the centreline of the joint between interconnecting members. The action of the restraint may constrain one, or any combination of, the three displacements, three rotations and warping deformation at that joint. Hence, the frame stiffness matrix [K] may be reduced by equating the relevant joint displacements to zero.

However, in some framed structures, the restraint positions are eccentric from the centreline of the joint. A simple example of an eccentric restraint is the support offered to an industrial portal frame by the purlins and girts located around it. For this type of structure the purlins and girts are fixed eccentrically to the member. They provide out of plane lateral support to the member at the point of attachment. They may also provide some degree of rotational restraint depending upon the fixity of the purlin to the cleat.

In general, an eccentric restraint can be located at a distance e_X, e_Y

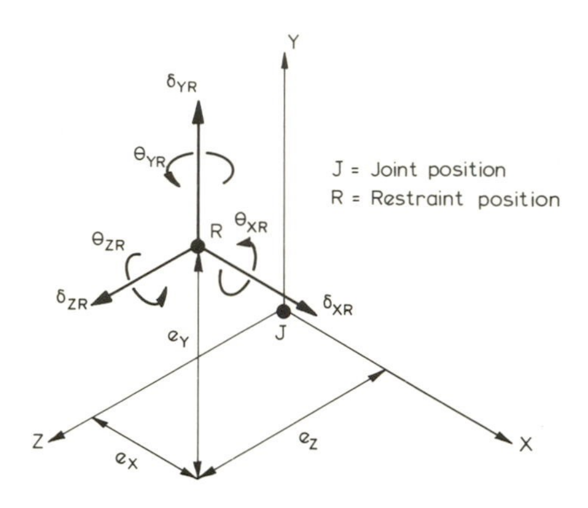

FIG. 8.6. Eccentric restraint.

and e_Z in the X, Y and Z directions from the joint origin, as shown in Fig. 8.6. To allow for the effect of this restraint on a matrix displacement analysis, the displacement degrees of freedom δ_{XR}, δ_{YR}, δ_{ZR} in the X, Y, Z directions, respectively, at the restraint point R can simply be transformed to those at the origin J of the joint, using rigid body kinematics. The resulting transformations are given by

$$\delta_X = \delta_{XR} - e_Z \, . \, \theta_{YR} + e_Y \, . \, \theta_{ZR} \tag{8.28}$$

$$\delta_Y = \delta_{YR} + e_Z \, . \, \theta_{XR} - e_X \, . \, \theta_{ZR} \tag{8.29}$$

$$\delta_Z = \delta_{ZR} - e_Y \, . \, \theta_{XR} + e_X \, . \, \theta_{YR} \tag{8.30}$$

The rotations θ_{XR}, θ_{YR}, θ_{ZR} are simply the rigid body rotations unaffected by the eccentric position of the restraint. These relationships, along with eqns (8.28), (8.29) and (8.30), have been assembled as an eccentric restraint transformation matrix $[ER]$, given in Appendix 3 and expressed in matrix notation by

$$\{\delta\} = [ER]\{\delta_R\} \tag{8.31}$$

where $\{\delta_R\}$ is the vector of joint displacements at the eccentric restraint. The corresponding set of force transformations is expressed in matrix notation by

$$\{W_R\} = [ER]^T\{W\} \tag{8.32}$$

where $\{W_R\}$ is the vector of joint loads at the eccentric restraint points.

Elimination of $\{W\}$ and $\{\delta\}$ from eqn (8.25) results in

$$\{W_R\} = [ER]^T[K][ER] . \{\delta_R\} \tag{8.33}$$

By equating the relevant eccentric joint displacements to zero, the effect of eccentric restraints is incorporated in the analysis.

8.3 STUDY OF PITCHED ROOF PORTAL FRAMES COMPOSED OF CHANNEL- AND Z-SECTIONS

8.3.1 Frame Geometry and Loading

Experimental work at the University of Sydney described by Baigent and Hancock (1980, 1982*b*) involved testing of pitched-roof portal frames composed of cold-formed sections. The structures tested were made from single-channel-sections connected at the eaves and apex, using stiffened plates bolted to the webs of adjacent sections. The same frame geometry is used in the study described in this chapter. However, both channel- and Z-section portal frames are considered.

The test frames consisted of pinned-base pitched-roof portals with the overall geometry as shown in Fig. 8.7(a). Lateral restraint consisted of two types. External restraint simulated the effect of purlins and girts and involved prevention of movement normal to the plane of the frame at the 16 external locations shown in Fig. 8.7(a). External plus internal restraint simulated the effect of purlins and girts with fly bracing attached to prevent lateral movement of the internal flange of the frame. The internal restraints were located opposite the third external restraint position in each stanchion and opposite the first and third restraint positions in each rafter. In the tests, the three load cases of dead and live load, transverse wind load, and longitudinal wind load producing suction were studied. However, for simplicity, in this chapter only the first case of vertical dead and live load is considered. These loads are assumed to be applied to the frames at the restraint points in the rafters, since these are the positions of the purlins which transmit the dead load of the roof, and the live load upon it, to the frame. Vertical point loads of $W/4$ are assumed to be applied equally at the four restraint points on each rafter shown in Fig. 8.7(a).

In the test frames, each of the monosymmetric channels were bolted through their webs to joints which consisted of stiffened plates, as shown in detail in Fig. 8.7(b). The web of each channel was rigidly attached to the joint. However, the nature of the connection allowed

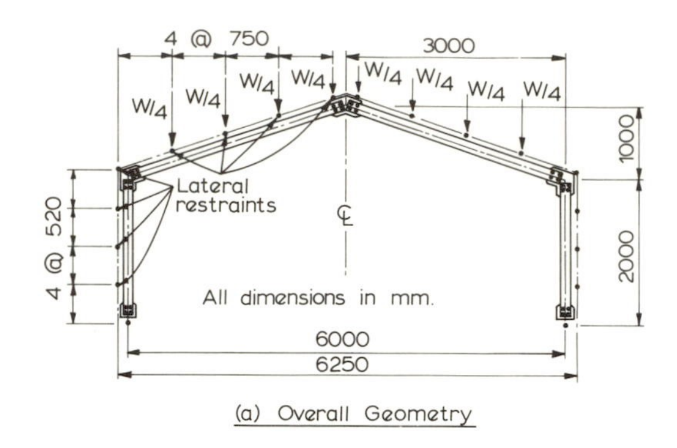

(a) Overall Geometry

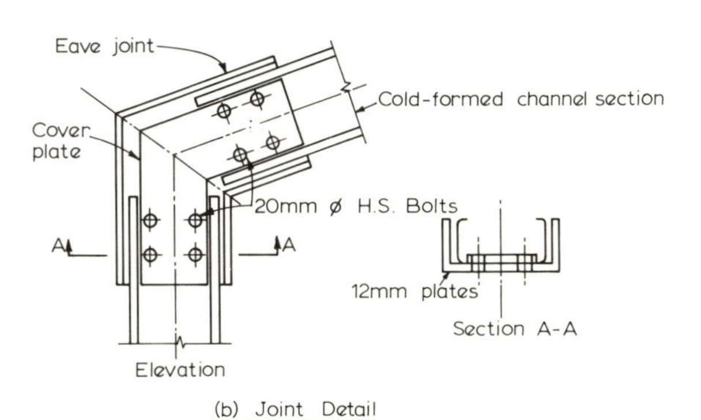

(b) Joint Detail

FIG. 8.7. Frame geometry.

freedom of movement of the flanges. The construction of the joint is similar to the diagrammatic representation shown in Fig. 8.5(b). The apex joint was of a similar construction to the eaves joint. In this chapter a similar type of joint has been assumed for connecting Z-sections such that only the webs of the sections are connected to the joint with the flanges free.

The channel section used in the tests had an overall depth of 153 mm, an overall flange width for both the top and bottom flanges of 79 mm, a plate thickness of 1·86 mm, an overall lip stiffener depth of 15 mm and internal corner radii of 10 mm. The Z-section used in this study is assumed to have the same dimensions, with only the direction of the outside flange reversed.

8.3.2 Computer Models

The computer models of the channel- and Z-section portal frames studied are shown in Figs. 8.8(a) and (b), respectively. The structures are each subdivided into 24 thin-walled elements with the nodal positions located at the boundaries between different member types, the restraint and loading points and the eaves, apex and base joints. It is not necessary to further subdivide the members, since the stiffness matrices of the thin-walled elements are based on an exact solution of the differential equations, as described in Section 8.2.1, and not on an approximate energy analysis. The lateral restraint points are 50 mm beyond the flange of the members and are located on the centreline of these flanges, as shown in Fig. 8.8.

The eaves and apex joints are assumed to be effectively rigid bodies linking the webs of two adjacent sections. The joints are treated as prismatic members with their shear centre and centroidal axes along the centreline of the plate. They are considered as having no warping torsion capability but transmit all torque by Saint–Venant torsion. Hence, using the results in eqns (8.26) and (8.27), a bimoment calculated with respect to the shear centre axis is applied on the end of

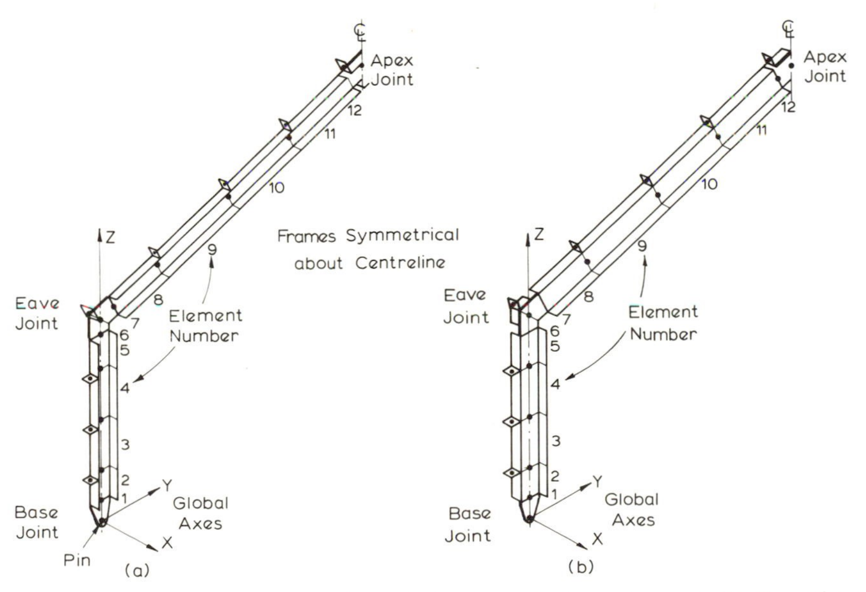

FIG. 8.8. Computer models of portal frames: (a) channel portal; (b) Z-section portal.

the channel (and is equal to the major axis moment multiplied by the distance from the shear centre to the channel web) and also on the end of the Z-section (and is equal to the axial force multiplied by the sectorial coordinate at the centre of the web).

As a consequence of the warping freedom of the thin-walled sections at the joints, the ends of the thin-walled members are assumed to be free to warp as described in Section 8.2.1.

8.3.3 Stress Resultants and Stresses in Model Portals

8.3.3.1 Channel Portal

Distributions of lateral deflection of the inside flange (δ_X), bending moments about both principal axes (M_x, M_y), axial force (F_z) and bimoment (B_z) are plotted in Fig. 8.9, for one-half of the channel portal subjected to vertical loading with $W = 1$ kN. The stress distributions resulting from the separate components (M_x, M_y, F_z, B_z) in a channel are given in Fig. 8.10. The analysis has been performed for

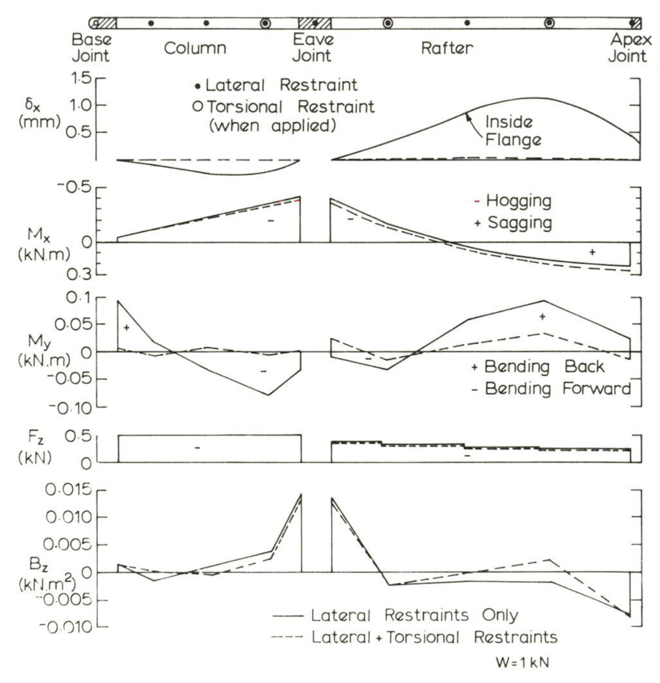

FIG. 8.9. Channel portal stress resultants and displacements.

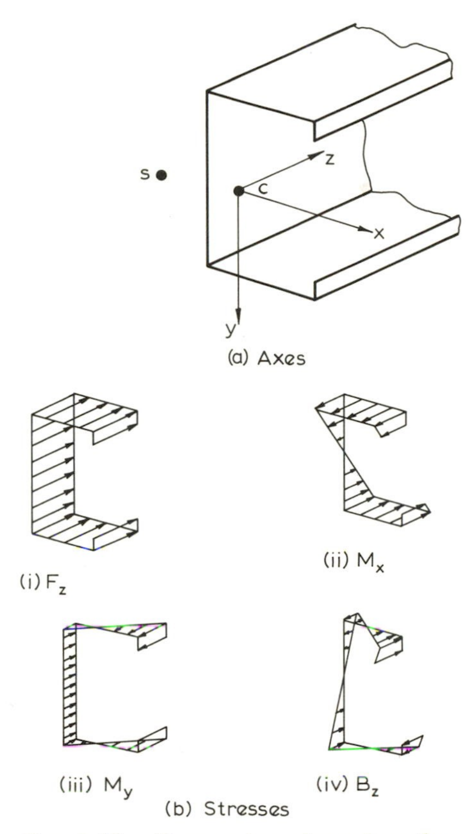

FIG. 8.10. Stresses in a channel-section.

lateral restraint only (solid lines in Fig. 8.9) and lateral plus torsional restraint (dashed lines in Fig. 8.9).

The most notable feature of these distributions is the large value of the bimoments adjacent to the eaves and apex joints. The bimomental values decay rapidly with distance from the joint. The bimoments are induced by the application of the major axis bending moment (M_x) in the joint onto the plane of the web of the channel-section, as described in Section 8.2.4. The longitudinal stresses, created principally by the major axis bending moment and bimoment, are plotted in Fig. 8.11 for the critical cross-section just below the eaves joint. The longitudinal stress values for the case of lateral plus torsional restraint are not significantly less than for the case of lateral restraint only, since the

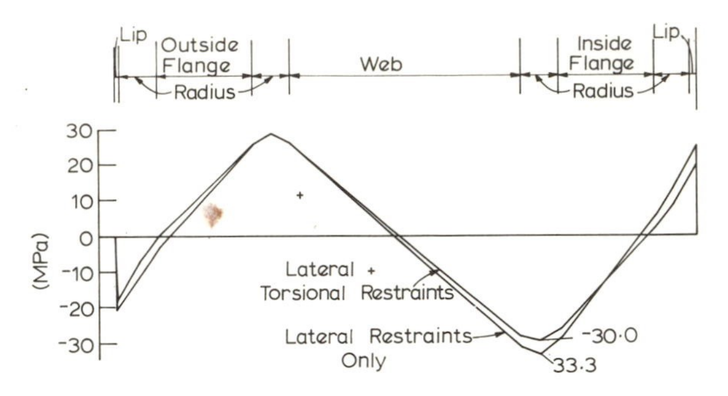

FIG. 8.11. Longitudinal stress below eaves of channel portal.

major axis moment, and hence bimoment, is not significantly altered by the torsional restraints. It is interesting to note that the minor axis moment (M_y) within the rafters and columns is altered significantly by the torsional restraints. However, its value adjacent to the eaves joint is small and does not seriously influence the longitudinal stress immediately adjacent to the eaves.

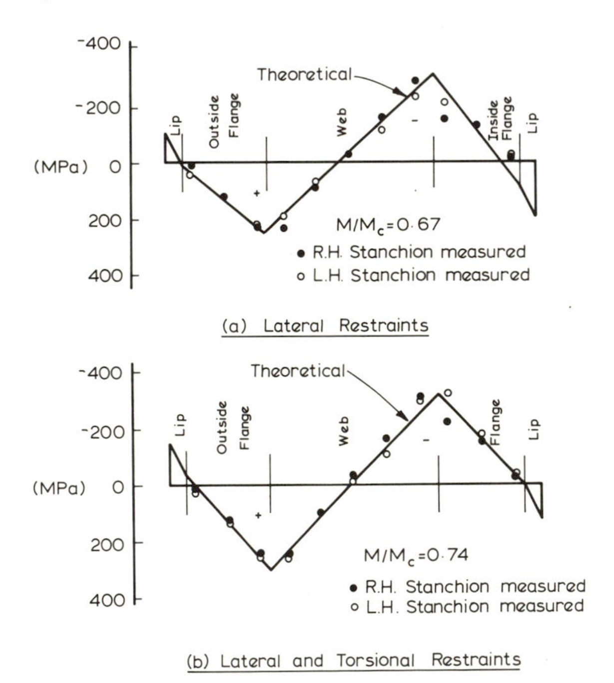

FIG. 8.12. Measured stress distributions below eaves of channel portal.

The stress distributions measured in the test frames at the critical cross-sections in the stanchions below both the right hand and left hand eaves are shown in Fig. 8.12(a) for the frame with lateral restraints only and in Fig. 8.12(b) for the frame with lateral and torsional restraints. The stresses are plotted at 67 per cent and 74 per cent of the frame collapse loads, respectively. They are compared with the theoretical estimates using the methods just described. The comparison is good, confirming the theoretical model in the vicinity of the joints.

8.3.3.2 Z-Portal

Distributions of lateral deflection of the inside flange (δ_X), bending moments about both principal axes (M_x, M_y), axial force (F_z) and bimoment (B_z) are plotted in Fig. 8.13 for one-half of the Z-portal subjected to a vertical loading with $W = 1$ kN. The stress distributions corresponding to the separate components (M_x, M_y, F_x, B_z) in a

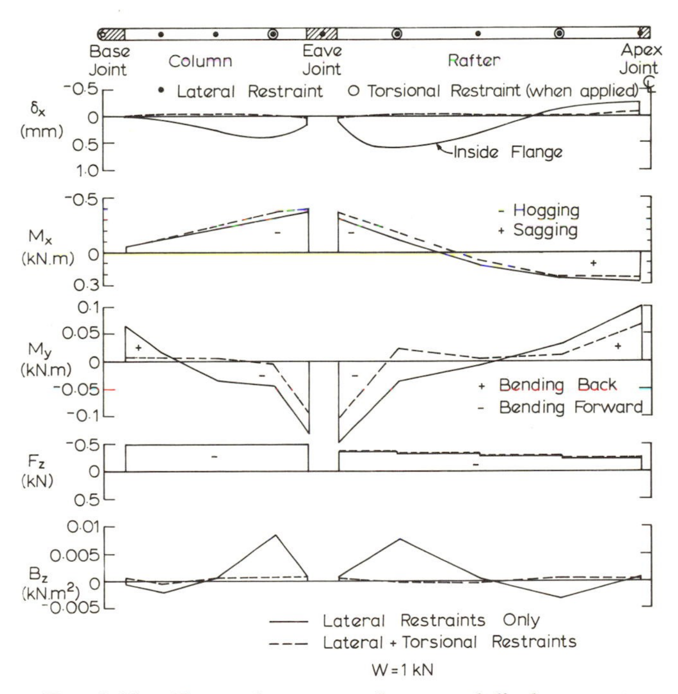

FIG. 8.13. Z-portal stress resultants and displacements.

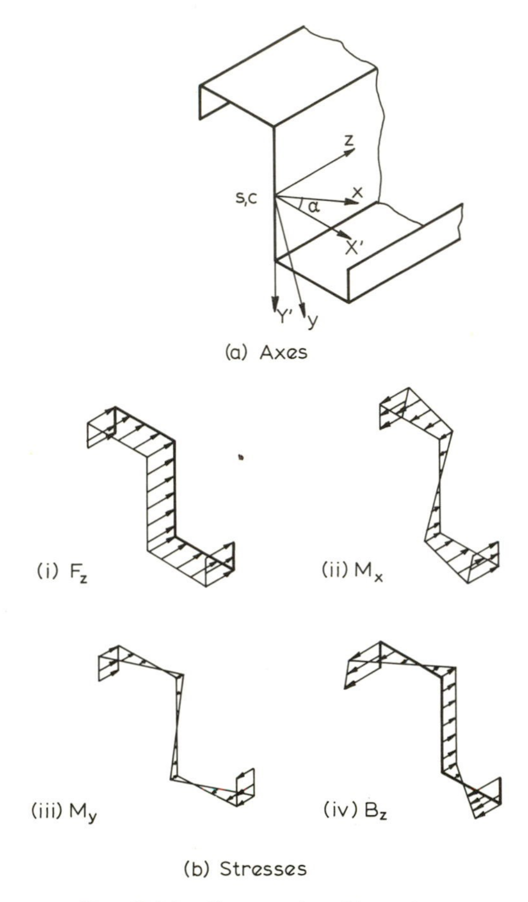

FIG. 8.14. Stresses in a Z-section.

Z-section are given in Fig. 8.14. The frame analysis has been performed for lateral restraint only (solid lines in Fig. 8.13) and lateral plus torsional restraint (dashed lines in Fig. 8.13).

The distributions of major axis moment and axial force around the frame are similar for both the Z- and channel-portals. However, the distributions of minor axis moment and bimoment are distinctly different. The minor axis moment distribution in Fig. 8.13 for the Z-portal is similar in pattern to the bimoment distribution for the channel-portal in Fig. 8.9. In both cases, there are large values adjacent to the eaves and apex joints with fairly rapid decay away from the joints. Of

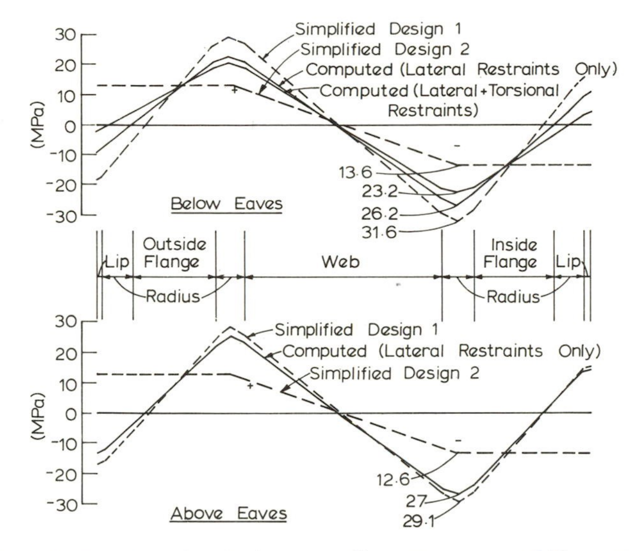

FIG. 8.15. Longitudinal stress adjacent to eaves of Z-portal.

interest, the bimoment distribution in the Z-portal has a similar pattern to the minor axis moment distribution in the channel-portal. In both cases, the values adjacent to the joints are small but have significant values within the span.

The stress distributions computed above and below the eaves joint are shown in Fig. 8.15. The most interesting feature of these stress distributions is their similarity around the section to those of the channel-portal shown in Figs. 8.11 and 8.12. Apparently, the combination of major and minor axis bending moment in a Z-section produces a similar stress distribution to the combination of major axis moment and bimoment in a channel. As for the channel-portal, the effect of the torsional restraints is to produce only a slight decrease in the stresses adjacent to the eaves.

8.3.3.3 Frame Deformations

The in-plane displacements of the eaves and apex, and twisting rotations of the rafters, are summarised for both portal types in Table 8.1. They have been compared with those for a prismatic-section portal frame with the same second moment of area of the section about a horizontal axis (X'-axis) as that for the channel- and Z-section.

TABLE 8.1
FRAME DEFORMATIONS ($W = 1$ kN)

Frame section	*Restraint*	δ_Z *(apex)* *(mm)*	δ_Y *(eave)* *(mm)*	θ_Y *max (rafter)* *(degrees)*
Channel	L	−2·16 (+61%)	−0·69	0·32
Channel	L+T	−1·85 (+38%)	−0·58	0·008
Z	L	−2·28 (+70%)	−0·68	−0·16
Z	L+T	−1·55 (+16%)	−0·48	−0·027
Prismatic	Either	−1·34	−0·42	0·0

L = lateral, T = torsional. Figures in brackets are per cent increase above prismatic values.

The in-plane deflections for both types of thin-walled portals, without torsional restraint, are significantly (60–70 per cent) in excess of those for the prismatic-portals. The additional deflections of the channel-portals have been confirmed in the tests described by Baigent and Hancock (1980, 1982*b*). When torsional restraints are included, the in-plane deflections are reduced to exceed the deflections of a prismatic-portal by 38 per cent and 16 per cent for the channel- and Z-portals, respectively.

The maximum twisting deformations of the rafters of the channel-portal are approximately double those of the Z-portal. This phenomenon is demonstrated by comparing δ_X in Figs. 8.9 and 8.13.

8.4 ELASTIC AND INELASTIC LOCAL BUCKLING OF COLD-FORMED SECTIONS

8.4.1 Finite Strip Method

As described in the introduction, failure of cold-formed members generally occurs when inelastic local buckles develop at critical cross-sections. To calculate theoretically the inelastic buckling load, a method is required which accounts for the particular geometry of the cross-section, the longitudinal stress distribution and the progression of yielding. Yoshida (1975) described a method in which he used the finite strip method of analysis developed by Cheung (1976) which was applied to local buckling by Przmieniecki (1973) and to membrane displacements as well as plate flexural displacements in the buckling mode by Plank and Wittrick (1974).

Yoshida extended the elastic analysis of Przmieniecki by allowing for yielding in I-section columns. He achieved this by reducing the effective moduli of yielded strips, according to the theory of plastic stability of thin-walled plates described by Bijlaard (1947). In the case of I-sections, progressive yielding of the web and flanges occurs with increasing load, as a result of gradients of residual stress being superimposed with the applied stress. For cold-formed sections of portal frames, progressive yielding of the web and flange will occur mainly as a result of the gradients of stress caused by superimposing bending and torsional effects to produce stress distributions of the type shown in Figs. 8.11, 8.12 and 8.15. A similar method to that described by Yoshida (1975) can be applied to cold-formed sections undergoing bending and twisting and is described in this chapter.

8.4.2 Elastic Buckling Analysis

A finite strip subdivision of the channel is shown in Fig. 8.16. The detailed analytical method of elastic buckling was described by Hancock (1978). The method involves performing a buckling analysis of the section subjected to the appropriate longitudinal stress distribution, shown in Fig. 8.17 for an assumed range of buckle half-wavelengths.

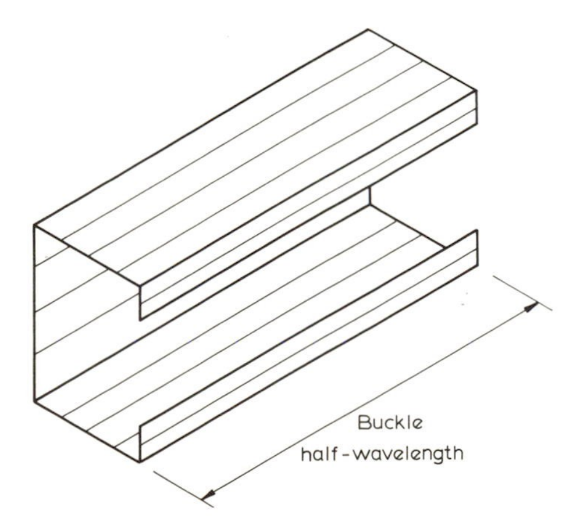

FIG. 8.16. Finite strip subdivision of channel-section for buckling analysis.

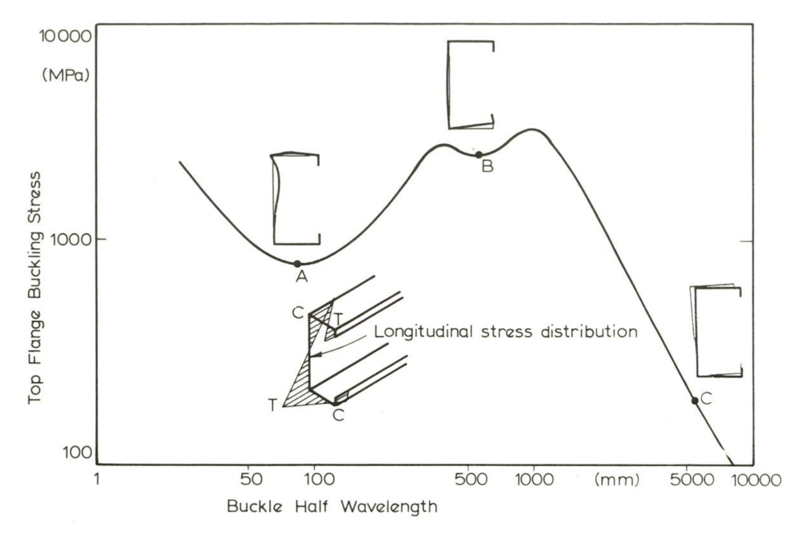

FIG. 8.17. Buckling curve for channel section.

The resulting critical stresses are plotted against the buckle half-wavelengths, as shown in Fig. 8.17. The modes corresponding to certain half-wavelengths are also shown in Fig. 8.17.

The mode shown at A is a local buckle involving the flange and web and occurs at a half-wavelength of 90 mm. A higher minimum occurs at B at a half-wavelength of 550 mm and corresponds to a stiffener buckle with the lip stiffener undergoing in-plane bending in the buckling mode. For long wavelengths of laterally unrestrained sections, a lateral buckle of the type shown at C occurs. However, for portal frames such as those tested, lateral restraints prevent this mode and so the local mode shown at A predominates.

8.4.3 Inelastic Buckling Analysis

Following initial yielding at a cross-section, increasing load causes yielding to progress along the thin-walled elements of the cross-section. In the case of the channel- and Z-sections with stress distributions shown in Figs. 8.11, 8.12 and 8.15, yielding commences at the flange–web junction and penetrates into both the web and flange. To calculate the progression of yielding, it is assumed that the stress resultant ratios calculated at first yield remain constant so that a monotonic load increase can be applied. This assumption is reasonable for small increases in load beyond first yield, where the localised yielding does not alter significantly the overall structural response.

Santathadaporn and Chen (1972) developed a tangent stiffness method for the biaxial bending analysis of column sections. However, their method did not include yielding resulting from the bimoment in thin-walled members. Hancock (1977) extended the method to include yielding resulting from warping torsion. The method has been applied by Baigent and Hancock (1982*c*) to study the channel-section portal frames as follows.

First, the axial force, bending moments and bimoment are increased monotonically beyond yield by applying a load factor λ to their values. Based on the elastic section rigidities (EA, EI_x, EI_y, EI_w), the resulting strain distribution is calculated. The yielded zones are determined and the consequent stress distribution, assuming the yield stress in the yielded zones, is integrated to calculate the net section stress resultants. Before convergence, these stress resultants will differ slightly from the applied values at load factor λ. The axial strain dw/dz, curvatures (d^2u_s/dz^2, d^2v_s/dz^2) and rate of change of twist ($d^2\theta_z/dz$) are adjusted using a tangent stiffness matrix based on the effective section rigidities. The effective rigidities are those of the elastic core ignoring yielded zones. A new strain distribution, and hence yield distribution, is calculated and the process repeated until the resulting stress distribution is in equilibrium with the applied stresses. By continuing this process at increasing load factors, the progression of yielding in a thin-walled cross-section can be calculated.

To account for yielding in the finite strip buckling analysis, the analytical process is performed with the Young's and shear moduli and Poisson's ratio reduced, to allow for plasticity. Two approaches have been developed for the purpose. The original approach developed by Bijlaard (1947) used the deformation (or total strain) theory of plasticity. The theory was further developed by Ilyushin (1947) and Stowell (1948) and has been found to correlate well with plastic buckling tests. The alternative flow (or incremental) theory of plasticity developed by Prager (1949) is more rationally based, but was not found to correlate well with plastic buckling tests of imperfect specimens. However, modifications of the application of the flow theory by Onat and Drucker (1953) and Haaijer (1957), to account for imperfections, produced results which correlate well with plastic buckling tests.

In the study of local buckling of partially yielded sections, the reduced value of the shear modulus (G) in the strain-hardening range

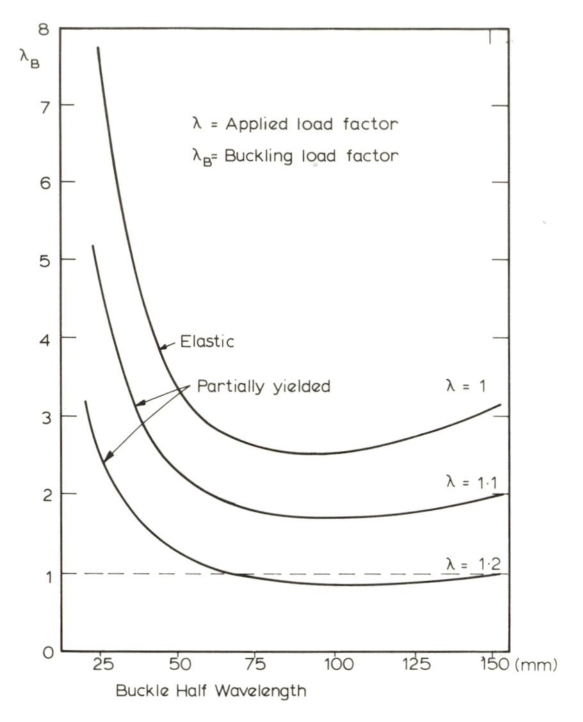

FIG. 8.18. Local buckling curves for partially yielded sections.

is important. Haaijer (1957) and Lay (1965) have concluded that a shear modulus based on mild steel in the strain-hardening range produces reasonable results. Accordingly, in this chapter, G_y/G has been taken as 0·25, based on the work of Lay. For simplicity, conservative values of E_t/E equal to 1/33 are taken in both directions and a Poisson's ratio of 0·5 has been used in the yielded zones.

The results of the inelastic local buckling analysis at increasing load factors are shown in Fig. 8.18. The curves are seen to drop with increasing load factor without a significant change in the local buckle half-wavelength. When the minimum on the buckling curve is equal to the load factor, failure is assumed to occur.

In the tests of the channel portals reported by Baigent and Hancock, this method always produced conservative estimates of the strengths of the channel section portal frames. The theoretical failure loads varied from 17 to 28 per cent greater than the theoretical first yield loads, whilst the experimental failure loads varied from 25 to 77 per cent greater than the theoretical first yield loads. The experimentally observed failure loads of the portal frames under uplift load were 75 and 77 per cent greater than the theoretical first yield loads since, although yielding occurred under the eaves, failure of the frame did not occur until plastic buckling in the rafters.

8.5 SIMPLIFIED DESIGN

8.5.1 Z-Portals

A common procedure in the design of portal frames is to perform a plane–frame linear elastic analysis, either by a manual method such as moment distribution or more recently a computer analysis using the matrix displacement approach. A possible simplified procedure for the design of a Z-portal would be to use the results of such an analysis to calculate first yield, without recourse to the three-dimensional analysis described in Section 8.2. To investigate this approach, the stresses at the critical cross-sections were computed by Hancock (1983) for three different load cases by applying the planar bending moments and axial force to the Z-section. In this chapter, the results for the vertical load case are summarised.

Two possible approaches exist for calculating the stresses from the planar bending moment. The first would be to resolve the planar moment applied about the horizontal axis (X'-axis in Fig. 8.14) into its components about the principal axes (x, y axes in Fig. 8.14). The superposition of the stresses resulting from the two principal axis moments is taken as the resultant stress distribution and is shown as 'Simplified Design 1' in Fig. 8.15. For all load cases studied, the simplified design method 1 produces slightly higher (more conservative) stress estimates than the three dimensional analysis, with the maximum value conservative by 17 per cent.

The second method for calculating stresses from the planar bending moment would be to simply divide the moment by the section modulus about the horizontal (X') axis. This method assumes that the section is fully restrained laterally and torsionally throughout its length and results in a uniform stress distribution in the flanges. The stress distribution using this approach is shown in Fig. 8.15 as 'Simplified Design 2'. The method is significantly unconservative as a result of the assumption regarding full lateral and torsional restraint, which is not satisfied in the frames studied. Hence the simplified procedure 2 should not be used. The simplified design procedure 1 appears suitable for estimating first yield.

8.5.2 Channel Portals

A similar procedure of using a plane frame linear elastic analysis could be applied to the channel-section portals. However, in this case, the bimoment produced by major axis bending would need to be calculated using eqn (8.26).

The maximum compressive stress, computed half way around the radius between the flange and web of the channel-section, is 30·3 MPa, consisting of 13·0 MPa from the major axis bending moment, 16·4 MPa from the bimoment, using eqns (8.3a) and (8.26), and 0·9 MPa from axial compression. This is less than the value of 33·3 MPa in Fig. 8.11 for the frame with lateral restraints only, but approximately equal to the value of 30·0 MPa for the frame with lateral and torsional restraints. This latter agreement is to be expected, since the minor axis moment has been ignored in the simplified calculation and is effectively zero below the eaves in Fig. 8.9 when lateral and torsional restraints are included.

There does not appear to be a simple way to produce a conservative estimate of the longitudinal stress in a channel portal without torsional restraints. The full three dimensional analysis appears to be necessary in this case.

8.6 CONCLUDING REMARKS

A systematic approach to the analysis and design of portal frames composed of cold-formed channel- and Z-section members has been described. To accurately understand the behaviour of such structures it has been demonstrated that a full three dimensional analysis, including non-uniform torsion and cross-section monosymmetry and asymmetry, is required. In addition, the specific nature of the connection of thin-walled members to joints is important and must be accurately modelled if the stiffness and strength of such structures is to be assessed. Simplified methods for calculating first yield based on plane frame linear elastic analyses have been described. However, these methods have only been investigated for the types of portals described and they should not be used for other joint types and frame geometries without further investigation.

A method for determining the strength of structures composed of cold-formed members has been presented. The method assumes that the mode of failure is inelastic local buckling. However, other failure modes such as elastic instability of members may occur if insufficient lateral and torsional restraints are attached to a frame. Further research in this area is required for a complete understanding of structures of this type.

REFERENCES

AMERICAN IRON AND STEEL INSTITUTE (1980) *Specification for the design of cold-formed members*, Washington, DC.

BAIGENT, A. H. and HANCOCK, G. J. (1980) The behaviour of portal frames composed of cold-formed members. *Thin-Walled Structures* (Ed. by J. Rhodes and A. C. Walker), Granada, London.

BAIGENT, A. H. and HANCOCK, G. J. (1982*a*) Structural analysis of assemblages of thin-walled members. *Engineering Structures*, **4**(3), 207–16.

BAIGENT, A. H. and HANCOCK, G. J. (1982*b*) The stiffness and strength of portal frames composed of cold-formed members. *Civil Engineering Transactions, Institution of Engineers, Australia*, **CE24**(3), 278–83.

BAIGENT, A. H. and HANCOCK, G. J. (1982*c*) The strength of portal frames composed of cold-formed channels. *Proceedings, Sixth International Speciality Conference on Cold-Formed Steel Structures*, University of Missouri-Rolla, November, 1982, Department of Civil Engineering, University of Missouri-Rolla, pp. 321–47.

BAKER, J. F., HORNE, M. R. and HEYMAN, J. (1956) *The Steel Skeleton—Volume II—Plastic Behaviour and Design*, Cambridge University Press, Cambridge.

BARSOUM, R. S. and GALLAGHER, R. H. (1970) Finite element analysis of torsional and torsional–flexural stability problems. *International Journal for Numerical Methods in Engineering*, **2**, 335–52.

BAZANT, Z. P. and EL NIMEIRI, M. (1973) Large deflection spatial buckling of thin-walled beams and frames. *Journal of the Engineering Mechanics Division, ASCE*, **99**, EM6 (Proc. paper 10247), 1259–81.

BIJLAARD, P. P. (1947) Some contributions to the theory of elastic and plastic stability. *International Association for Bridge and Structural Engineering*, **8**, 17–80.

CHEUNG, Y. K. (1976) *Finite Strip Method in Structural Analysis*, Pergamon Press, Oxford.

HAAIJER, G. (1957) Plate buckling in the strain-hardening range. *Journal of the Engineering Mechanics Division, ASCE*, **83**, EM2 (Proc. paper 1212), 1212-1–1212-47.

HANCOCK, G. J. (1977) Elastic–plastic analysis of thin-walled cross-sections. *Proceedings, Sixth Australasian Conference on the Mechanics of Structures and Materials*, Christchurch, August 1977, University of Canterbury, Christchurch, New Zealand, pp. 35–42.

Hancock, G. J. (1978) Local, distortional and lateral buckling of I-beams. *Journal of the Structural Division, ASCE*, **100,** ST11 (Proc. paper 14155), 2205–22.

Hancock, G. J. (1983) The behaviour and design of Z-section portal frames. *Civil Engineering Transactions, Institution of Engineers, Australia*, **CE25**(4), 268–78.

Harrison, H. B. (1980) *Structural Analysis and Design—Some Minicomputer Applications*, Pergamon Press, Oxford.

Ilyushin, A. A. (1947) The elasto-plastic stability of plates. NACA Technical Memorandum, TM 1188.

Krahula, J. L. (1967) Analysis of bent and twisted bars using the finite element method. *Journal American Institute Aeronautics and Astronautics*, **5** (6), 1194–7.

Krajcinovic, D. (1969) A consistent discrete elements technique for thin-walled assemblages. *International Journal of Solids and Structures*, **5,** 639–62.

Lay, M. G. (1965) Flange local buckling in wide-flange shapes. *Journal of the Structural Division, ASCE*, **91,** ST6 (Proc. paper 4554), 95–116.

Livesley, R. K. (1956) The application of an electronic digital computer to some problems of structural analysis. *Structural Engineer*, **34** (1), 1–12.

Morrell, P. J. B. (1980) The influence of joint detail on the torsional behaviour of thin-walled structures having an axial discontinuity. *Thin-Walled Structures* (Ed. by J. Rhodes and A. C. Walker), Granada, London, 539–52.

Onat, E. T. and Drucker, D. C. (1953) Inelastic instability and incremental theories of plasticity. *Journal of the Aeronautical Sciences*, **20** (3), 181–6.

Plank, R. J. and Wittrick, W. H. (1974) Buckling under combined loading of thin, flat-walled structures by a complex finite strip method. *International Journal for Numerical Methods in Engineering*, **8**(2), 323–39.

Prager, W. (1949) Recent developments in the mathematical theory of plasticity. *Journal of Applied Physics*, **20**(3), 235–9.

Przmieniecki, J. S. (1973) Finite element structural analysis of local instability. *Journal American Institute of Aeronautics and Astronautics*, **11**(1), 33–9.

Rajasekaran, S. (1977) Finite element method for plastic beam-columns. *Theory of Beam-Columns, Vol. II* (Ed. by W. F. Chen and T. Atsuta) Chapter 12, McGraw-Hill, New York.

Renton, J. D. (1962) Stability of space frames by computer analysis. *Journal of the Structural Division, ASCE*, **88,** ST4 (Proc. paper 3237), 81–103.

Santathadaporn, S. and Chen, W. F. (1972) Tangent stiffness method for biaxial bending. *Journal of the Structural Division, ASCE*, **98,** ST1 (Proc. paper 8637), 153–63.

Stowell, E. A. (1948) A unified theory of plastic buckling of columns and plates. NACA Technical Note, TN 1556.

Timoshenko, S. P. (1945) Theory of bending, torsion and buckling of thin-walled members of open cross-section. *Journal of the Franklin Institute*, **239** (3), 201–19; (4), 249–68; (5), 343–61.

Vacharajittiphan, P. and Trahair, N. S. (1974) Warping and distortion at I-section joints. *Journal of the Structural Division, ASCE*, **100,** ST3 (Proc. paper 10390), 547–64.

VLASOV, V. Z. (1959) *Thin-Walled Elastic Beams*, Moscow. (English Translation, Israel Program for Scientific Translations, Jerusalem, 1961.)
YOSHIDA, H. (1975) Coupled strength of local and whole bucklings of H-columns. *Proceedings of the Japan Society of Civil Engineers*, No. 243, 19–32.

APPENDIX 1

Member stiffness matrices
(a) Axial stiffness matrix

$$\begin{bmatrix} F_{zA} \\ F_{zB} \end{bmatrix} = \frac{EA}{L} \begin{bmatrix} 1 & -1 \\ -1 & 1 \end{bmatrix} \cdot \begin{bmatrix} \delta_{zA} \\ \delta_{zB} \end{bmatrix}$$

(b) Flexural stiffness matrices

$$\begin{bmatrix} F_{yA} \\ M_{xA} \\ F_{yB} \\ M_{xB} \end{bmatrix} = \frac{EI_x}{L^3} \begin{bmatrix} 12 & 6L & -12 & 6L \\ 6L & 4L^2 & -6L & 2L^2 \\ -12 & -6L & 12 & -6L \\ 6L & 2L^2 & -6L & 4L^2 \end{bmatrix} \cdot \begin{bmatrix} \delta_{yA} \\ \theta_{xA} \\ \delta_{yB} \\ \theta_{xB} \end{bmatrix}$$

$$\begin{bmatrix} F_{xA} \\ M_{yA} \\ F_{xB} \\ M_{yB} \end{bmatrix} = \frac{EI_y}{L^3} \begin{bmatrix} 12 & 6L & -12 & 6L \\ 6L & 4L^2 & -6L & 2L^2 \\ -12 & -6L & 12 & -6L \\ 6L & 2L^2 & -6L & 4L^2 \end{bmatrix} \cdot \begin{bmatrix} \delta_{xA} \\ \theta_{yA} \\ \delta_{xB} \\ \theta_{yB} \end{bmatrix}$$

(c) Torsional stiffness matrix

$$\begin{bmatrix} M_{zA} \\ B_{zA} \\ M_{zB} \\ B_{zB} \end{bmatrix} = -\frac{EI_w\lambda}{g} \begin{bmatrix} \lambda^2 l & -\lambda m & -\lambda^2 l & -\lambda m \\ -\lambda m & n & \lambda m & -p \\ -\lambda^2 l & \lambda m & \lambda^2 l & \lambda m \\ -\lambda m & -p & \lambda m & n \end{bmatrix} \begin{bmatrix} \theta_{zA} \\ -\theta'_{zA} \\ \theta_{zB} \\ -\theta'_{zB} \end{bmatrix}$$

$\lambda = \sqrt{(GJ/EI_w)}, \; l = \sinh \lambda L$
$g = 2(\cosh \lambda L - 1) - \lambda L \sinh \lambda L$
$n = \lambda L \cosh \lambda L - \sinh \lambda L$
$p = \lambda L - \sinh \lambda L$
$m = \cosh \lambda L - 1$

APPENDIX 2

Cross-section transformation matrix $[G]$

$$\begin{bmatrix} F_{X'} \\ F_{Y'} \\ F_{Z'} \\ M_{X'} \\ M_{Y'} \\ M_{Z'} \\ B_{Z'} \end{bmatrix} = \begin{bmatrix} \cos\alpha & -\sin\alpha & & & & & \\ \sin\alpha & \cos\alpha & & & & & \\ & & 1 & & & & \\ & & b_{Y'} & \cos\alpha & -\sin\alpha & & \\ & & -b_{X'} & \sin\alpha & \cos\alpha & & \\ -a_y & a_x & & & & 1 & \\ & & C & a_x & a_y & & 1 \end{bmatrix} \begin{bmatrix} F_x \\ F_y \\ F_z \\ M_x \\ M_y \\ M_z \\ B_z \end{bmatrix}$$

For channel in Fig. 8.1, $\alpha=0$, $C=0$, $a_y=0$, $b_{Y'}=0$.
For Z-section in Fig. 8.1, $a_x=0$, $a_y=0$, $b_{X'}=0$, $b_{Y'}=0$, $C=\alpha_w$.

APPENDIX 3

Eccentric restraint transformation matrix $[ER]$

$$\begin{bmatrix} \delta_X \\ \delta_Y \\ \delta_Z \\ \theta_X \\ \theta_Y \\ \theta_Z \\ -\theta'_Z \end{bmatrix} = \begin{bmatrix} 1 & & & & -e_Z & e_Y & \\ & 1 & & e_Z & & -e_X & \\ & & 1 & -e_Y & e_X & & \\ & & & 1 & & & \\ & & & & 1 & & \\ & & & & & 1 & \\ & & & & & & 1 \end{bmatrix} \begin{bmatrix} \delta_{XR} \\ \delta_{YR} \\ \delta_{ZR} \\ \theta_{XR} \\ \theta_{YR} \\ \theta_{ZR} \\ -\theta'_{ZR} \end{bmatrix}$$

Chapter 9

BRACED STEEL ARCHES

SADAO KOMATSU
Department of Civil Engineering, Osaka University, Japan

SUMMARY

This chapter deals with the ultimate strength of parabolic steel arches with bracing systems for the following cases; (1) when they fail by inelastic in-plane instability under uniformly distributed vertical loads, (2) when they fail by inelastic out-of-plane instability under vertical loads or combined vertical and uniform lateral loads. Design formulae are presented for each case.

NOTATION

A	Average cross-sectional area over the whole length of the arch rib with variable cross section, or cross-sectional area of arch rib with uniform cross section
A_1, A_2	Cross-sectional areas of bracing members in the end and next panel, respectively
A_j	Cross-sectional area of arch rib at the jth segment
A_s	Cross-sectional area at springings of arch rib
a	Distance of twin arch ribs
$\bar{a}$	Slenderness parameter
$\mathbf{B}$	Coefficient matrix including reduced bending stiffness and elongation stiffness in elastic–plastic range
C_d	Drag coefficient of arch rib, which is 2·19 for box cross-section with height-to-width ratio of 2
D_i	Axial force of bracing member in the ith panel
E	Young's modulus

f	Rise of arch rib
h	Height of arch rib
I	Second moment of area of arch rib about horizontal axis
I_0, I_1	Second moments of area of arch rib in the portal frame and the end panel of bracing system, respectively
$I_b(=I_3), I_y$	Second moments of area of cross beam and arch about vertical axis, respectively
K	Effective-length factor
$\bar{K}$	Coefficient included in eqn. (9.11)
$\mathbf{K}_e, \mathbf{K}_p$	Stiffness matrices corresponding to the elastic and plastic parts of the whole structure, respectively
$\mathbf{K}_g$	Initial stress matrix
L	Total length of arch
L_j	Length of the jth segment of arch rib
l	Span length of arch
l_0	Panel length of portal frame
l_p	Panel length of bracing system
l_t	Effective buckling length of bracing
M	Vertical bending moment
M_0	Vertical bending moment at the end of arch
M_1, M_3, M_5	Lateral bending moments of arch given by analysis, as shown in Fig. 9.13
M_b	Lateral bending moment of individual rib for the integrated whole structure
M_c	Lateral bending moment at the springings of arch rib
M_s	Lateral bending moment of individual rib computed by elastic analysis for load intensity $q/q_0=1$
M_y	Yield moment at springings of arch rib
m	Force ratio given by formula (9.29)
$\bar{m}$	Non-dimensional vertical bending moment
N	Axial force at springings of arch produced by vertical load, p
$N_{cr,s}, N_{cr}$	Critical axial forces at the quarter point of span length under symmetric and asymmetric loading, respectively
N_E	Euler buckling load
N_s	Axial force of leeward rib computed by elastic analysis for load intensity $q/q_0=1$
N_u	Ultimate axial force at springings of arch rib
$\bar{n}$	Non-dimensional load normal to original arch axis
P	Incremental load vector

p	Uniform live load
p_n, p_t	Loads normal and tangential to original arch axis, respectively
Q_0	Shear force at the end of arch
$Q_i (i = 1, 2)$	Lateral shear forces in the ith panel of idealised plane frame under lateral loading
Q_r	Lateral shear forces of idealised plane frame under lateral loading
q	Wind load
q_0	Reference wind load
$\bar{q}$	Non-dimensional wind load
q_u	Ultimate lateral load
R_0	Radius of finite circular arc-element
r	Radius of gyration about horizontal centroidal axis of arch cross-section
r_t	Radius of gyration of bracing member
r_y	Radius of gyration about vertical centroidal axis of arch cross-section
$\bar{t}$	Non-dimensional load tangential to original arch axis
$\mathbf{u}$, u	Displacement vector and displacement normal to original arch axis
V_0	Reference wind velocity, 50 m/s
v	Specified design wind velocity
W_{cr}	Critical load producing in-plane instability
W_y, W_z	Section moduli about vertical and horizontal centroidal axes at the springings of arch rib, respectively
w	Dead load
$\mathbf{X}$	State vector at the end of arch
$\mathbf{Y}$	Load vector
α	Angle between diagonal member and a transverse line in idealised plane frame, as shown in Fig. 9.13
β	Ratio of the length of braced portion to the total length of arch
θ	Polar coordinates for finite circular arc-element
κ	Buckling coefficient given by Stüssi (1935)
λ	Generalised slenderness ratio about horizontal axis of arch rib
λ_{cr}	Buckling coefficient for in-plane instability
λ_y	Slenderness ratio about vertical axis of arch rib

$\bar{\lambda}_y$	Non-dimensional slenderness parameter defined by eqn (9.16)
ν	Load factor
ρ	Density of air, 0·125 kg/m^3
$\sigma_n = N/A_s$	Normal stress at the springings of arch rib due to normal force
$\sigma_u, \bar{\sigma}$	Ultimate stress and non-dimensional stress of braced arch subjected to vertical load, respectively
$\sigma_s, \bar{\sigma}_s$	Maximum elastic fibre stress and non-dimensional stress at the springings of leeward rib, respectively
σ_{sa}	Theoretical value of σ_s
σ_{se}	Approximate value of σ_s
σ_y	Yield stress
Φ	Reduction factor
ψ_0	Slope of deflection curve at the end of arch

9.1 INTRODUCTION

Currently the concept of ultimate state design is gaining international acceptance. Such a trend in structural design emphasises the importance of understanding the true characteristics of the ultimate behaviour and predicting accurately the instability of metal compression members and structures, when they include some initial imperfections, which influence their load carrying capacity.

Few studies have been carried out on the inelastic instability of arches, so far. Unfortunately the elastic buckling theory only permits us to estimate the critical loads of extremely slender arches which are hardly ever used as components of a bridge structure.

Plastic analysis based on the assumption of formation of plastic hinges is not realistic for the arch, because the plastic zone always spreads along the arch axis under vertical overloading. In such a situation, it appears that the elasto-plastic analysis taking account of finite displacement is an indispensable tool.

9.2 IN-PLANE INSTABILITY OF ARCHES UNDER VERTICAL LOADS

9.2.1 Theoretical Background

Details of theoretical investigations can be found in the thesis of Shinke (1977). As this and the relevant papers were written only in

Japanese, a brief explanation of the theoretical procedure is given below.

The structural model treated in the analysis is composed of finite circular arc-elements, being able to fit for any shape of arch axis as a whole.

In the equilibrium equation between the stress resultants and the external forces acting on the element, account is taken of the deformation and the elongation of the arch axis; from this, simultaneous differential equations for the two components of in-plane displacement u and v were derived.

Eliminating one component v from these equations, the following sixth-order fundamental differential equation has been obtained.

$$\frac{1}{1+\bar{m}}\frac{d^6u}{d\theta^6}+(2+\bar{n})\frac{d^4u}{d\theta^4}+\bar{t}\frac{d^3u}{d\theta^3}+(1+\bar{m}+2\bar{n})\frac{d^2u}{d\theta^2}+\bar{t}\frac{du}{d\theta}+\bar{n}u=-\bar{t}R_0 \quad (9.1)$$

where $\bar{m}=MR_0/\kappa EI$, nondimensional bending moment; $\bar{n}=p_nR_0^3/\kappa EI$, nondimensional load normal to the original arch axis; $\bar{t}=p_tR_0^3/\kappa EI$, nondimensional load tangential to the original arch axis.

By dividing the cross-section of each element into a sufficient number of sub-elements, the relation betwen the bending moment, the normal force, the curvature of arch axis and the strain, the so-called 'M–N–ϕ–ε relationship', in the elasto-plastic range was numerically determined taking into consideration both residual stress and strain-hardening.

Equation (9.1) is transformed into simultaneous differential equations of the first-order which can be solved by means of numerical integration, such as by the Runge-Kutta method. Thus, a field transfer matrix determines the relationship between the two state vectors at both ends of the element.

Then, the following matrix equation for a vector **X** including three unknown mechanical quantities as its components is derived by means of a general procedure of transfer matrix method, as follows

$$\mathbf{BX}=\mathbf{Y} \quad (9.2)$$

where $\mathbf{X}=\{\psi_0; Q_0; N_0\}$ for two-hinged arch, $\mathbf{X}=\{M_0; Q_0; N_0\}$ for fixed arch.

Thus, the inelastic stability of the arch occurs when the following equation is satisfied

$$|\mathbf{B}| = 0 \tag{9.3}$$

The nonlinear equation (eqn (9.3)) for the characteristic value K_{cr} (referred to as buckling coefficient) can be found by using iteration procedure.

Finally, the critical load W_{cr} producing in-plane instability can be given by

$$W_{cr} = K_{cr} \frac{EI}{l^3} \tag{9.4}$$

9.2.2 Experimental Investigation

Tests have been performed on 20 models having various combinations of parameters, such as rise-to-span ratio, end condition, slenderness ratio, load condition, grade of steel, initial deflection and residual stress, to investigate some important characteristics of arches and also to examine the rationality of the analysis.

A typical example indicating the effect of initial deflection on the ultimate strength of arches is shown in Fig. 9.1. For two cases of two-hinged and fixed arches, the initial deflection, assumed to be sinusoidal, affects distinctly the ultimate strength of the arches under the load ratio of $p/w = 0{\cdot}05$.

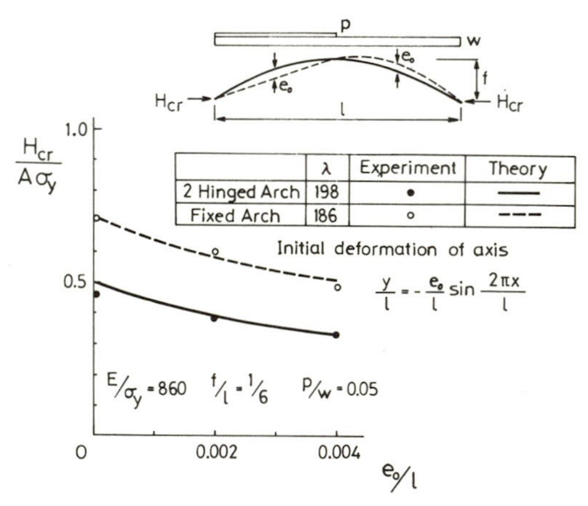

FIG. 9.1. Critical horizontal reaction versus initial deformation curves (Model series II).

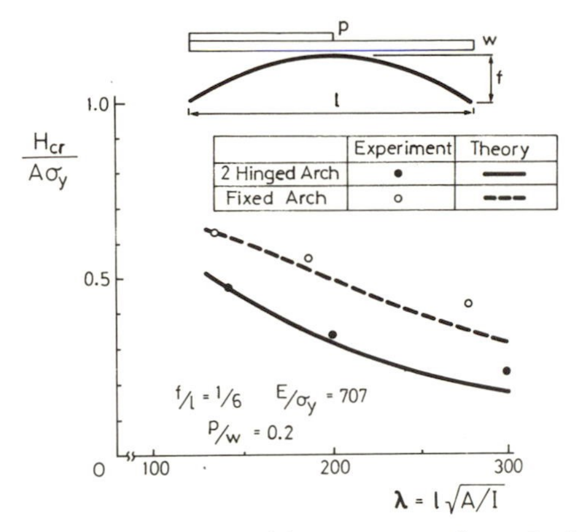

FIG. 9.2. Relationship between ultimate strength and slenderness ratio (Model series I).

The effect of residual stress on the reduction of the ultimate strength was also noticeable, especially for the small value of load ratio p/w.

Good agreement is seen between computed and experimental results for both types of arches, as shown in Fig. 9.2.

Similar results have been obtained in the experiments on other models.

9.2.3 Design Formulae

Based upon a large number of parametric computations for the appropriate range of all the parameters involved in existing arch bridges, the following design formulae for steel arches under symmetrical and asymmetrical loading, shown in Fig. 9.3, were obtained by Komatsu and Shinke (1977).

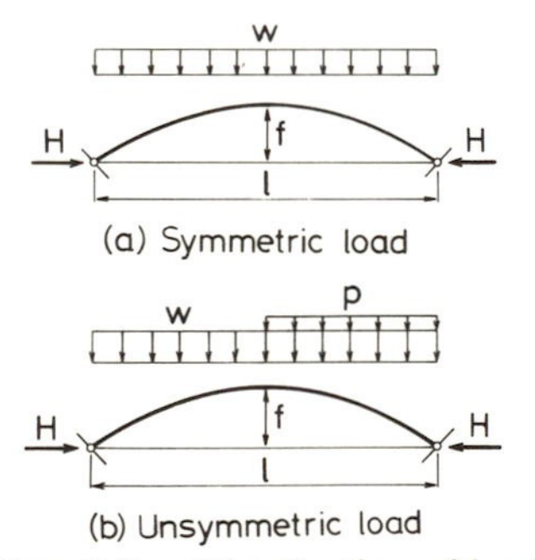

FIG. 9.3. Distribution of load.

9.2.3.1 Symmetric Loading

$$\left.\begin{aligned} \frac{N_{cr,s}}{A\sigma_y} &= 1 - 0{\cdot}136\bar{a} - 0{\cdot}3\bar{a}^2 \qquad && \bar{a} \leqslant 1 \\ \frac{N_{cr,s}}{A\sigma_y} &= \frac{1}{0{\cdot}773 + \bar{a}^2} && \bar{a} > 1 \end{aligned}\right\} \tag{9.5}$$

where

$$\bar{a} = \frac{1}{\sqrt{(\gamma\kappa(E/\sigma_y))}}\frac{l}{r} \tag{9.6}$$

$$\gamma = \sqrt{\left(1 + 4\left(\frac{f}{l}\right)^2\right)} \tag{9.7}$$

The values of buckling coefficient κ in eqn (9.6), given by Stüssi (1935), are given in Table 9.1.

TABLE 9.1
BUCKLING COEFFICIENT κ GIVEN BY STÜSSI (1935)

Type	*f/l*			
	0·1	0·15	0·2	0·3
Two-hinged arch	36·0	32·0	28·0	20·0
Fixed arch	76·0	69·5	63·0	48·0

Figure 9.4(a) and (b) show how the proposed formulae give an adequate estimation for the ultimate strength of two-hinged and fixed arches under symmetrical loading. In the Figures, SS41 and SM58 show the mild steel and high-strength low-alloy steel with tensile strengths of 402 N/mm^2 and 569 N/mm^2 specified according to the Japanese Industrial Standard, respectively.

9.2.3.2 Asymmetric Loading

As is well known, the critical axial force N_{cr} of arches under asymmetric loading is considerably small compared with $N_{cr,s}$ under symmetric loading, even though all other parameters are unchanged.

From the results of parametric studies, the following formulae for estimating the critical axial force N_{cr} could be obtained.

$$N_{cr} = \Phi N_{cr,s} \tag{9.8}$$

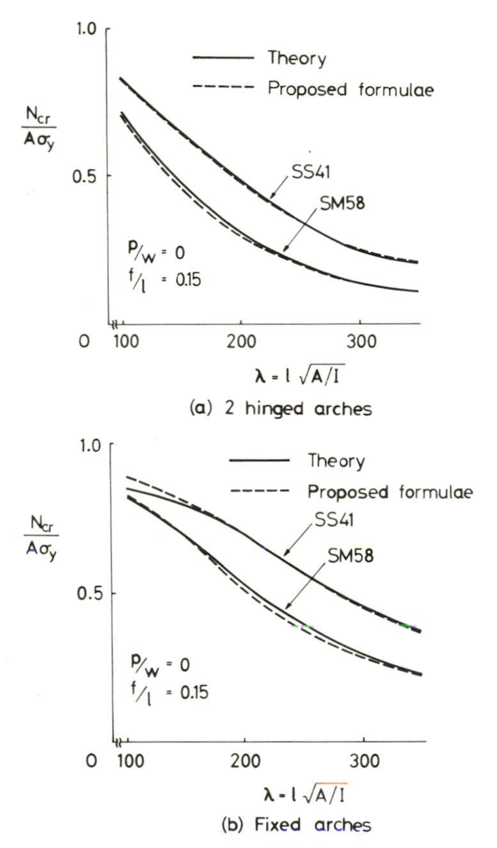

FIG. 9.4. Comparison between theory and the proposed formulae for symmetric loading.

where

$$\Phi = 1 - C\sqrt{\left(\frac{P}{W}\right)}, \text{ reduction factor} \tag{9.9}$$

$$C = C_1 + C_2$$

$$C_1 = 2{\cdot}2\frac{f}{l} + 0{\cdot}018\sqrt{\left(\frac{E}{\sigma_y}\right)} - 0{\cdot}190 \tag{9.10}$$

$$C_2 = -4\bar{K}\left(\frac{l}{r} - 6\bar{K}\sqrt{\left(\frac{E}{\sigma_y}\right)}\right)^2 \times 10^{-6} \tag{9.11}$$

$$\left.\begin{aligned} \bar{K} &= 1 \quad \text{for two-hinged arch} \\ \bar{K} &= 1{\cdot}7 \quad \text{for fixed arch} \end{aligned}\right\} \tag{9.12}$$

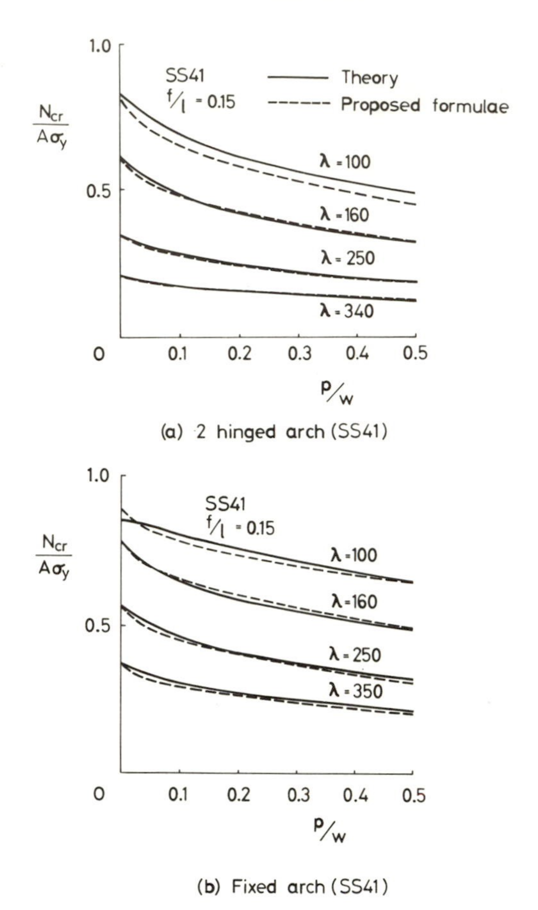

FIG. 9.5. Comparison between theory and the proposed formulae for asymmetric loading.

For the two-hinged and fixed arches made of mild steel or high-strength low-alloy steel, the critical axial forces N_{cr} calculated by the theory mentioned previously are compared with those of the proposed formulae, in Fig. 9.5(a) and (b).

It will be observed from these figures that the proposed formulae will serve as a convenient tool in design practice.

9.2.3.3 Arches With Variable or Hybrid Cross-Section

Arch bridges are usually designed in such a way that the cross-sectional properties vary along the arch axis; sometimes, two or three grades of steels are used in various parts of the span. It is necessary to employ an engineering approach to the estimation of the ultimate strength for such bridges. For this purpose, the following treatment may be useful.

From a number of numerical results obtained from typical models, it has been realised that an arch with a variable cross-section has approximately the same ultimate strength as an ideal arch with a uniform cross-section averaged over the whole length. In a similar way, an arch composed of various grades of steels may be regarded as a homogeneous arch having an idealised yield stress averaged over the whole length.

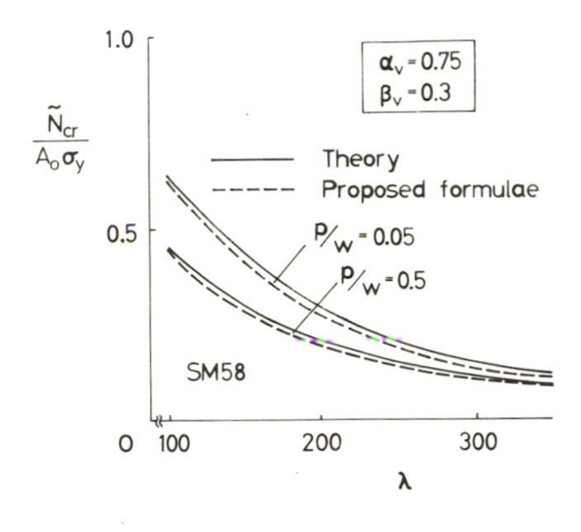

(a) 2 hinged arches

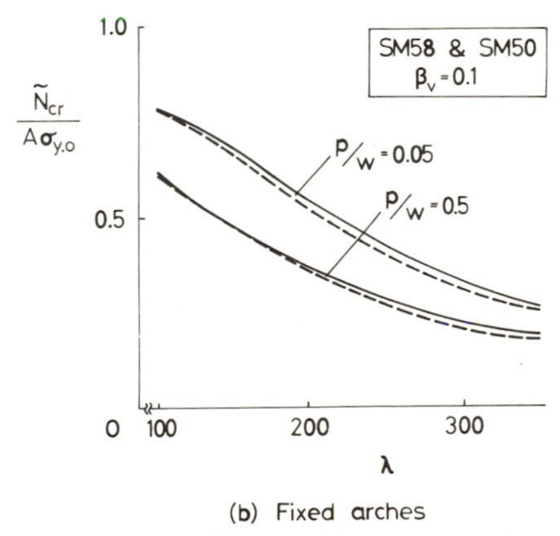

(b) Fixed arches

FIG. 9.6. Ultimate strength of (a) variable cross-sections; (b) hybrid sections.

Typical examples of numerical comparison between theoretical results and the proposed approaches can be seen in Fig. 9.6(a) for the case of variable cross-section. Similar examples for the case of hybrid arches display good correlation, as shown in Fig. 9.6(b) for variations of materials.

9.3 OUT-OF-PLANE INSTABILITY OF ARCHES WITH BRACING SYSTEMS

9.3.1 Theoretical Background

A method evaluating both torsional and bending rigidities of a partially yielded closed cross-section has been presented by Komatsu and Sakimoto (1975) in a form suitable for matrix analysis, and examined experimentally to find its validity.

Introducing it into a matrix stiffness method, the tangential stiffness matrix was derived by Komatsu and Sakimoto (1976) for the spatial frames subjected to nonproportional loads, considering the finite displacement and the spanwise and transverse spread of plastic zones.

The formulation was carried out using the following assumptions and idealisations;

(1) symmetric closed cross-sections of arch ribs made of elastic/perfectly plastic materials,
(2) Von Mises yield criterion,
(3) small strains and Bernoulli–Navier hypothesis,
(4) no local buckling of plate elements and no cross-sectional distortion induced,
(5) uniform shear flow in elasto-plastic element,
(6) immovability of shear centre after partial yielding in the cross-section, i.e. an almost double-symmetric configuration, and
(7) Prandtl–Reuss stress–strain relations.

The arch rib is divided into a number of finite bar elements of which the cross-sections are further divided into 47 sub-elements. The bracing members are also divided into 24 cross-segments like the arch rib.

The principle of minimum potential energy for each element is applied to derive the incremental equilibrium equation in terms of the local coordinates on the basis of up-dated Lagrangian formulation.

After transforming the coordinate system and assembling it for the whole structural system, the governing equation for the increment $\mathbf{u}$ of nodal displacements under given load increments $\mathbf{p}$ can be obtained in the global coordinate system as follows

$$(\mathbf{k}_e + \mathbf{K}_p + \mathbf{K}_g)\mathbf{u} = \mathbf{P} - (\mathbf{T}\bar{\mathbf{f}} - \bar{\mathbf{P}}) \tag{9.13}$$

The last term $\mathbf{T}\bar{\mathbf{f}} - \bar{\mathbf{P}}$ expresses the unbalanced forces due to the linearisation in the formulation for each incremental stage as well as yielding of the material during the incremental loading process. This force can be cancelled by the iterative Newton–Raphson procedure for each incremental loading stage.

The convergence criterion in the iterative computation is such that the increment of every nodal displacement becomes less than 10^{-3} times the up-to-date total displacement at the same node.

The small residual unbalanced forces still remaining just after ending the iteration process for each incremental load are added to the next incremental load so as to make the computing time short. The maximum load is determined as the average of the last two load values, i.e. the ultimate equilibrium load and the next divergent one. Every arch studied herein reaches the collapse caused by unbounded increase of lateral deflections without forming any plastic hinge.

9.3.2 Structural Models

The effects of the stiffness, location and type of lateral bracing system on the inelastic lateral stability of the arch are mainly investigated in the following paragraphs. The other structural parameters, such as the magnitude and distribution of residual stresses, the amplitude and variation of the initial crookedness, the configuration of the arch, the rise-to-span ratio, the height-to-width ratio of box cross section, the loading direction, and the grade of steel, are assumed somewhat conservatively on the basis of the previous study performed by Sakimoto and Komatsu (1977*a*, *b*), in which the influences of these parameters on the lateral instability of unbraced arches were reported upon quantitatively. The range of parameters treated here is listed in Table 9.2, on the basis of the justification for its choice as examined by that study. Moreover, the typical models are representatives of what might actually be found in existing arches, and are shown in Fig. 9.7.

The notations included in Table 9.2 are described below:

The λ_y values of existing bridges constructed in Japan lie somewhere within the range of 100–600. The λ_t value varies from 70 to 150 in existing arch bridges.

TABLE 9.2
PROPERTIES OF STANDARD MODELS

Items	*Properties*
Section modulus ratio, W_z/W_y	2·0
Axial variation of cross-section	Uniform
Slenderness ratio, $\lambda_y = L/r_y$	100–800
Grade of steel, E/σ_y	875, homogeneous mild steel
Maximum initial crookedness, $\bar{W}_0$	0, $L/1000$
Slenderness ratio of truss member, $\lambda_t = l_t/r_t$	70, 140, 280
Stiffness ratio of cross beam, $n = I_b/I_y$	0·01, 0·1, 1
Number of finite elements in individual arch rib	14 (D6X, D7B) 16 (D12X)

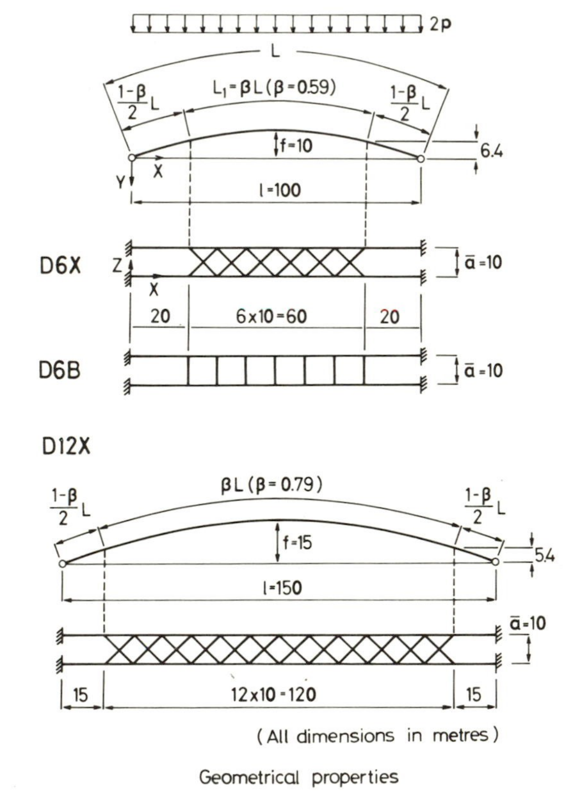

FIG. 9.7. Typical models

The bending stiffness ratio I_b/I_y of the cross beam to the arch rib in the frame type of bracing system is denoted by n.

For numerical analysis, each rib is divided into 14 and 16 member elements for the cases D6X, D6B, and D12X shown in Fig. 9.7, respectively.

Notations D6X and D12X denote twin arches having the double Warren type of bracing system with six and twelve panels, respectively, while D6B and D12B denote twin arches having the frame type of bracing system with six and twelve panels, respectively.

Each of the diagonal members and the cross beams constructing the bracing system is treated as a single element and the intersecting points of the diagonal members in the double Warren truss type of bracing system are regarded as loose, for the sake of simplicity.

The cross-sections of the arch rib element and the bracing member element are subdivided into 48 and 24 segments, respectively, which are adequate to model the development of plastic zone in the cross-sections.

9.3.3 Ultimate Strength of Braced Arches Under Vertical Loads

The non-dimensional ultimate strength $\bar{\sigma} = \sigma_u/\sigma_y$ of the model D6X is shown in Fig. 9.8, where σ_u is the ultimate stress due to axial force at

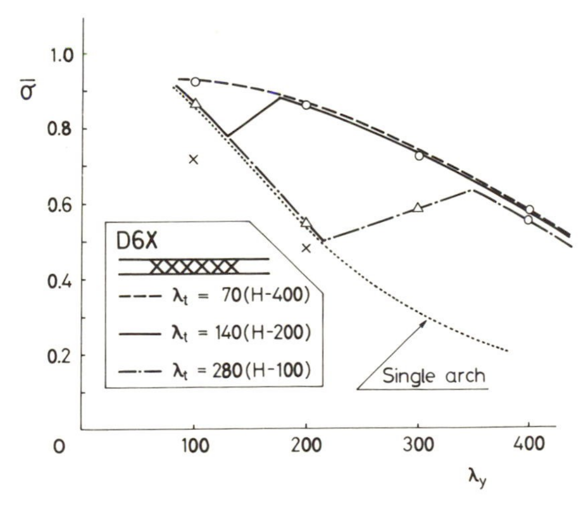

FIG. 9.8. Ultimate stresses of model D6X: (○), collapse without buckling of bracings; (△), collapse with premature buckling of bracings; (×), initiation of buckling of bracings.

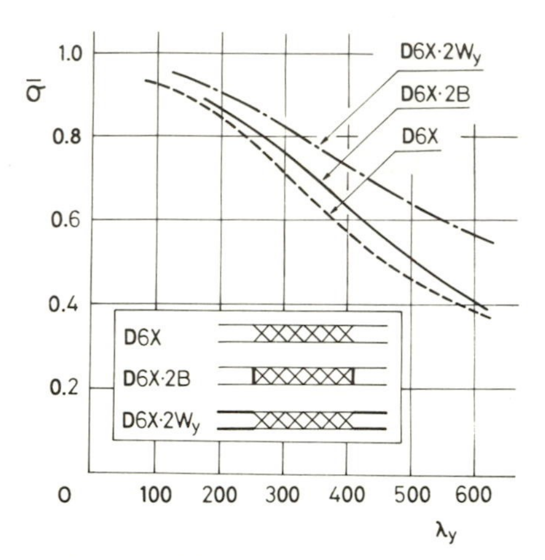

FIG. 9.9. Comparison of ultimate strength between several models.

springings (Sakimoto and Komatsu, 1977*b*). The $\bar{\sigma}$ value of the corresponding single-rib arch is also displayed in the figure.

It is worth noting that the ultimate strength of the twin arches braced with lateral members is reduced considerably due to the premature buckling of the end bracing member, if its slenderness ratio is considerably large. Such an important characteristic was also seen for other types of bracing systems in the numerical studies.

The ultimate strength curves for D6X . $2W_y$ and D6X . $2B$ are shown in Fig. 9.9. The end portions of arch rib of the model D6X . $2W_y$ have a heavy cross-section, of which the radius of gyration is twice as large as that of the standard model. It is remarkable that the ultimate strength of the model D6X . $2W_y$ exceeds that of the D6X by a significant amount. This suggests how the slenderness ratio of the unbraced end portion of the arch rib itself plays an important role in strengthening the through arch bridges with twin arch ribs.

D6X . $2B$ is the model braced by the end cross beams of $I_b/I_y = 1$, in addition to the double Warren truss with six panels. However, the increase of the ultimate strength by the addition of such comparatively stiff cross beam is not significant, compared with the model D6X.

The ultimate stress σ_u of braced arches subjected only to the vertical loads can be obtained by the following formulae, which are based on the results of theoretical and experimental investigation previously described for lateral instability (Sakimoto, Yamao and Komatsu (1979)),

$$\left.\begin{aligned}\bar{\sigma} &= 1-0\cdot136\bar{\lambda}_y-0\cdot300\bar{\lambda}_y^2 && \text{for} \quad \bar{\lambda}_y \leq 1\\ \bar{\sigma} &= 1\cdot276-0\cdot888\bar{\lambda}_y+0\cdot176\bar{\lambda}_y^2 && \text{for} \quad 1 \leq \bar{\lambda}_y \leq 2\cdot52\\ \bar{\sigma} &= \frac{1}{\bar{\lambda}_y^2} && \text{for} \quad 2\cdot52 \leq \bar{\lambda}_y\end{aligned}\right\} \tag{9.14}$$

where

$$\bar{\sigma} = \frac{\sigma_u}{\sigma_y} = \frac{N_u}{A\sigma_y} \tag{9.15}$$

$$\bar{\lambda}_y = \frac{1}{\pi}\sqrt{\left(\frac{\sigma_y}{E}\right)}\frac{KL}{r_y} \tag{9.16}$$

$$r_y = \sqrt{\left(\frac{I_y}{A}\right)} \tag{9.17}$$

$$K = K_e K_\beta K_l, \qquad K_\beta = 1-(1-C)\beta, \qquad C = \frac{2r_y}{K_e a} \tag{9.18}$$

$$K_e = \begin{cases} 0\cdot5 \text{ for fixed ends about lateral bending} \\ 1\cdot0 \text{ for hinged ends about lateral bending} \end{cases} \tag{9.19}$$

$$K_l = \begin{cases} 0\cdot65 \text{ for hanger-loaded arches} \\ 1\cdot0 \text{ for vertically-loaded arches} \end{cases} \tag{9.20}$$

$$A = \frac{1}{L}\sum_j A_j L_j, \qquad L = \sum_j L_j \tag{9.21}$$

9.3.4 Ultimate Strength of Braced Arches Subjected to Combined Vertical and Lateral loads

In the analysis, the lateral load q is increased step-by-step under the constant vertical load p until an instability occurs. The longitudinal distributions of stress resultants and deflection in the ultimate state are similar to those observed in the case of vertical loading, except for the torsional behaviour as shown in Fig. 9.10.

The difference in the in-plane bending moment diagrams between the two ribs, as shown in Fig. 9.10(d), may be caused by the torsional action of the lateral load about the bridge axis passing through both ends of the arch. The most important characteristic in the lateral

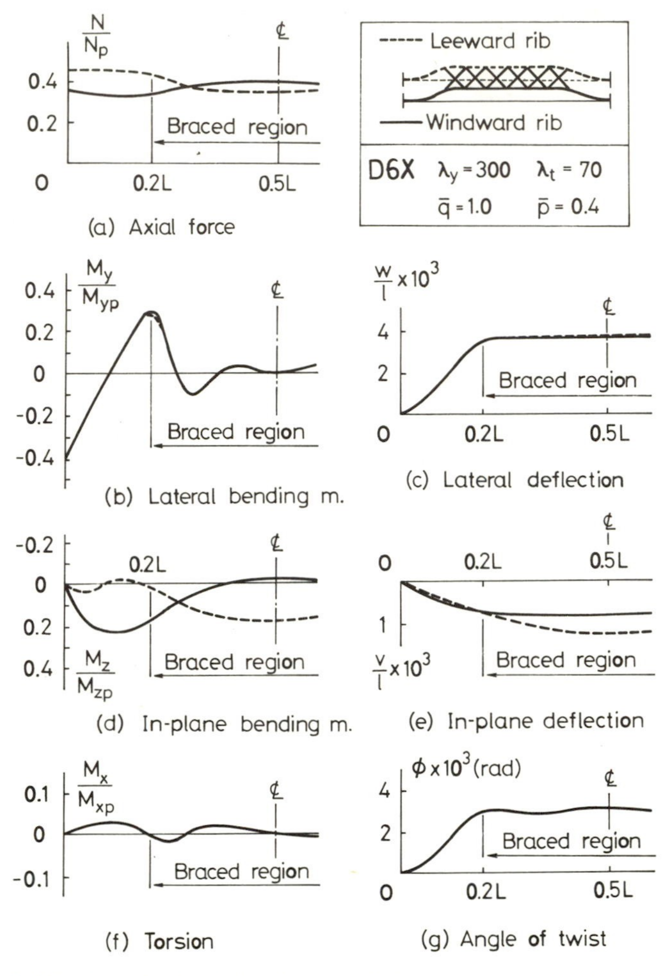

FIG. 9.10. Longitudinal distribution of stress resultants and deflections.

bending moment diagram of the individual rib is that it has a large negative gradient in the end panels of the bracing system, where the negative shear forces acting in the panel produce an additional axial force in the bracing members, as shown in Fig. 9.10(b). Similar behaviour was also seen for the case of pure vertical loading.

After a large number of parametric studies, it has been concluded that the ultimate strength of the models D6X and D12X under lateral loading may be classified by the following three categories according to the slenderness ratio of arch rib λ_y.

(1) Stocky arches having slenderness ratios in the range of $\lambda_y < 180$ can have sufficiently large ultimate strengths under lateral loads even if there are no bracing systems. For this reason, the central-arch-girder bridge can be useful for this range of slenderness ratio from economical and aesthetic standpoints.

(2) The bridge consisting of twin arches with medium slenderness ratios ranging from 180 to 300 needs an appropriate bracing member, rationally designed by the procedure described later.

(3) If the slenderness ratio of arch rib is in the range of $\lambda_y > 300$, a sufficient margin against collapse cannot be ensured even by means of a rigid lateral system, because they collapse owing to a remarkable decrease in lateral rigidity by the yielding of the arch rib in the neighbourhood of the springings. Therefore, it is necessary to strengthen the unbraced parts of the individual ribs to prevent yielding there.

9.3.5 Relation Between the Ultimate Strength and the Stresses at Springings

In order to ascertain if the ultimate strengths of the slender arches with $\lambda_y > 300$ subjected to lateral load may be controlled by the stresses at the arch springings, the ultimate strengths of four different models were compared with each other, as shown in Fig. 9.11. The unbraced portion of model D6X.2W_y has twice the section modulus of the standard model D6X. The ultimate strength of the model D6X.2W_y under lateral load is comparatively large even for the range of $\lambda_y >$ 300, shown in Fig. 9.11.

These facts show that there exists a close relationship between the ultimate strength of braced arches and the stresses at the springings.

In design practice, the lateral load q is represented by the wind load. So, the ultimate strength under lateral load can be defined by the following non-dimensional form

$$\bar{q} = \frac{q_u}{q_0} \tag{9.22}$$

where

$$q_0 = \tfrac{1}{2}\rho V_0^2 C_d h \tag{9.23}$$

Both windward and leeward ribs are supposed to be subject to an identical wind load q as shielding effect does not appear for parallel twin arches.

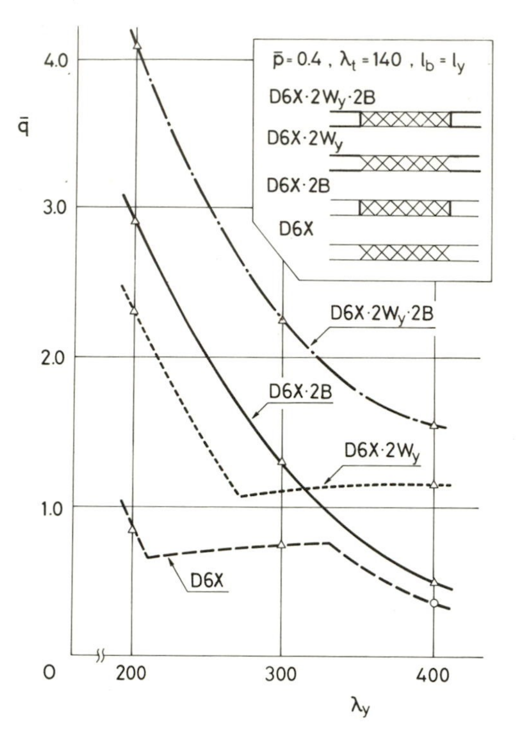

FIG. 9.11. Effects of double section modulus and portal frame.

Thus the value of $\bar{q}$ may be considered alternatively as a reservation factor for a nominal load corresponding to the wind velocity V_0.

On the other hand, the maximum elastic fibre stress σ_s at the springings of the leeward rib can be given by the following formula in non-dimensional form under the assumption that the lateral bracing system never buckles.

$$\bar{\sigma}_s = \frac{\sigma_s}{\sigma_y} = \frac{1}{\sigma_y}\left(\frac{N_s}{A_s} + \frac{M_s}{W_y}\right) \tag{9.24}$$

$\bar{\sigma}_s$ can be regarded as a measure of the maximum load effect in the braced arch. Since it may be a fictitious stress for the completely elastic arch, the stress resultants N_s and M_s should be estimated, multiplying q_0/q by their values computed under a certain load q lower than both the elastic limit load and the critical load at which the arch will collapse elastically.

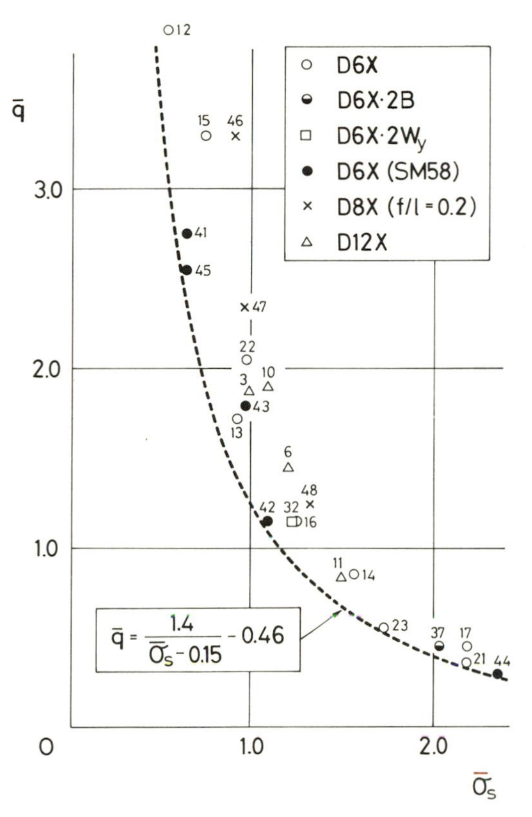

FIG. 9.12. Ultimate strength $\bar{q}$ versus maximum stress at the arch springings $\bar{\sigma}_s$.

The theoretical values of $\bar{q}$ and $\bar{\sigma}_s$ for various models are plotted in Fig. 9.12, which shows that there is a distinctive correlation between the ultimate strength $\bar{q}$ and the maximum stress $\bar{\sigma}_s$ at springings. From the figure, the conservatively approximate formula can be found as follows

$$\bar{q} = \frac{1{\cdot}4}{\bar{\sigma}_s - 0{\cdot}15} - 0{\cdot}46 \tag{9.25}$$

If the arch must be designed against a specified design wind velocity v m/s, the safety criterion as to the ultimate strength can be formulated as follows

$$\left(\frac{50}{v}\right)^2 \bar{q} > \nu \tag{9.26}$$

9.3.6 Derivation of Design Formulae

9.3.6.1 Stress Resultants of Framed Structure

Let us consider a laterally-loaded plane frame, which is obtained by straightening the twin arch with bracing members in the horizontal plane, as shown in Fig. 9.13. Since this frame is an indeterminate structure of high-order, it certainly is cumbersome to solve it analytically except by matrix analysis relying on a computer. However, it is advantageous for practical purposes in design to solve it approximately by giving attention to its lateral bending characteristics shown in Fig. 9.10(b). As illustrated schematically in this figure, the lateral bending moment of the individual rib abruptly decreases along the span in the end panel of the bracing system, so that the bending moment in the central part may be disregarded.

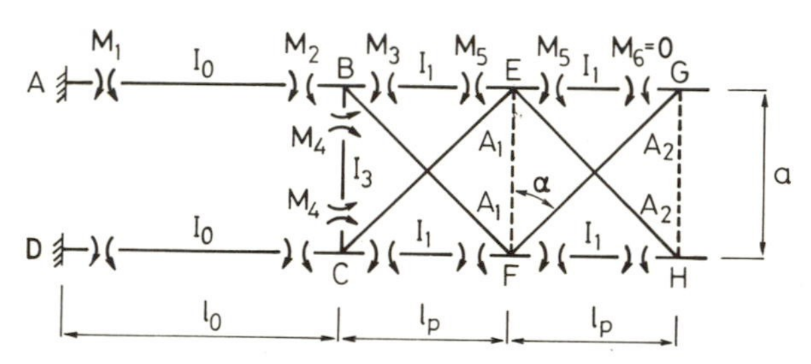

FIG. 9.13. Plane frame analysis for lateral bending moments of arch.

Therefore, the lateral bending moments of an individual rib can be easily determined by analysing the sub-structure consisting of a portal frame and two end panels of the bracing system.

According to the compatibility conditions of the slopes at points B and E, unknown bending moments M_3 and M_5 can be readily determined and then the portal frame can be analytically solved, as follows

$$M_1 = \frac{1}{6k_3+1} M_3 - \frac{3k_3+1}{12k_3+2} Q_1 l_0 - \frac{4k_3+1}{12k_3+2} q l_0^2 \tag{9.27a}$$

$$M_3 = \frac{1}{7+24k_1+42k_3} \left[2k_1(3Q_1+2ql_0)l_0 - \frac{3(6k_3+1)}{\sin\alpha\cos^2\alpha} \left(\frac{5Q_1}{A_1} - \frac{Q_2}{A_2} \right) \frac{I_1}{l_p} \right] \tag{9.27b}$$

$$M_5 = \frac{3}{4\sin\alpha\cos^2\alpha} \left(\frac{Q_1}{A_1} - \frac{Q_2}{A_2} \right) \frac{I_1}{l_p} - \frac{1}{4} M_3 \tag{9.27c}$$

$$k_1 = \frac{I_1 l_0}{I_0 l_p}, \qquad k_3 = \frac{I_3 l_0}{I_0 a} \tag{9.28}$$

$$m = \frac{N_E}{N_E - N} \tag{9.29}$$

N is the axial force at the springings of the arch rib produced by the vertical load p according to the linear theory.

$$N_E = \frac{\pi^2 E I_0}{l_0^2} \tag{9.30}$$

$$Q_1 = q(L - 2l_0), \qquad Q_2 = Q_1 - 2ql_p \tag{9.31}$$

9.3.6.2 Axial Forces of Bracing Members

The axial force D_i of a bracing member in the ith panel is given by

$$D_i = (\tfrac{1}{2}Q_i - Q_r) \sec \alpha \quad (i = 1, 2) \tag{9.32}$$

$$Q_r = \frac{m}{l_p}(M_5 - M_3) \tag{9.33}$$

The shear force of the individual rib in the central panel of $i > 2$ may be regarded as negligibly small and approximately equal to zero.

Substituting eqns (9.31) and (9.33) into eqn (9.32) gives a fairly approximate axial force of bracing members. It should be noted that some iterative procedure is needed for the solution of eqns (9.27) to (9.33) for purposes of design, because the axial forces from which the cross-sectional areas of bracing members should be decided cannot be computed without assuming the cross-sectional area. The lateral bracing members can be dimensioned, however, by using the standard column strength curve, after estimating the axial force D_i by means of trial and error.

Since the validity of the concept on which these formulae are based is unaffected by the different types of bracing systems, these formulae might be applicable to the other types of trussed bracing systems (e.g. K-truss) after some modification, if necessary.

In conventional design practice, the shear force of arch rib Q_r has been neglected through judging mistakenly that such a treatment will give the result only on the safety side for the bracing members. In fact, Fig. 9.14 clearly shows that this assumption underestimates their axial force.

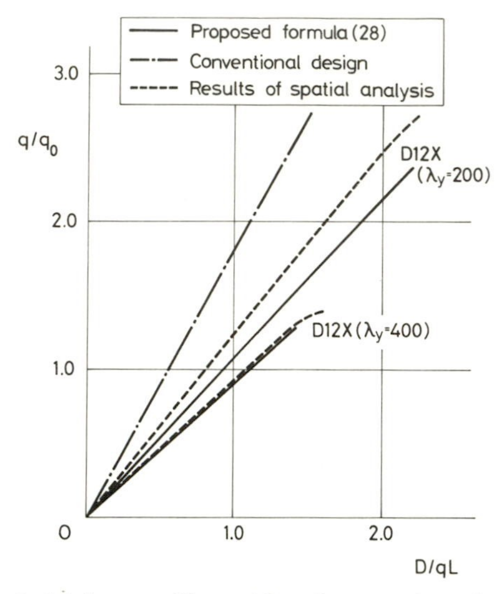

FIG. 9.14. Axial forces of lateral bracing members in end panel.

9.3.6.3 Maximum Stress at Springings

The maximum stress at the springings of the arch ribs can be approximately estimated by the following simple formula

$$\sigma_s = \frac{N}{A_s} + \frac{M_c}{A_s a}\beta + \frac{M_b}{W_y}m \tag{9.34}$$

where M_c is the lateral bending moment at the springings of arch rib, regarded as a curved beam, under lateral load $2q$.

M_b is given by the following formula

$$M_b = \frac{3k_3+1}{12k_3+2}Q_1 l_0 + \frac{4k_3+1}{12k_3+3}ql_0^2 - \frac{1}{6k_3+1}M_3 \tag{9.35}$$

The approximate values σ_{se} of σ_s calculated by formula (9.34) are compared with the theoretical ones σ_{sa} for various structural models in Fig. 9.15. The general validity of the proposed formula (eqn (9.34)) is demonstrated by such a comparison shown in this figure, so that σ_{se} can be substituted into eqn (9.25) to find the ultimate strength $\bar{q}_e$. In addition, the ultimate strengths $\bar{q}_e$ estimated by the formula (eqn (9.25)) are compared with the values $\bar{q}_a$, calculated by the elasto-plastic finite displacement theory described in Section 9.2.1, in Fig. 9.16 for

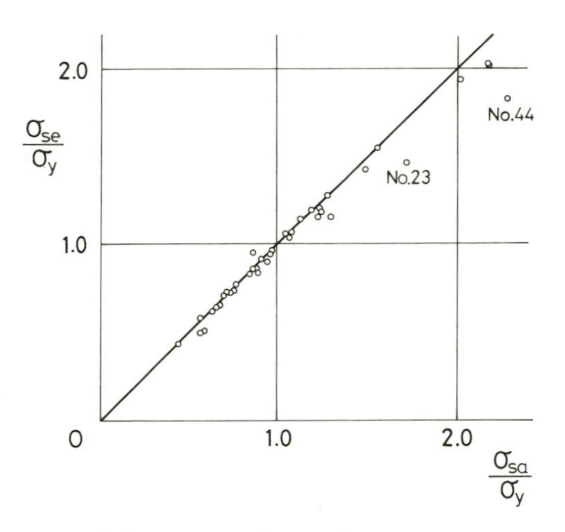

FIG. 9.15. Accuracy of the stress estimated by proposed formula (eqn (9.34)).

various structural models showing that the former gives fairly conservative predictions.

Before applying the formula (eqn (9.25)), it should be checked up if the stress σ_n included in the first term of the stress σ_s is smaller than the value σ_u at which the arch fails due to lateral instability under pure vertical load p.

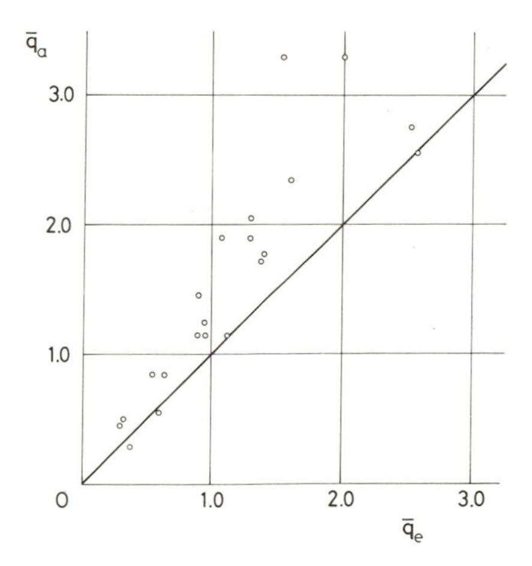

FIG. 9.16. Applicability of proposed formula (eqn (9.25)).

9.3.6.4 Interaction Formula

It has been found from comparison with theoretical results that the following interaction formula, for the ultimate strength of steel arches subjected to combined in-plane and lateral loads, is fairly conservative if the rigidity of bracing members is sufficient to prevent its premature buckling

$$\frac{N}{N_u} + m\frac{M_b}{M_y} = 1 \qquad (9.36)$$

9.4 CONCLUSIONS

This Chapter describes the strength characteristics of arches with bracing system and some design formulae have been presented for estimating their ultimate strengths for the following cases;

(1) when in-plane instability occurs due to in-plane load,

(2) when lateral instability occurs due to in-plane load,

(3) when lateral instability occurs due to combined in-plane and lateral load.

These formulae can be applied to the design of braced steel arch bridges with span lengths less than about 300 m, which was the limit covered in the parametric study. A detailed description of Section 9.3. can be found in the papers by Sakimoto and Komatsu (1979, 1982).

REFERENCES

KOMATSU, S. and SAKIMOTO, T. (1975) Elasto-plastic behavior of thin-walled steel tubes under combined forces. *Proc. Japan Soc. Civ. Engrs*, **235,** 125–36.

KOMATSU, S. and SAKIMOTO, T. (1976) Nonlinear analysis of spatial frames consisting of members with closed cross sections. *Proc. Japan Soc. Civ. Engrs*, **252,** 143–57.

KOMATSU, S. and SAKIMOTO, T. (1977) Ultimate load carrying capacity of steel arches. *Proc. Am. Soc. Civ. Engrs*, **103,** ST12, 2323–36.

KOMATSU, S. and SHINKE, T. (1977) Practical formulas for in-plane load carrying capacity of arches. *Trans. Japan Soc. Civ. Engrs*, **9,** 92–5.

SAKIMOTO, T. and KOMATSU, S. (1977*a*) Ultimate load carrying capacity of steel arches with initial imperfections. *Prel. Rept. 2nd. Int. Coll. on Stability of Steel Structures*, Liege, Imprimerie B. Nelissen, Liege, pp 545–50.

SAKIMOTO, T. and KOMATSU, S. (1977*b*) A possibility of total lateral bracings.

Final Rept. 2nd. Int. Coll. on Stability of Steel Structures, Liege, Imprimerie B. Nelissen, Liege, pp. 299–301.

SAKIMOTO, T. and KOMATSU, S. (1979) Ultimate strength of steel arches under lateral loads. *Proc. Japan Soc. Civ. Engrs*, **292,** 83–94.

SAKIMOTO, T. and KOMATSU, S. (1982) Ultimate strength of arches with bracing systems. *Proc. Am. Soc. Civ. Engrs*, **108,** ST5, 1064–76.

SAKIMOTO, T., YAMAO, T. and KOMATSU, S. (1979) Experimental study on the ultimate strength of steel arches. *Proc. Japan Soc. Civ. Engrs*, **286,** 139–49.

SHINKE, S. (1977) *A method of calculating the in-plane load carrying capacity of steel arches and its application to design*. Dissertation, Osaka University (in Japanese).

STÜSSI, F. (1935) Aktuelle baustatische Probleme des Konstruktionspraxis. *Schweiz. Bauzeitung*, **106,** 132–5.

Chapter 10

MEMBER STABILITY IN PORTAL FRAMES

L. J. MORRIS

Simon Engineering Laboratories, University of Manchester, UK

and

K. NAKANE

University of Canterbury, Christchurch, New Zealand

SUMMARY

One of the more interesting design problems associated with portal frame construction is member stability and the location of lateral supports. A review of existing procedures for checking both unrestrained and restrained members, together with the conditions to which they apply, is undertaken. These stability checks are discussed with reference to both uniform and haunched members. New proposals for determining the limiting slenderness ratio for the compression flange are introduced and then the theoretical failure loads, as predicted by the appropriate methods, are compared with known experimental evidence. Guidance is also given on the design of lateral supports.

NOTATION

A Cross-sectional area; stability parameter ($\sqrt{k}\ (l/r_y)$)
a Distance from centroidal axis of member to axis of restraint
a_y Extreme fibre distance about minor axis
B Width of flange
c Shape factor
D Depth of uniform section

D^* Largest depth of haunch
d Distance between centroid of flanges $(= D - t_f)$
E Modulus of elasticity
f_0 Bending stress about minor axis
f_x Major axis bending stress due to M_x
G Shear modulus
I_y Second moment of area about minor axis
K St Venant's constant
k Equivalent uniform moment distribution factor
L_{ch} Maximum unsupported length of haunched member
L_{cr} Elastic critical length
L'_{cr} Critical buckling length
L_m Allowable design length
l Distance between lateral restraints to compression flange
M_{pr} Reduced plastic moment due to axial load
M_x Larger end moment
M'_x Equivalent uniform major axis moment
M^*_x Equivalent major axis moment, including effects of axial load
M_y Yield moment $(= Z_x p_y)$
m Equivalent moment factor; sub-lengths (l/s)
n Number of flanges
P Axial thrust
P_E Euler buckling load
p Mean axial stress
p_y Yield stress; design strength
q Ratio of haunched length to haunch length
r Ratio of maximum and minimum depths $(= D^*/D)$
r_0 Polar radius of gyration
r_y Radius of gyration (minor axis)
s Spacing between tension flange restraints
T Torsion constant $(= AGK/Z_x^2)$
t_f flange thickness
x D/t_f ratio
Z_x Elastic modulus (major axis)

α Regression factor
β Ratio of end moments
ρ P/P_E
ϕ Rotation
ϕ_0 Initial imperfection

10.1 INTRODUCTION

Following the innovative research undertaken at Cambridge (Baker *et al.*, 1956) structural engineers were provided with a powerful new design method known as plastic design, with the result that, since the mid-fifties, portal (gable) frame construction in the UK has been almost exclusively designed by the plastic theory. Such designs produce quite slender rafter members with fairly long shallow haunches at the eaves (see Fig. 10.1), the normal practice being to make the haunch depth twice that of the basic rafter section (see enlarged detail of Fig. 10.1). At least one plastic hinge of those required for the collapse mechanism is usually assumed to develop at either the column/haunch intersection or the haunch/rafter intersection. Secondary members such as purlins and sheeting rails provide restraint on one flange against lateral displacement.

A basic assumption of plastic theory is that the frame is able to collapse by the rotation of the plastic hinges before member or frame instability can develop. That is, plastic deformation can take place without the geometry of the structure changing to such an extent that the conditions of equilibrium are significantly modified. Such changes of geometry are associated with problems of local, frame and member instability, i.e.

(a) *overall* change of geometry of the structure, causing joints to displace in-plane relative to each other (e.g. sway deformation),

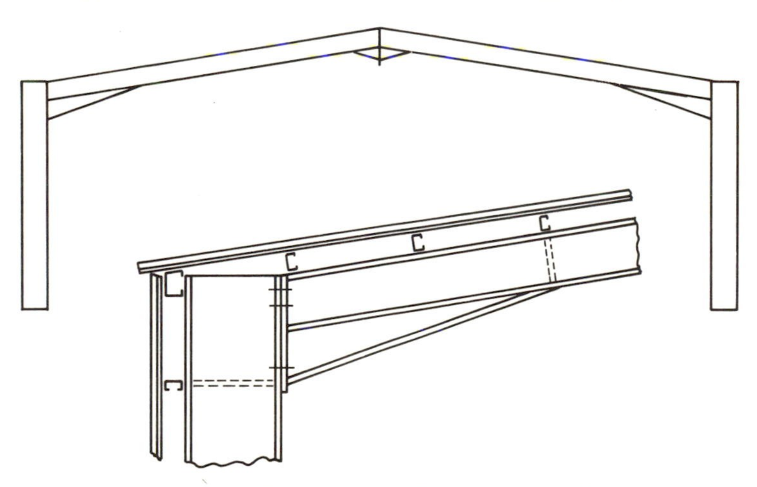

Fig. 10.1. Typical portal frame construction.

(b) out-of-plane displacements within the length of a *member* relating to straight lines drawn between adjacent lateral supports which define the length being considered (due to bending and/or twisting of the member), and
(e) deformation within the cross-section of a member (resulting from *local* buckling in the plate elements constituting the web or flanges).

Though the preceding comments are applicable to any framed structure, this chapter will concentrate on criteria by which the adequacy of portal frame construction can be checked against member instability. The problems of frame instability are discussed by Horne in Chapter 1.

Producing realistic design guidance with respect to the instability problems is beset with difficulties, not only because of the complexity, but because the theoretical solutions are not always supported by experiments on a level approaching full-scale conditions. Despite the difficulties, a great deal of effort has gone into studying these problems in recent years.

At present, there are no current recommendations in the British, European and American codes of practice for structural steelwork for determining the maximum safe unsupported length for lateral stability of *haunched* members. Though a design procedure for checking the adequacy of *uniform* members against stability has been available since the mid-sixties (Horne, 1964*a*, *b*), it is only in the last decade that tentative design proposals based on theoretical considerations have been suggested for *non-uniform* members, e.g. haunched members. Also, it has become important economically to assess the influence of the 'middle' flange, which is a feature of British portal frame construction.

Previous work at Manchester University has produced two different approaches for checking the adequacy of two-flanged members, one of which has been extended to include three-flanged members. The approach of Horne *et al.* (1979*a*), which covers the design of tapered and haunched two-flanged members, appears to give realistic solutions, but the method needs to be extended to include three-flanged members. Horne and Morris (1977) modified a safe stress method to check the adequacy of the three-flanged members against lateral instability. These methods are further developed to account, where possible, for the beneficial effect of the middle flange. Also, there is a need for any tentative conservative design proposal to be checked against experimental evidence.

Due to the definite gap in the research knowledge on the behaviour of haunched rafter members, a series of tests was undertaken, which has provided useful information (Morris and Nakane, 1983). This, together with other published research evidence, is used in the comparison of existing and new theoretical treatments for member stability.

10.2 THE PROBLEM

A steel I-section is very efficient and strong when loaded through its shear centre, but it is inherently weak when subject to lateral bending (minor axis) or torsion, particularly the universal beam sections. The engineer must, therefore, guard against the possibility of failure caused by lateral movement of a member combined with a twisting action of the cross-section. Such a phenomenon is known as lateral–torsional buckling—a condition which can arise when in-plane bending of an I-section causes both sudden sideways deflection and twisting at a lower load than designed. Thus, lateral–torsional stability is an important criterion in the design of steel I-section members. In the case of single-storey construction, there is the added complication that the rafter is usually haunched within the most highly stressed length. Deep haunches can be more susceptible to overall member instability by twisting about the purlin/rafter connections.

The complex subject of lateral–torsional stability has attracted the attention of many researchers over a number of years and various design formulae or charts have been proposed for a wide range of problems. Thus majority of these solutions have dealt with the stability of symmetrical I-sections having a uniform depth—few have examined the stability of members having a varying depth, i.e. tapered (Vickery, 1962; Nethercot, 1973). When viewed in terms of real structures, it is inevitable that approximate solutions only have been derived because of the numerous 'practical' factors that need to be considered, such as plasticity, imperfections, residual stresses and effectiveness of restraints.

The particular problem being considered involves lateral–torsional buckling of a haunched member subjected to large in-plane moments, the magnitude of which may be sufficient to cause a plastic hinge to develop at the designated positions assumed in analysis, e.g. haunch/rafter intersection. The engineer needs to define the critical length over which the outstand (compression) flange is stable under factored loading. Important parameters include the effect of lateral restraint on the tension flange and/or compression flange, the ability of the member to develop the necessary moment–rotation capacity to allow a plastic hinge to form without instability, the influence of local buckling on lateral stability and the effect of non-uniform members.

10.3 STABILITY OF UNRESTRAINED UNIFORM MEMBERS

In elastic design, the conditions for member stability are assessed simply by reference to the forces and corresponding stresses derived from an elastic analysis of the frame. A limited acknowledgement in the proposed British code (to be published as BS5950) is made of the influence of redistribution of forces after partial yielding by allowing a redistribution of moments up to 10% of the peak moments.

On the other hand, in plastic design account must be taken of plastic rotation requirements, and these affect design for member instability in the following ways.

(a) Inelastic deformations due to plasticity in bending about the major axis of an I-section causes the warping resistance to decrease more drastically than St Venant torsional resistance.
(b) Rotation requirements are less severe in regions of low moment gradient (e.g. sagging moment regions in beams) than in regions of high moment gradient (e.g. at rigid or continuous supports) because of the greater elastic rotation capacity of the former.
(c) A modest reduction in the full plastic capacity is not deleterious when this occurs in a region of uniform or near-uniform moment, since such a region will contain the 'last hinge to form'. The effect of such a reduction on the collapse load of the structure will be small.

For these reasons, special treatment procedures of member stability requirements are needed in plastic design; the following are among those suitable for the treatment of such problems in single-storey frames.

10.3.1 Elastic Stability of Unrestrained Uniform Member

Apart from localised plastic hinge action at designated locations in a portal frame (necessary to produce a failure mode) the frame behaves elastically. Therefore, the stability of the *uniform* members or parts of uniform members, not containing plastic hinges, can be readily checked by using conventional allowable stress limitations as given in the current British code (BS449) or the design charts given in the BCSA publication No. 23 (Horne, 1964*b*) using factored loading. The latter method for checking elastic stability is based on earlier studies by the Cambridge team (Baker *et al.*, 1956; Horne, 1956). The procedure is best suited for members not carrying high axial loads and can deal with any combination of end moments (including minor axis bending). The treatment of stability is based on the calculation of an equivalent uniform major axis moment M'_x given by $M'_x = mM_x$, where m depends on β (ratio of end moments) (see Fig. 10.2). The values of m are presented graphically for different moment ratios β. For members with slenderness greater than a certain limiting value, buckling occurs at loads only slightly greater than these causing yield in a extreme fibre near mid-length. The design charts therefore facilitate a stability check

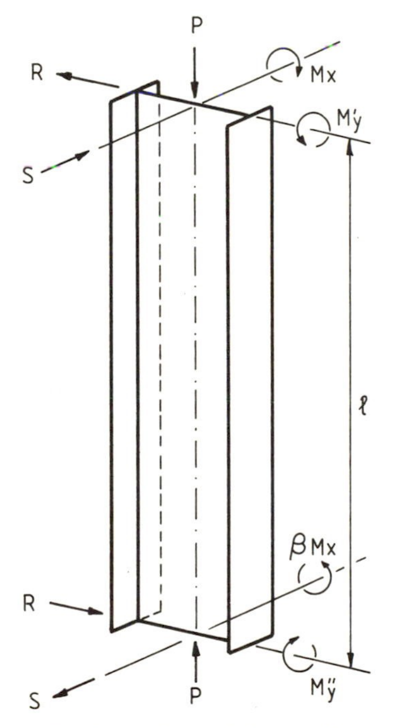

FIG. 10.2. Unrestrained member subject to linear moment gradient.

based on the criterion that the extreme fibre stress does not exceed yield anywhere along the length of the member.

10.3.2 Plastic Stability of Unrestrained Uniform Member

A design procedure for checking the stability of *uniform* members with plastic hinges at the ends has been available since the mid-sixties and also takes the form of design charts, the basis of which is given elsewhere (Horne, 1964*a*). In using the charts, the uniform member is assumed to be subjected to end moments that act about the major axis only, the larger moment causing a plastic hinge to form at one end. The BCSA publication No. 23 (Horne, 1964*b*) gives a direct checking procedure for such members subject to a *linear moment gradient.* That is, plastic moment occurs at one end, M_{pr} $(=Z_{pr} \cdot p_y)$, with βM_{pr} at the other end (where $-1 \leq \beta \leq +1$). To use the charts it is necessary to know the slenderness ratio l/r_y, where l is the length of member being considered, and the torsion constant T $(\simeq AGK/Z_x^2)$ a property of the cross-section of the chosen member. Values of T are given as a

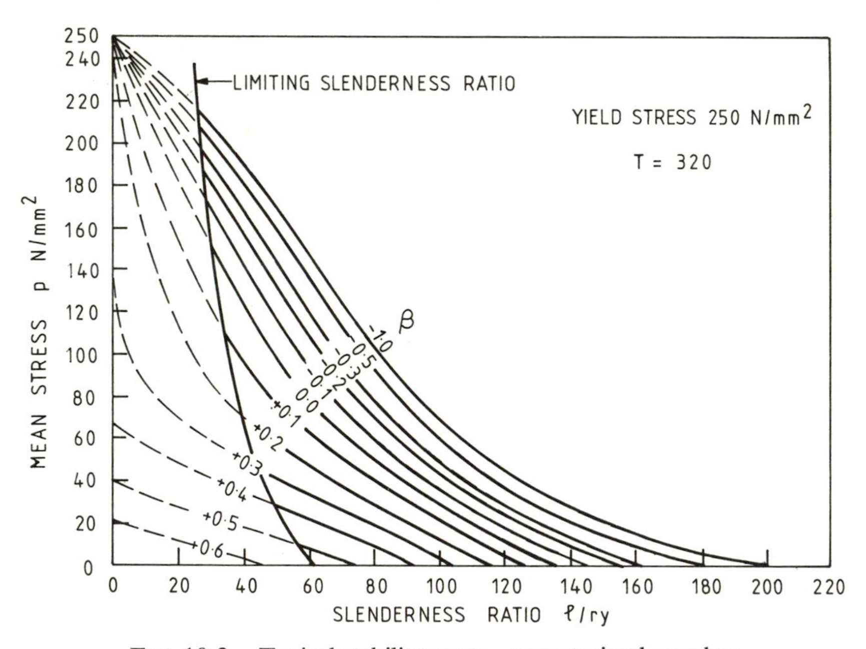

FIG. 10.3. Typical stability curve—unrestrained member.

property for both the universal sections and the RSJs in Table C of *Plastic Design* (Morris and Randall, 1979). Alternatively, it may be assumed to be given (with sufficient accuracy) by $T \simeq 12\,000(1/x)$, where $x = D/t_f$. Charts are plotted in terms of slenderness ratio and permissible axial stress for different values of β and are reproduced for both grade 43 and 50 steels and specific values of T: Fig. 10.3 shows a typical chart. Knowing the end moment ratio β and permissible axial stress p, the engineer can rapidly assess whether or not the design slenderness ratio is acceptable.

10.3.2.1 Limiting Slenderness Curve for Unrestrained Members

The various charts given in BCSA publication No. 23 (Horne, 1964*b*) for checking plastic instability indicate the maximum axial stresses allowed while ensuring the attainment of full plastic action at one or both ends of a member. It would appear from these charts that plastic action is not possible, whatever the slenderness ratio, when a condition of uniform moment is approached, e.g. $\beta > 0{\cdot}6$ in the chart shown in Fig. 10.3. However, at low slenderness ratios, plastic collapse mechanisms involving assumed 'plastic hinges' in such members will give satisfactory estimates of carrying capacity of continuous structures, provided

(a) the peak of the moment–rotation curve is not more than a few per cent below the theoretical plastic moment for the given axial load, and
(b) the curve of moment versus in-plane hinge rotation is sufficiently flat-topped, i.e. adequate rotation capacity.

The theoretical treatment of this problem is difficult since it involves following the post-buckling behaviour in the elastic–plastic range. This has been done for the lateral instability of uniform beams in Chapter 12 of Baker *et al.* (1956) and is the basis for the design recommendations given in the preceding section. The recommended maximum unsupported length is given by

$$L_m = \frac{43 r_y}{\sqrt{\left[\dfrac{p}{100} + \left(\dfrac{p_y}{240}\right)^2 \left(\dfrac{x}{36}\right)^2\right]}} \tag{10.1}$$

This equation gives the limiting slenderness curve indicated on the design charts, as illustrated by Fig. 10.3. At slenderness below this limiting curve, full plastic action may be assumed in the member for design purposes, irrespective of ratio of end moments. That is, eqn (10.1) gives the safe permissible spacing of supports to the compression flange whatever the ratio of end moments and the degree of plasticity, *provided there is no destabilising force acting on the compression flange between the supports.* For example, the effect of the compression flange of the haunch trying to move laterally at the eaves connection introduces a destabilising lateral force into the inner flange of the portal leg, thereby requiring adequate restraint at that position.

As the axial stresses in a portal frame are generally low, and taking account of the safe assumptions involved in its derivation, eqn (10.1) can be simplified to $L_m/r_y = 1500/x$ for grade 43 steel. Alternatively, the safe spacing of supports for a *uniform* member containing a plastic hinge has been tabulated for each section in Table B of *Plastic Design* (Morris and Randall, 1979). However, if the length of members being considered is elastic and/or contains the last hinge to form, then the maximum spacing can be relaxed as rotational capacity is not a requirement and a value of $2500/x$ for grade 43 steel is recommended.

If the axial stress in a member exceeds the value given by the appropriate chart, the member may be rendered safe by the introduction of lateral supports to the compression flange or by changing the section size.

10.3.2.2 Maximum Spacing of Restraints on Tension Flange

All lengths of any member between lateral restraints to the *tension flange*, i.e. maximum spacing between secondary members such as purlins, must satisfy the stability requirement for an unrestrained member of that effective length. A simple conservative estimate is to ensure the spacing of these secondary members does not exceed the value given by the limiting slenderness values in Section 10.3.2.1.

10.4 STABILITY OF UNIFORM MEMBERS RESTRAINED ALONG ONE FLANGE

Since the early seventies, researchers have been attempting to take account of the effect of restraint to *uniform* members, normally supplied by secondary members. Considering the particular case of portal frame construction, the columns are usually restrained at intervals along one (tension) flange by cold-formed sheeting rails. The rafters are also restrained by similar secondary members (purlins). Composite action with the cladding by the purlins and sheeting rails is sufficient to ensure that the particular flange involved (at its point of attachment to a rail) is fully restrained against displacement perpendicular to the web of the rafter or column under consideration, i.e. positional lateral restraint is provided.

The effective torsional restraint given to the main member by the purlins or sheeting rails depends on many factors, such as the length and cross-section of the rails, the moment–rotation characteristics of the joints between the rails and the main members, the local deformation of the main member and the nature of the cladding. However, standard connections between purlins and the tension flange of the rafter are generally not markedly moment resistant, i.e. as the connections occur on one flange of the main member only, local deformation of the web and the outstand flange can occur. Therefore, researchers in recent years have assumed there is adequate lateral restraint, but any torsional restraint (as might be present) has been ignored.

To develop a design procedure for the general case, the design of restrained symmetrical I-section members subject to uniform moment is considered initially. The modification required to deal with any distribution of moments is then considered.

10.4.1 Elastic Stability of Restrained Uniform Member

The elastic buckling of members, laterally supported at intervals along one flange, was presented by Horne and Ajmani (1971, 1972). The loading condition considered was for a symmetrical I-section member (Fig. 10.4(a)) supported against lateral deflection in *both* flanges at its ends (distance l apart) and subjected to an axial thrust P, together with major axis end moments M_x and βM_x (Fig. 10.4(b)). The ends are assumed free to rotate about the minor axis of the section with no restraint against warping. Also, the section is assumed to be restrained by the rails at intervals of s along the axis AB. As these secondary members are attached to one of the flanges, the effective axis of restraint is assumed to act at some distance a from the centroidal axis of the main member. Due to the finite size of connections, a is somewhat larger than half the depth of the main member. If d is the distance between the centroids of the flanges, the ratio a/d will tend to lie between the values of 0·5 and 0·75, i.e. the distance from the axis of restraint to the flange lies between zero and $0{\cdot}25d$. To be on the safe side a high value should be taken and after due consideration of constructional details it is assumed that the ratio $a/d = 0{\cdot}75$.

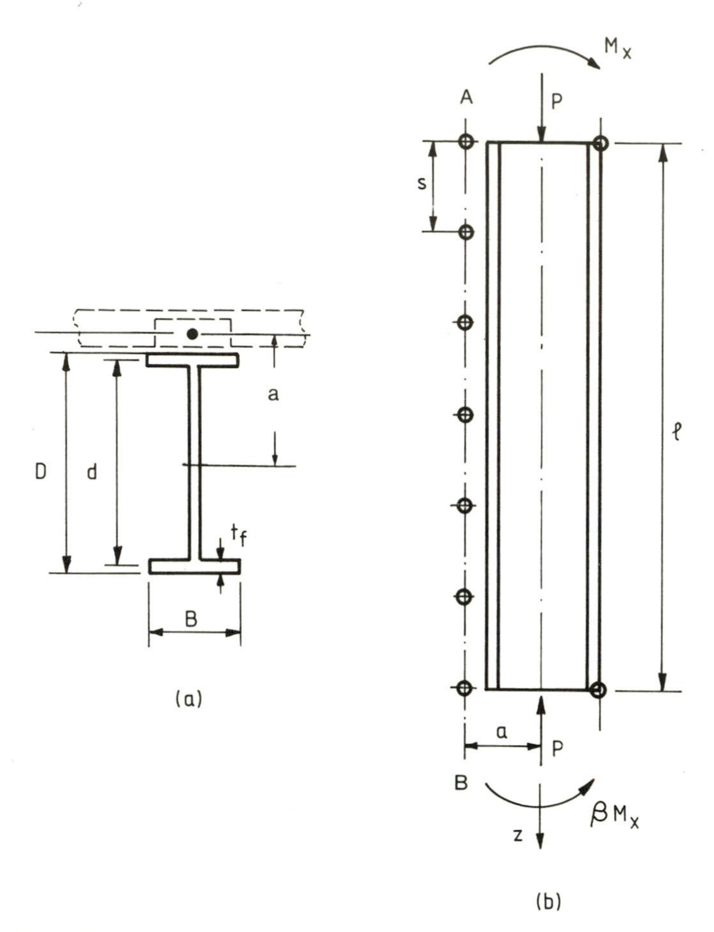

FIG. 10.4. Restrained member subject to linear moment gradient.

It is shown by Horne and Ajmani (1971) that, for sufficiently small values of s, a member (loaded as indicated in Fig. 10.4) will buckle by twisting about the restrained axis AB ('overall failure', Fig. 10.5(a)), but that for s larger than some critical value, buckling will occur between supports ('failure between supports', Fig. 10.5(b)). The critical value of s depends only slightly on the ratio of end moments β, provided that the larger end moment M_x produces compression in the unrestrained flange, as illustrated in Fig. 10.4(b). It is therefore sufficient to consider the case of uniform moment.

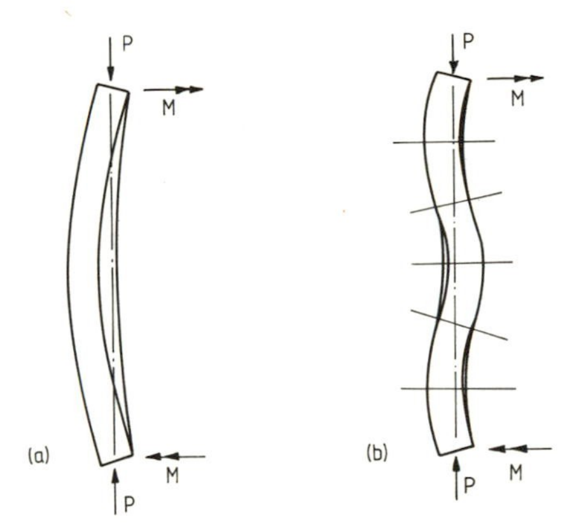

FIG. 10.5. Different modes of member instability: (a) overall buckling; (b) buckling between supports.

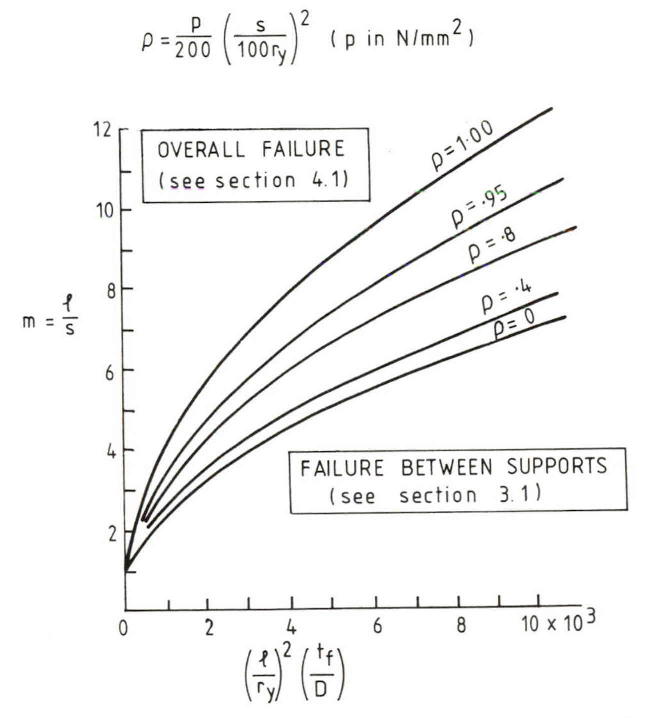

FIG. 10.6. Determination of mode of failure when $a/d = 0{\cdot}75$.

Some results obtained by Horne and Ajmani (1971) for the uniform moment condition are given in Fig. 10.6 for $a/d = 0{\cdot}75$. The vertical scale gives the number of sub-lengths $m(=l/s)$ into which the member is divided by the lateral restraints. The horizontal scale is the value of $(l/r_y)^2(t_f/D)^2$. The curves are plotted for various values of $\rho = P/P_E = (p/\pi^2E)(s/r_y)^2$, where P_E is the Euler buckling load for the member as a pin-ended strut over a length s and p is the mean axial stress $(=P/A)$. Failure occurs by overall instability for cases lying above the appropriate curve, and between supports for cases lying below. When failure occurs between supports, design methods appropriate to unrestrained members should be used (see Section 10.3) and should be applied to the most critically loaded length between supports, see Horne (1956). However, in the haunched region of portal frames, overall failure (by twisting about the axis of restraint) tends to be more critical and is a particular problem where deep haunches are used, as they are more susceptible to instability of the outstand compression flange.

10.4.1.1 *Restrained Uniform Member Subject to Uniform Moment*

Taking the member shown in Fig. 10.4 which is subject to uniform moment, $M_x = \beta M_x$ (i.e. $\beta = 1$), the differential equation governing the rotation ϕ at a distance z from one end, assuming the initial imperfection ϕ_0 is given by

$$\phi_0 = \Phi_0 \sin \frac{\pi z}{l} \tag{10.2}$$

is

$$GK \, . \, \frac{\mathrm{d}(\phi - \phi_0)}{\mathrm{d}z} - EI_y\left(a^2 + \frac{d^2}{4}\right) . \, \frac{\mathrm{d}^3(\phi - \phi_0)}{\mathrm{d}z^3} = (Pr_0^2 + 2M_x a) \, . \, \frac{\mathrm{d}\phi}{\mathrm{d}z} \tag{10.3}$$

Here, GK is the St Venant torsional rigidity, EI_y the flexural rigidity about the minor axis and r_0 the polar radius of gyration about the axis of restraint. Assuming the ends of the members to be free about the minor axis and free from warping restraint, the solution to eqn (10.3) is

$$\phi = \Phi \sin \frac{\pi z}{l} \tag{10.4}$$

where

$$\Phi = \frac{GK + \dfrac{\pi^2 EI_y}{l^2}\left(a^2 + \dfrac{d^2}{4}\right)}{GK + \dfrac{\pi^2 EI_y}{l^2}\left(a^2 + \dfrac{d^2}{4}\right) - Pr_0^2 - 2M_x a} \, \Phi_0 \tag{10.5}$$

Following the Perry–Robertson approach for struts, it is assumed that the initial lateral deflection at mid-span of the outstand flange is $0{\cdot}0015(lr_y/a_y)$ where a_y is the extreme fibre distance about the minor axis, e.g. $B/2$ for the section shown in Fig. 10.4. Hence

$$\Phi_0 = 0{\cdot}0015\left[\frac{lr_y}{a_y\left(a+\dfrac{d}{2}\right)}\right] \tag{10.6}$$

By considering the bending of the outstand flange, it is readily shown from eqns (10.5) and (10.6) that the bending stress about the minor axis at mid-length is given by f_0 where

$$f_0 = 0{\cdot}0015\pi^2 E\left(\frac{r_y}{l}\right)\left[\frac{Pr_0^2+2M_x a}{GK+\dfrac{\pi^2 EI_y}{l^2}\left(a^2+\dfrac{d^2}{4}\right)-Pr_0^2-2M_x a}\right] \tag{10.7}$$

In order to simplify eqn (10.7), various approximations to section properties may be made, in this instance based on the British range of universal beam sections. It is found that, approximately,

$$K = 0{\cdot}32\left(\frac{t_f}{D}\right)^2 Ad^2$$

If r_x is the radius of gyration about the major axis then

$$r_0^2 = a^2 + r_x^2 + r_y^2 \simeq a^2 + \frac{d^2}{4} = \tfrac{13}{16}d^2 \qquad \text{if} \quad a = 0{\cdot}75d$$

Hence

$$Pr_0^2 + 2M_x a = Ar_0^2 p + \tfrac{48}{13}\frac{r_x^2}{d}\frac{d}{D}f_x$$

where f_x is the major axis bending stress due to M_x. Inspection of typical universal sections shows that the second term inside the square brackets approximates to $0{\cdot}65f_x$. Taking $E = 200\ \text{kN/mm}^2$ and $G = 80\ \text{kN/mm}^2$, eqn (10.7) then reduces to

$$f_0 = 0{\cdot}3\left(\frac{l}{r_y}\right)\left[\frac{p+0{\cdot}65f_x}{200-\left[(p+0{\cdot}65f_x)10^{-4}-3\left(\dfrac{t_f}{D}\right)^2\right]\left(\dfrac{l}{r_y}\right)^2}\right] \tag{10.8}$$

The design criterion is that the maximum stress, given by $p + f_x + f_0$, should not exceed the yield stress p_y, i.e.

$$p + f_x + f_0 \leqslant p_y \tag{10.9}$$

When $p = 0$, it follows that $f_x \leqslant p_y$, so that a safe (high) value of f_0 may be derived from eqn (10.8) by putting $f_x = p_y$. Substituting the resulting value of f_0 in eqn (10.9) gives, approximately, the design criterion

$$(p_y - f_x) \geqslant \frac{2p_y\left(\dfrac{l}{100r_y}\right)}{20 - \left[0{\cdot}065p_y - 3000\left(\dfrac{t_f}{D}\right)^2\right]\left(\dfrac{l}{100r_y}\right)^2} \tag{10.10}$$

It is intended to produce a set of design charts for checking the elastic stability of restrained uniform members, based on the theoretical treatment by Horne and Ajmani (1971). These charts, similar to the charts (34–36) for unrestrained members given in BCSA publication No. 23 (Horne, 1964*b*), are intended for publication in a companion volume to the Constrado monograph (Horne and Morris, 1981).

10.4.1.2 Restrained Uniform Member Subject to Non-Uniform Moment

It has been shown by Singh (1969) that a satisfactory method of calculating the critical elastic buckling conditions for a restrained uniform member, subjected to *non-uniform moment*, in the absence of axial load (which is the usual design case) is as follows. Let f_1, f_2, f_3, f_4 and f_5 be the elastic extreme fibre stresses (based on *elastic* moduli) due to the applied moments M_1, M_2, M_3, M_4 and M_5, where M_1 and M_5 are the moments at the ends (Fig. 10.7(b)), M_3 is the moment at mid-length and M_2 and M_4 are the quarter point moments. A factor k is calculated where

$$k = \frac{1}{12p_y}\left[f_1 + 3f_2 + 4f_3 + 3f_4 + f_5 + 2(f_{S\,max} - f_{E\,max})\right] \tag{10.11}$$

The elastic stresses $f_{S\,max}$ and $f_{E\,max}$ are the maximum span and end stresses, respectively. The stresses f_1 to f_5, $f_{S\,max}$ and $f_{E\,max}$ are positive if they correspond to compression in the outstand flange, otherwise they are zero. Similarly, the quantity $(f_{S\,max} = f_{E\,max})$ is only included if it is positive. If L_{cr} is the critical buckling length for a member subjected to a uniform moment producing extreme fibre stress of p_y, then the

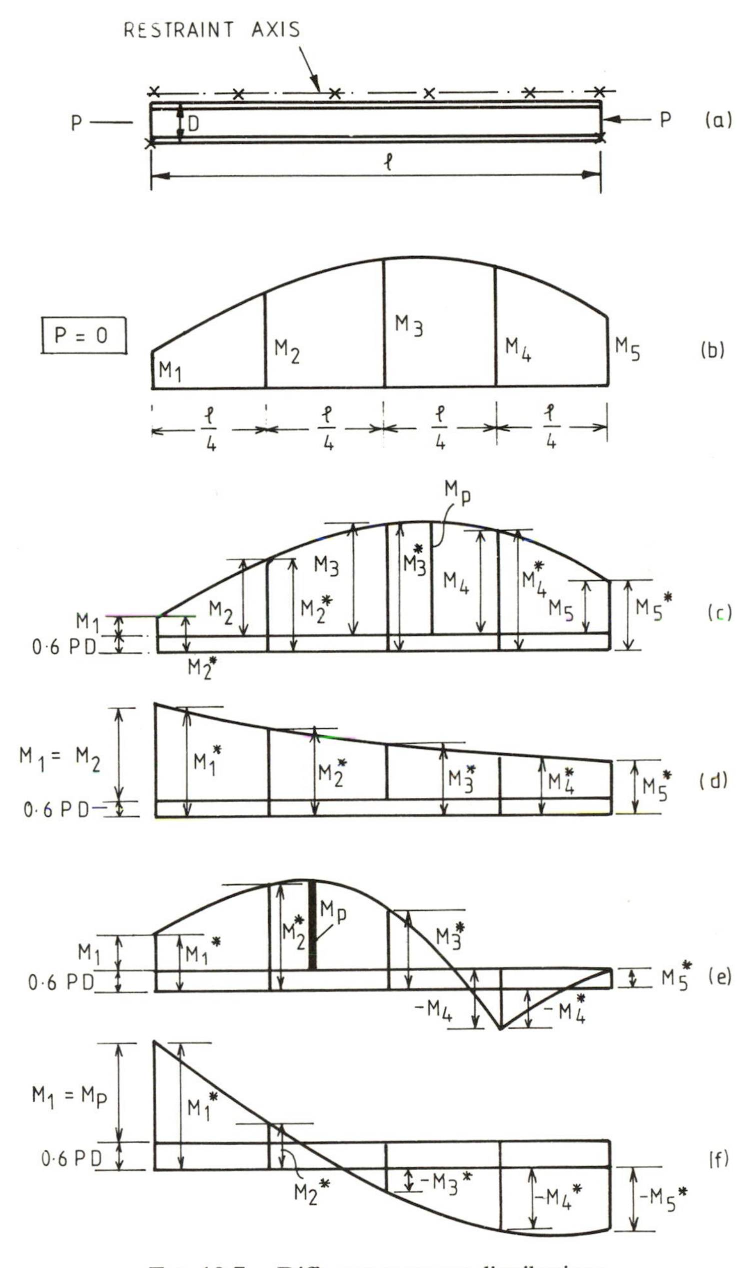

FIG. 10.7. Different moment distributions.

critical buckling length for the given moment distribution is L'_{cr} where

$$L'_{cr} = \frac{L_{cr}}{\sqrt{k}} \tag{10.12}$$

It follows that, to obtain the equivalent length for buckling under a uniform moment, the actual length should be multiplied by $\sqrt{k}$. Hence the term $l/100r_y$ in eqn (10.10) should be replaced by $\sqrt{k}l/100r_y$ giving as the design criterion

$$(p_y - f_x) \geqslant \frac{2p_y\sqrt{k}\left(\frac{l}{100r_y}\right)^2}{20 - k\left[0{\cdot}065p_y - 3000\left(\frac{t_f}{D}\right)^2\right]\left(\frac{l}{100r_y}\right)^2} \tag{10.13}$$

The stress condition (10.13) aims to limit the maximum stress in the member to the yield value. The bending stress about the minor axis f_0 occurs only between the points of lateral support at sections 1 and 5, and may be assumed to occur equally at sections 2, 3 and 4. Hence f_x on the left-hand side of eqn (10.12) refers to the greatest positive stress anywhere within the restrained length, which might not coincide with the quarter-points 2, 3 or 4. Note, if the right-hand side of inequality (10.13) is negative, then the member is *not* stable. There is no necessity to check the stability if $D/t_f < 220/\sqrt{p_y}$.

10.4.1.3 Effect of Axial Load

The effect of an axial load P may be derived by noting from eqn (10.7) that a moment M_x in the presence of a force P has the same effect as a moment M'_x where

$$M^*_x = M_x + \frac{r_0^2}{2a}P$$

It has been shown that when $a = 0{\cdot}75d$ then r_0^2 approximates to $13d^2/16$, hence

$$M^*_x = M_x + \tfrac{13}{24}Pd$$

In order always to produce a safe result it may be assumed that

$$M^*_x = M_x + 0{\cdot}6PD$$

The stresses f in eqn (10.11) are therefore replaced by corresponding stresses f^* due to the moments M^*_x (see Fig. 10.7(c)). The calculation of the modified moments to allow for the effect of axial load is illustrated by Figs. 10.7(c)–(f), inclusive. Because of the safe approximations made to derive eqn (10.10) it may be safely used for members carrying loads up to a mean factored stress of 10% of the yield stress.

10.4.2 Plastic Stability of Restrained Uniform Member

The research work by Horne and Ajmani (1971, 1972) describes a method of checking a member subject to plastic end moments in the presence of axial thrust when the tension flange is laterally restrained at intervals, not exceeding that recommended in Section 10.3.2.2. Failure occurs due to torsional buckling about the restrained axis and the theoretical treatment is similar to that used for unrestrained members (Section 10.3.2). Lower limiting slenderness curves applicable to loading in the plastic range with any ratio of end moments β are also derived by considering post-buckling behaviour. In producing design charts it is assumed that a/d is 0·75. Adopting this value, then the limiting slenderness between restraints to compression flange while

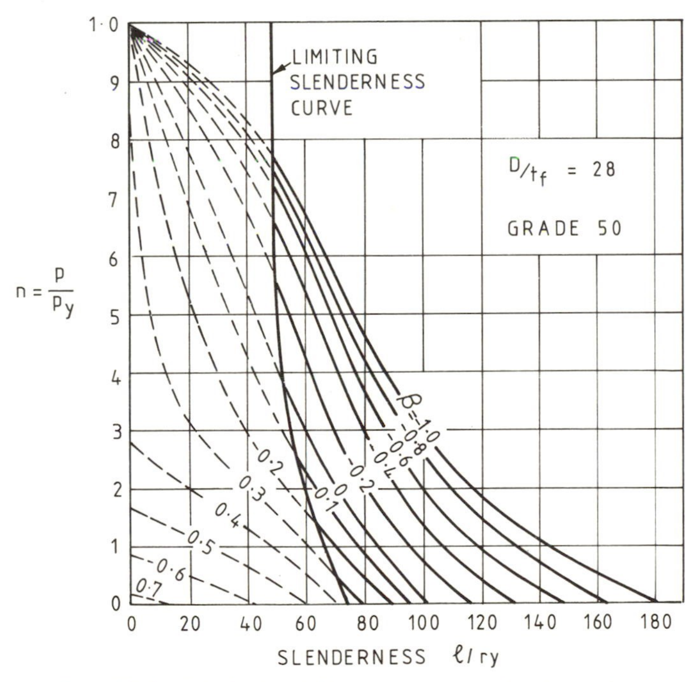

FIG. 10.8. Typical stability curve—restrained member.

allowing plastic action is given by

$$\frac{L_m}{r_y}=\frac{\left[5{\cdot}4+0{\cdot}7\left(\frac{p_y}{240}\right)\right]x}{\sqrt{\left[\left(\frac{x}{12{\cdot}3}\right)^2\left(\frac{p_y}{240}\right)-1\right]}} \tag{10.14}$$

Assuming grade 43 steel, then eqn (10.14) becomes

$$\frac{L_m}{r_y}=\frac{6{\cdot}1x}{\sqrt{\left[\left(\frac{x}{12{\cdot}3}\right)^2-1\right]}} \tag{10.15}$$

A typical design chart, similar to the charts (1–32) for unrestrained members given in the BCSA publication No. 23 (Horne, 1964*b*), is reproduced in Fig. 10.8. A complete set of design curves is intended for publication in a companion volume to the Constrado monograph (Horne and Morris, 1981).

10.5 STABILITY OF RESTRAINED NON-UNIFORM MEMBERS

The preceding sections and comments have dealt only with members having a constant cross-section throughout their length. The method outlined in Section 10.4.1 needs to be modified to take account of the varying depth of a haunched member and, when necessary, the 'middle' flange. Also, the theoretical approach of Horne *et al.* (1979*a*) is discussed due to its particular relevance to tapered and haunched members.

10.5.1 Elastic Stability of Non-Uniform Members Restrained Along One Flange

Initially, design methods dealing with the stability of two-flanged, tapered and haunched members are described, followed by a discussion of procedures concerned with three-flanged members.

10.5.1.1 Two-Flanged Haunched Members

(a) Horne and Morris (1977) approach. In the design of non-uniform members, such as the two-flanged haunched rafter shown in

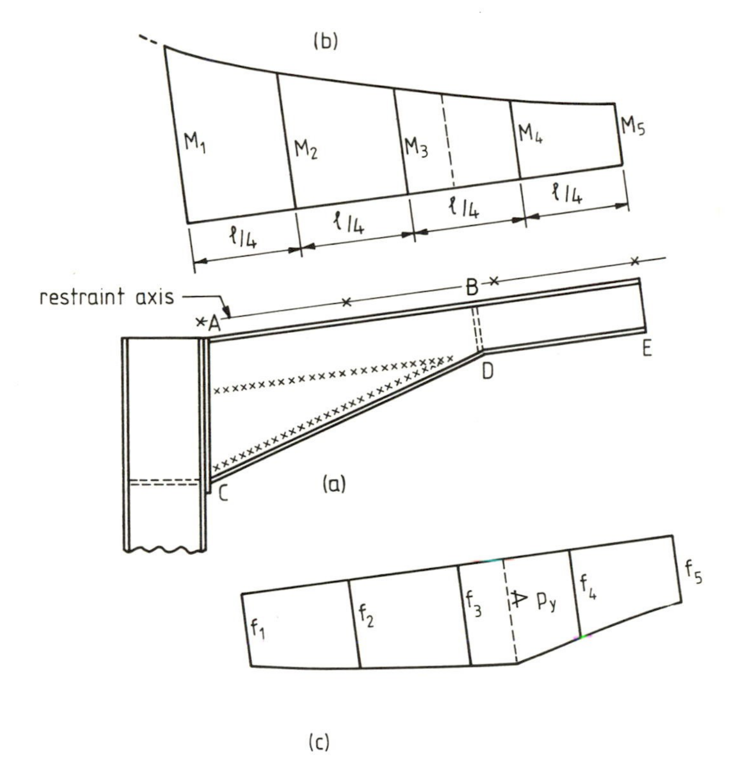

FIG. 10.9. Two-flanged haunched rafter: (a) two-flanged haunch member; (b) bending moment diagram; (c) typical stress distribution.

Fig. 10.9(a), it is assumed that restraint is provided at intervals along the flange AB as indicated (see enlarged detail in Fig. 10.1). The outstand flange is assumed to be laterally restrained at the points C and D, or point E. The design condition, eqn (10.13), can be applied, provided D^*, t_f and r_y all refer to the properties at the deepest section of the haunched rafter and k is calculated from eqn (10.11), based on the induced compression stresses in the outstand flange. This approach involves numerous assumptions. The most important is that eqn (10.11) gives the factor k to be applied to the critical length L_{cr} for a uniform member subject to a uniform yield moment (based on properties of the deepest section) in order to obtain the elastic critical length L'_{cr} $(=L_{cr}/\sqrt{k})$ of the haunched member under its actual load conditions.

(*b*) *Horne et al.* (*1979a*) *method*. The general approach adopted by Horne *et al*. is to express the maximum stable length L_s' of a non-uniform (tapered or haunched) I-section rafter in the form

$$L_s' = L_s / c\sqrt{k} \tag{10.16}$$

where L_s is the maximum stable length of a uniform member (properties based on minimum section) subject to a uniform moment. In the case of elastic instability, the reference length L_s is in fact the critical length L_{cr} for a uniform member subject to a uniform moment just sufficient to cause yield in the extreme fibres, i.e. M_y. The factor k allows for the arbitrary distribution of stresses along its length based on elastic moduli (see eqn (10.11)) while the shape factor c allows for the varying depth of the tapered or haunched member.

Using the expression derived by Horne and Ajmani (1969) for the critical length L_{cr}, Horne *et al.* investigated the influence of various parameters affecting the elastic critical length in the context of the British range of universal beam sections, and proposed an empirical expression for the elastic critical slenderness ratio

$$\frac{L_{cr}}{r_y} = \frac{(8{\cdot}0 + 150 p_y/E)(x)}{\sqrt{[4{\cdot}4(p_y/E)x^2 - 1]}} \tag{10.17}$$

which reduces to the following equation for grade 43 steel

$$\frac{L_{cr}}{r_y} = \frac{8{\cdot}2x}{\sqrt{[(x/13{\cdot}5)^2 - 1]}} \tag{10.18}$$

where $x = D/t_f$ of the basic section. The parameter c represents an estimate of the ratio of the elastic critical length of a haunched member, subject to a moment distribution that would just cause yield in all extreme fibres, to the critical length L_{cr} of a uniform member, i.e.

$$c = L_{ch}/L_{cr}$$

Subsequent investigation, with respect to the more 'economic' universal beam sections, resulted in c being defined as

$$c = 1 + \frac{3}{(x-9)}(r-1)^{2/3}\sqrt{q} \tag{10.19}$$

where $r = D^*/D$ (ratio of maximum and minimum depths) and q is the ratio of the haunched length to the total length of member between lateral supports. A slightly more accurate evaluation of c is given in

Horne *et al.* (1979*a*). However, it should be noted that this method deals only with elastic critical length and does not include for the effect of imperfections as other methods do. Therefore, it will tend to give an overestimate of the allowable design length.

10.5.1.2 Three-Flanged Haunched Members

(a) Horne and Morris (1977) method. Practical haunched members (Morris and Packer, 1977, 1984) tend to differ from the arrangement given in Fig. 10.9(a) in that the basic section is continued through into the haunched portion. This produces within the haunch region a 'middle flange' (see Fig. 10.10(a)). Considering the buckling of a uniform three-flanged member of the cross-section indicated in Fig. 10.11, restrained about the axis R, then the differential eqn (10.3) remains

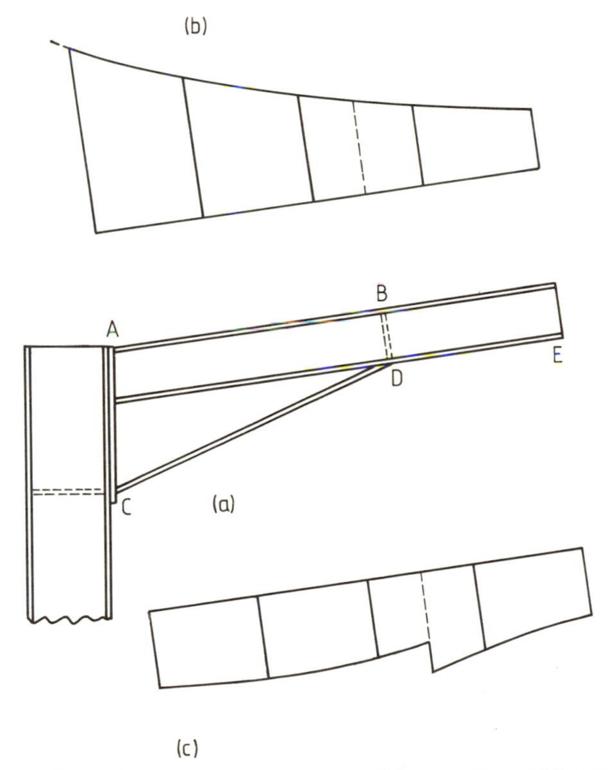

FIG. 10.10. Three-flanged haunched rafter: (a) three-flanged haunch member; (b) bending moment diagram; (c) typical stress distribution.

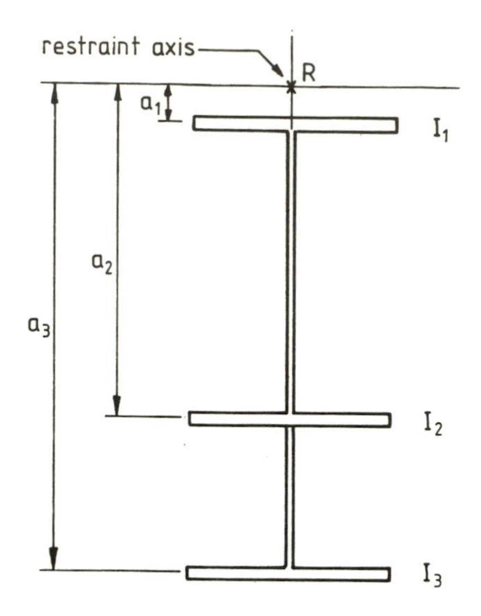

FIG. 10.11. Cross-section of three-flanged member.

the same, except that the term $EI_y(a^2+d^2/4)$ is replaced (to a sufficiently close approximation) by $E(I_1 \,.\, a_1^2+I_2 \,.\, a_2^2+I_3 \,.\, a_3^2)$ where I_1, I_2 and I_3 are the second moments of area of the flanges about the minor axis of the section, and a_1, a_2 and a_3 are their distances from the axis of restraint. Thus, eqn (10.7) is amended accordingly.

When the necessary modifications to eqns (10.8) and (10.10) including approximations to the value of r_0 are explored numerically, it is found that the net effect is to replace $3(t_f/D)^2$ in eqn (10.8) by $4{\cdot}2(t_f/D^*)^2$, and $3000(t_f/D)^2$ by $4200(t_f/D^*)^2$ in eqn (10.10). This corresponds to the increase in the St Venant term GK in eqn (10.7) to allow for the torsional resistance of the third flange. It may be noted that the axis restraint is still assumed to be a distance from the tension flange of a quarter of the depth of basic section. A further refinement is suggested by Horne (1983) to cover the case when the haunch flange thickness is different from the flange of the basic rafter section. That is, a universal beam (UB) cutting different from the basic section is deliberately used to improve the stability of the haunch.

The use of the modifying factor $\sqrt{k}$ by which l in eqn (10.7) is multiplied to derive the effect of varying extreme fibre stresses has not been investigated theoretically for three-flanged haunched members. Tentatively, however, it may be applied in the same way as for two-flanged tapered members, the stresses f being based on *elastic* moduli, for cross-sections perpendicular to the axis of the basic rafter section. Equation (10.13) becomes

$$(p_y - f_x) \geqslant \frac{2p_y\sqrt{k}\left(\frac{l}{100r_y}\right)^2}{20 - k\left[0{\cdot}065p_y - 3000\left(\frac{t_f}{D^*}\right)^2 - 1200\left(\frac{t_h}{D^*}\right)^2\right]\left(\frac{l}{100r_y}\right)^2} \tag{10.20}$$

where t_h is the thickness of haunch compression flange.

(b) Horne et al. method (1979a). It is suggested that a conservative estimate for the three-flanged, haunched member using the method outlined in Section 10.5.1.1(b) would be obtained if c is made equal to unity, i.e.

$$L'_{cr} = L_{cr}/\sqrt{k}$$

where L_{cr} is defined by eqn (10.17) or (10.18).

(c) Morris and Nakane approach. One of the comments occasionally made by engineers regarding empirical design formulae is that they should be simple (Morris, 1980, 1983*a*, *b*). In an attempt to simplify eqns (10.13) and (10.20), it is necessary to know the influences of each variable parameter, i.e. k, l/r_y, p_y and D/t_f or D^*/t_f, on the equations. The relationship between each variable and each equation could be shown graphically by means of scatter diagrams, using a simple random sampling technique for the variables. Although it is possible to obtain some indication from these diagrams of the relationship between each variable within the context of a given equation, this method is not sufficiently precise for most statistical purposes. It becomes necessary to compute a quantitative index of each relationship; one such index is the Pearson product–moment correlation coefficient, or simply the Pearson r coefficient. A basic assumption dependent on the use of the Pearson r technique is that the variables and equations have a linear relationship. However, though the variable l/r_y exhibits a curvilinear relationship it has a dominant influence when compared with k and D/t_f or D^*/t_f and is retained in a non-linear form. Also, k plays an important role in reflecting the stress distribution along the member, and is retained in its present form.

Equation (10.13) can be expressed in a simple form, i.e.

$$f(k, l/r_y, D/t_f, p_y) = \frac{A}{4 - \alpha\sqrt{\left(\frac{D}{t_f}\right)}A^2}$$

where $A = \sqrt{k}(l/r_y)$ and α is the regression coefficient. A regression analysis was undertaken to evaluate α, based on the following ranges of the variables for the two-flange case, assuming grade 43 steel,

$$0{\cdot}5 \leqslant k \leqslant 1{\cdot}0; \qquad 20 \leqslant d/t_f \leqslant 44; \qquad 40 \leqslant l/r_y \leqslant 150; \qquad p_y = 250$$

The resulting value of r was 0·997, with $\alpha = 0{\cdot}000047$. Similar analyses were carried out for the three-flanged case. Further study revealed that a general, though slightly less accurate, expression can be used for both cases, i.e.

$$p_y - f_x \geqslant \frac{A}{4 - \frac{0{\cdot}0001}{n}\sqrt{(x)}A^2} \tag{10.21}$$

where x is dependent on n, the number of flanges in haunched region, i.e. $x = D/t_f$ when $n = 2$ and $x = D^*/t_f$ when $n = 3$. It is suggested that k does not exceed a value of 0·75–0·80, otherwise there is a strong possibility of plasticity occurring in the haunched region, which could lead to premature failure. Equation (10.21) can be easily used for grade 50 steel, by replacing the constant 4 by 3.

Extending the concept of the shape factor c to include three-flanged members, both tapered and haunched, the following approximate empirical expressions have been developed.

(i) *tapered* members (depth varies continually within design length, i.e. $q = 1$)

$$c = (2 - r)[0{\cdot}007\sqrt{(x - 20)} + 0{\cdot}9158] + r - 1 \tag{10.22}$$

(ii) *haunched* members (part of design length has constant depth, i.e. $q \neq 1$)

$$c = 1 + q[(44 - x)^3 \times 10^{-5} - 0{\cdot}0364] \tag{10.23}$$

where $x = D/t_f$.

10.5.2 Plastic Stability of Non-Uniform Members Restrained Along One Flange

10.5.2.1 Two-Flanged Haunched Members

Horne *et al.* (1979*a*) also gives a theoretical treatment, based on Horne and Ajmani (1969, 1971), for the derivation of maximum possible slenderness ratios appropriate to two-flanged haunched rafters laterally restrained on the top flange by purlins, when the rafter is assumed to contain a plastic hinge at the haunch/rafter intersection. A convenient expression similar to eqn (10.14), for the limiting slenderness ratio, is

$$\frac{L_m}{r_y}=\frac{(5{\cdot}4+600p_y/E)(x)}{\surd[(5{\cdot}4p_y/E)(x)^2-1]} \qquad (10.24)$$

The permissible length L'_m for the two-flanged haunched member is obtained by substituting $L_s=L_m$ in eqn (10.16). In the evaluation of k the stresses f are calculated, based on the *plastic* moduli of the appropriate cross-sections. The factor c is determined from eqn (10.19). For grade 43 steel, eqn (10.24) reduces to eqn (10.15).

Experimental confirmation of eqn (10.24) was limited to cases where plastic hinges formed at the haunch/rafter intersection, while the moment at the eaves connection was always less than 0·7 of the plastic moment at the deepest section. If plasticity occurs in the haunched region, other than local yielding near the eaves connection, then it becomes difficult to ensure satisfactory plastic rotation capacity. It is found that in order to develop a plastic hinge at the haunch/rafter intersection, full depth web stiffness is required at that position. This was confirmed later by the research of Morris and Nakane (1983).

10.5.2.2 Three-Flanged Haunched Members

An estimate of L'_m for three-flanged, haunched rafters can be made by using the Horne *et al.* method outlined in Section 10.5.2.1 and making $c=1$.

Alternatively, it may be assumed that the appropriate expression for c developed by Morris and Nakane (see eqns (10.22) and (10.23)) may be used with the Horne *et al.* method given in Section 10.5.2.1.

10.6 LATERAL RESTRAINTS

In checking for member stability the engineer assumes the positions of lateral restraints to the compression flange, based on practical considerations. For instance, members must be adequately braced against lateral and torsional displacements at hinge positions. In addition, a member may need to be braced away from the hinge position, depending on stability requirements. In all cases, the restraining brace should be sufficiently stiff so that the member is induced to buckle between braces, i.e. restraint must have adequate strength and stiffness.

Recent tests (Morris and Nakane, 1983) have indicated that the magnitude of the restraining force before instability occurs is of the order of 2% of the *squash load* of the compression flange, i.e. $0{\cdot}02Bt_fp_y$. Though the restraining force is relatively small, it is *essential* that such a force (in the form of a brace) be supplied. If a lateral restraint buckles (or is removed) then the buckling mode will change to a more severe condition, leading to premature failure as was noted recently (Morris and Packer, 1984). Adequate stiffness is therefore probably more important than strength and in view of the lack of sufficient experimental evidence a limiting slenderness ratio of 100 is recommended (Morris, 1980) for diagonal braces.

It is sometimes difficult to give lateral support exactly at a plastic hinge position. However, if the hinge position is to be regarded as being laterally restrained, then the point of support to the *compression flange* must not be more than $D/2$ from the hinge position.

10.7 COMPARISON OF METHODS

In an attempt to assess the different procedures outlined in this chapter it was decided to compare several of the methods against known experimental evidence. There are three main sources of test information: the research on two-flanged tapered and haunched members by Horne *et al.* (1979*a*, *b*), the test series on haunched members summarised in a paper by Morris and Nakane (1983), and more recently the reported behaviour of practical rafters by Morris and Packer (1984). Though each series of tests have differing support conditions at the 'column'/rafter intersection, they are, in simple terms, haunched rafter members cantilevered from a 'column' member and loaded at the free end to produce a linear moment gradient along the length of the member. Each test arrangement is such as effectively to restrain torsionally both ends of the member.

The experimental research used in the comparison has been broadly classified into groups, basically related to the geometry of the member and the amount of restraint applied to the member, i.e.

(a) Horne *et al.* (1979*b*)—two-flanged, tapered members ($q=1$) tension flange restraint (3 tests),

TABLE 10.1

PREDICTED/ACTUAL FAILURE LOAD RATIO FOR VARIOUS STABILITY CHECKS

Test reference (see text, Section 10.7)	*Mean predicted/actual failure load ratio*					
	Horne and Morris eqns (10.13), (10.20),	*Morris and Nakane eqn* (10.21)	*Morris and Nakane eqns* (10.21), (10.19), (10.22), (10.23)	*Horne et al. eqns* (10.24), (10.19)	*Horne et al. eqn* (10.17)	*Morris and Nakane eqns* (10.24), (10.23)
	exclude c *Method* (1)	$c = 1$ *Method* (2)	*include c* *Method* (3)	*include c* *Method* (4)	$c = 1$ *Method* (5)	*include c* *Method* (6)
Two-flanged members						
(a) Horne *et al.*	0·76	0·87	0·77	0·95	—	—
(b) Horne *et al.*	0·84	0·88	0·85	0·96	—	—
Three-flanged members						
(c) Morris and Nakane	0·60	0·75	0·78	—	0·88	0·92
(d) Morris and Nakane	0·56	0·71	0·74	—	0·80	0·86
(e) Morris and Nakane	0·83	0·85	0·86	—	>1·0	>1·0
(f) Morris and Packer, Morgan	0·84	0·94	0·94	—	0·88	0·92

(b) Horne *et al.* (1979*b*)—two-flanged, haunched members ($q<1$) with tension flange restraint (8 tests),
(c) Morris and Nakane (1983)—three-flanged, haunched members with no tension flange restraint (6 tests),
(d) Morris and Nakane (1983)—three-flanged, haunched members with only tension flange restraint (4 tests),
(e) Morris and Nakane (1983)—three-flanged, haunched members with tension flange restraint, coupled with lateral support to compression flange at the haunch/rafter intersection (3 tests), and
(f) Morris and Packer (1984), Morgan (1978)—represents practical conditions; three-flanged, haunched rafter with cold-formed purlins attached to tension flange and diagonal braces to outstand flange at haunch/rafter intersection (3 tests).

Table 10.1 gives the mean predicted/actual failure load ratio for each group of tests against the various methods being compared. The comparison indicates that method (2) is an improvement on the original basic method (1), giving a more accurate but safe prediction of failure load. In effect, this means an increase in the limiting allowable slenderness. Though method (3) produces similar results to procedure (2) it does involve the additional calculation of the factor c, and method (2) is preferred because of its simplicity. The method by Horne *et al.* (4) gives an extremely good estimate of failure load for the two-flanged condition. In the context of the three-flanged members, methods (5) and (6) produce fairly consistent predictions, apart from one unsafe result. An inspection of the estimates for the practical haunched members (*f*) shows that there is little to choose between the methods (2) and (6), apart from the calculation of c for the latter. On the other hand, method (6) gives a direct evaluation of the limiting slenderness ratio.

10.8 CONCLUSIONS

A preliminary assessment of both existing and new methods, used for checking the adequacy of haunched members against instability, is made on the basis of predicting the failure load for a range of test specimens, the actual failure loads of which are known. Though the two methods (2) and (6) appear to produce good results, the procedure by Morris and Nakane (2) might be preferred as it gives safe answers. However, further detailed examination of the test evidence needs to be undertaken in order to give better guidance as to the influence of geometry of the member, the amount of restraint and stress distribution factor k in the control of stability of haunched members.

REFERENCES

BAKER, J. F., HORNE, M. R. and HEYMAN, J. (1956) *The Steel Skeleton, Vol. II*, Cambridge University Press, Cambridge.

HORNE, M. R. (1956) The stanchion problem in frame structures designed according to ultimate carrying capacity. *Proc. ICE, Pt.* 3, **5,** 105.

HORNE, M. R. (1964*a*) Safe loads on I-section columns in structures designed by plastic theory. *Proc. ICE,* **29,** 137–50.

HORNE, M. R. (1964*b*) The plastic design of columns. BCSA Publication No. 23 (now supplement to *Plastic Design*, Constrado Publication, London).

HORNE, M. R. (1983) Contribution to: A commentary on portal frame design. *Structural Engineer,* **61A,** 188–9.

HORNE, M. R. and AJMANI, J. L. (1969) Stability of columns supported laterally by side-rails. *Int. J. Mech. Sci.,* **11,** 159.

HORNE, M. R. and AJMANI, J. L. (1971) Design of column restrained by side rails; and Post buckling behaviour of laterally restrained columns. *Structural Engineer,* **49,** 339–52.

HORNE, M. R. and AJMANI, J. L. (1972) Failure of columns laterally supported on one flange. *Structural Engineer,* **50,** 355–66.

HORNE, M. R. and MORRIS, L. J. (1977) The design against instability of haunched members restrained at intervals along the tension flange. Prel. Report, *3rd Int. Coll. on Stability and Dynamics of Steel Structures,* Washington, ASCE, New York, pp. 618–29.

HORNE, M. R. and MORRIS, L. J. (1981) *Plastic Design of Low-Rise Frames.* Constrado Monograph, Granada Publishing, London.

HORNE, M. R., SHAKIR-KHALIL, H. and AKHTAR, S. (1979*a*) The stability of tapered and haunched beams. *Proc. ICE, Pt. 2,* **67,** 677–94.

HORNE, M. K., SHAKIR-KHALIL, H. and AKHTAR, S. (1979*b*) Tests on tapered and haunched beams. *Proc. ICE, Pt. 2,* **67,** 845–50.

MORGAN, A. N. (1978) Interaction between local buckling and plastic hinge formation. MSc Dissertation, University of Manchester, UK.

MORRIS, L. J. (1980) A commentary on portal frame design. *Structural Engineer,* **59A,** 394–404.

MORRIS, L. J. (1983*a*) Discussion on: A commentary on portal frame design. *Structural Engineer,* **61A,** 181–9.

MORRIS, L. J. (1983*b*) Correspondence on: A commentary on portal frame design. *Structural Engineer,* **61A,** 212–21.

MORRIS, L. J. and NAKANE, K. (1983) Experimental behaviour of haunched members. *Int. Conf. on Instability and Plastic Collapse of Steel Structures*, Granada Publishing, London, pp. 547–59.

MORRIS, L. J. and PACKER, J. A. (1977) Stability of haunched rafter. *2nd Int. Coll. on Stability of Steel Structures*, Liege, ECCS/IABSE, pp. 539–44.

MORRIS, L. J. and PACKER, J. A. (1984) The behaviour and design of haunched steel portal frame knees. *Proc. ICE, Pt. 2*, **77,** 211–37.

MORRIS, L. J. and RANDALL, A. L. (1979) *Plastic Design*, Constrado Publication, (originally produced as BCSA publication No. 28, 1965) London.

NETHERCOT, D. A. (1973) Lateral buckling of tapered beams. *IABSE*, **33,** 173–92.

SINGH, K. P. (1969) Ultimate behaviour of laterally supported beams. PhD Thesis, University of Manchester, UK.

VICKERY, B. J. (1962) The behaviour at collapse of simple steel frames with tapered members. *Structural Engineer*, **40,** 365.

INDEX